Decision Making in a World of Comparative Effectiveness Research

Howard G. Birnbaum • Paul E. Greenberg
Editors

Decision Making in a World of Comparative Effectiveness Research

A Practical Guide

Editors
Howard G. Birnbaum
Analysis Group, Inc.
Boston
Massachusetts
USA

Paul E. Greenberg
Analysis Group, Inc.
Boston
Massachusetts
USA

ISBN 978-981-10-3261-5   ISBN 978-981-10-3262-2   (eBook)
DOI 10.1007/978-981-10-3262-2

Library of Congress Control Number: 2017941182

Printed on acid-free paper

This Adis imprint is published by Springer Nature
The registered company is Springer Nature Singapore Pte Ltd.
The registered company address is: 152 Beach Road, #21-01/04 Gateway East, Singapore 189721, Singapore

Acknowledgements

Howard G. Birnbaum and Paul E. Greenberg gratefully acknowledge our Analysis Group colleagues who were instrumental in effectively coordinating with the authors of the chapters in this book. We also want to thank all of the authors for their significant contributions. This book was made possible by their collective experience and detailed subject matter knowledge. Editing assistance was provided by Renata Cavalier, MA; Lucia Antras, PhD; Ana Bozas, PhD; and Cinzia Metallo, PhD, employees of Analysis Group, Inc. In particular, Renata Cavalier diligently worked through all the numerous issues in organizing and coordinating each of the chapters into a consistently formatted book.

Contents

Contributors

Husam Albarmawi (BS Pharm, University of Jordan; MS, University of Maryland, Baltimore) is a PhD candidate in the Department of Pharmaceutical Health Services Research at the University of Maryland School of Pharmacy. He specializes in pharmacoeconomics and health outcomes research.

Carl V. Asche (PhD, Economics, University of Surrey; MSc, Health Economics, University of York) is the Director of the Center for Outcomes Research and Research Professor of Medicine at the University of Illinois College of Medicine at Peoria. His research focuses on the use of comparative-effectiveness research and cost-effectiveness analysis in healthcare decision-making.

Carole Baas (PhD, Texas A&M University) is an Educational Consultant at Cancer Information and Support Network and a Cancer Research Advocate with Houston Methodist Research Institute and UT Southwestern Simmons Comprehensive Cancer Center. She is also one of the founding editors of the scientific journal *Convergent Science Physical Oncology*, representing the voice of the patient. Her expertise includes physical science approaches to cancer research, nanomedicine, breast cancer in young women, clinical trials, and advocacy education and program development.

Justin E. Bekelman (MD, Yale University) is Associate Professor in the Department of Radiation Oncology and the Department of Medical Ethics and Health Policy at the University of Pennsylvania Perelman School of Medicine and Principal Investigator of the Radiation Therapy Comparative Effectiveness (RadComp) Consortium's pragmatic randomized trial of proton vs. photon therapy for patients with nonmetastatic breast cancer, funded by the Patient-Centered Outcomes Research Institute (PCORI). Dr. Bekelman leads research programs in cancer comparative-effectiveness and delivery system and payment reform, integrating methods from the fields of epidemiology, clinical trials, health economics, and public policy.

Marc L. Berger (MD, Johns Hopkins University School of Medicine) is Vice President of Real World Data and Analytics in Global Health and Value at Pfizer Inc. He has coauthored more than 100 publications in the fields of outcomes research, health economics, health policy, and data analysis.

Howard G. Birnbaum (PhD, Economics, Harvard University) is a Principal at Analysis Group, Inc. He specializes in real-world comparative-effectiveness and models regarding the budget impacts and societal costs and benefits of pharmaceutical/biologic products and clinical interventions.

Michael W. Bonney (BA Economics, Bates College) is a Pharmaceutical Executive. Throughout his career, he has specialized in delivering impressive results for shareholders and important therapies to patients while simultaneously creating a dynamic company culture and building a talented, diverse leadership team. He most recently served as CEO of Cubist Pharmaceuticals Inc. He has also been employed at Biogen and Zeneca Pharmaceuticals.

Kalipso Chalkidou (MD, University of Athens; PhD, Newcastle University) is the Director of the Global Health and Development Group at the Institute of Global Health Innovation, Imperial College London. She helps governments build technical and institutional capacity for improving the value for money of their healthcare investment.

Joshua P. Cohen (PhD, Economics, University of Amsterdam) is an Associate Professor at the Tufts Center for the Study of Drug Development (Tufts University School of Medicine). He specializes in pricing and reimbursement of biopharmaceuticals and prescription drug policy in the private and public sectors.

Nick Dadson (PhD, Economics, Simon Fraser University) is an Economist at Analysis Group, Inc. He specializes in econometric analysis and data science, particularly in health, finance, and litigation.

Joseph A. DiMasi (PhD, Economics, Boston College) is a Research Associate Professor and Director of Economic Analysis of the Tufts Center for the Study of Drug Development at Tufts University School of Medicine. He specializes in the cost, risk, and length of new drug development, innovation incentives for biopharmaceutical R&D, and changes in the structure and performance of the biopharmaceutical industry.

Rebecca E. Dittrich (BS, Cornell University) is a JD candidate at Georgetown University Law Center and an MPH Candidate at Johns Hopkins Bloomberg School of Public Health. Her research focuses on the intersection of law and health policy.

Dean T. Eurich (PhD, BSP) is a Canada Research Chair in Prevention and Management of Chronic Diseases and Professor of the School of Public Health, University of Alberta, Canada. His research focuses on health outcomes and comparative-effectiveness research in patients with chronic diseases.

Marcy Fitz-Randolph (DO, University of North Texas Health Science Center at Fort Worth; MPH, University of Iowa) is Manager of Research Options at PatientsLikeMe. She specializes in patient-centered research, patient engagement in condition management, and patient, provider, and caregiver narratives in patient care.

Christian Frois (PhD, Economics, MIT) is a Vice President at Analysis Group, Inc. He specializes in the development of innovative pricing, market access, and evidence strategies for biopharmaceutical companies.

Mary E. Gately (JD, Harvard Law School) is a partner in the Washington DC office of DLA Piper US LLP where she is the co-head of the DC litigation practice. Her practice includes reputation management and defamation, products liability, consumer fraud law, complex civil litigation and class actions.

Paul E. Greenberg (MS, MIT Sloan School of Management; MA, University of Western Ontario) is the Director of Analysis Group's Health Care Practice. He specializes in health economics, both in health outcomes research and in complex business litigation.

Jens Grueger (PhD, Mathematical Statistics and Theoretical Medicine, TU Dortmund University) is a Vice President and Head of Global Pricing and Market Access at Roche Pharmaceuticals based in Basel, Switzerland. He leads a department of pricing specialists and health economists responsible for developing and implementing global pricing, reimbursement, and market access strategies for the company.

James P. Harnett (PharmD, Ernest Mario School of Pharmacy, Rutgers University; MS, Weill Cornell Graduate School of Medical Sciences) is a Senior Director in Real World Data and Analytics in Global Health and Value at Pfizer Inc. He specializes in rapid cycle and advanced analysis of Real World Data, the use of data visualization tools, health economics, epidemiology, and research partnerships.

Stephen Hippler (MD, FACP) is the Chief Clinical Officer of OSF Healthcare and Clinical Assistant Professor of Medicine at UICOMP. His focus is on creating high reliability and excellence in outcomes for the patients they serve.

Anna Hung (PharmD, MS, University of Maryland School of Pharmacy) is a PhD candidate at the University of Maryland School of Pharmacy. Her interests include pharmacoeconomics, comparative-effectiveness research, and patient-centered outcomes research.

Rahul Jain (PhD, Economics, SUNY–Buffalo) is Research Manager at HealthCore Inc. (Anthem Company). He specializes in health services research, health economics and outcomes research, and pharmacoeconomics/comparative-effectiveness research.

Kenneth I. Kaitin (PhD, Pharmacology, University of Rochester) is a Professor of Medicine and Director of the Tufts Center for the Study of Drug Development at Tufts University School of Medicine. He specializes in pharmaceutical R&D efficiency, strategic alliances, and academic-industry partnerships.

Paul Kalb (MD, Boston University School of Medicine; JD, Yale School of Law) is the Head of Sidley Austin LLP's national healthcare practice and serves as a coleader of its life sciences practice. Paul brings an uncommon clinical perspective, grounded in his experience as an attending physician at the Memorial Sloan-Kettering Cancer Center, to bear in his legal work for a wide range of healthcare clients.

Isao Kamae (MD, Kobe University; MPH, DrPH, Harvard School of Public Health) is a Professor at the University of Tokyo. He specializes in pharmacoeconomics, health technology assessment, and public policy, with a focus in value-based pricing methodologies.

Marie-Hélène Lafeuille (MA, Economics, and BSc, Economics, Sherbrooke University) is a Senior Economist at Analysis Group, Inc. She specializes in health economics and outcomes research.

Patrick Lefebvre (MA, Economics, and BSc, Mathematics, Laval University) is a Managing Principal at Analysis Group, Inc. He specializes in the application of biostatistics and economics of health outcomes research.

Genia Long (MPP, Harvard University, Kennedy School of Government; SB, Massachusetts Institute of Technology) is a Senior Advisor at Analysis Group, Inc., where she focuses on the business economics of innovation in the life sciences, including the effects of changes in policy, law and regulation, science and technology, and competition on the medical device and biopharmaceutical industries.

Rifaiyat Mahbub (BA, Economics, Wellesley College) is a Senior Program Associate at Results for Development (R4D), where she works on projects in the Health Analytics team. She specializes in tracking resources for health and assessing donor policies for health in low- and middle-income countries.

Cinzia Metallo (PhD, Tufts University School of Medicine; MS, University of Tennessee, Knoxville) is part of the healthcare practice at Analysis Group, Inc., where she supports health economics and outcomes research projects involving pharmaceuticals and medical technologies.

C. Daniel Mullins (PhD, Duke University) is Professor and Chair of the Pharmaceutical Health Services Research Department and Executive Director of The PATIENTS Program at the University of Maryland School of Pharmacy as well as co-editor in chief for *Value in Health*. His research and teaching focus on comparative-effectiveness and cost-effectiveness research, patient-centered outcomes research, and health disparities research.

Dave Nellesen (PhD, Biology, University of California, San Diego; MBA, University of California, Berkeley) is a Vice President at Analysis Group, Inc. He specializes in developing clinical and economic evidence of the value of new health technologies to support reimbursement.

Eberechukwu Onukwugha (PhD, Economics, Virginia Tech) is an Associate Professor in the Department of Pharmaceutical Health Services Research at the University of Maryland School of Pharmacy and Executive Director of the Pharmaceutical Research Computing center at UM School of Pharmacy. She specializes in pharmacoeconomics and health outcomes research with an emphasis on health disparities.

Catherine Tak Piech (MBA, Boston University Graduate School of Management; Health Policy Certificate, Jefferson School of Population Health) is Vice President of Health Economics and Outcomes Research at Janssen Scientific Affairs within the Janssen Pharmaceutical Companies of Johnson & Johnson. She has over 30 years of experience in pharmaceutical pricing, value assessment, and economic research and currently leads a multidisciplinary team of scientists engaged in the development and dissemination of comparative research studies to support informed decision-making by healthcare stakeholders.

Crystal T. Pike (MBA, MIT Sloan School of Management) is a Vice President at Analysis Group, Inc. She specializes in the analysis of complex healthcare data, in the context of business strategy, risk management, and litigation.

Lisa Pinheiro (MFin, Princeton University) is a Vice President at Analysis Group, Inc. She specializes in quantitative analysis and statistical modeling, which she applies to various practice areas, including biostatistics, finance, intellectual property, and antitrust.

Richard Price (MA, Johns Hopkins School of Advanced International Studies; BA, Johns Hopkins University) is Senior Vice President of Health Care Delivery Policy at the Advanced Medical Technology Association (AdvaMed), a trade organization of medical technology innovators. He focuses on Medicare delivery reform issues related to ACOs and bundled payment programs, Medicare's payments for inpatient hospital care, and competitive bidding for durable medical equipment.

Louis F. Rossiter (PhD, Economics, University of North Carolina, Chapel Hill) is Research Professor of Thomas Jefferson Program in Public Policy at the College of William and Mary, Williamsburg, VA. He specializes in hospital and health care antitrust, payment policy for Medicare and Medicaid, and issues affecting the pharmaceutical industry.

Jimmy Royer (PhD, Economics, Université Laval) is a Vice President at Analysis Group, Inc. He specializes in economics and statistics. He has applied his analytical and econometric skills in data science, predictive modeling, and health economics.

Paul D. Schmitt (PhD, University of Maryland; JD, Georgetown University; MA, History and English, University of Maryland; BA, English, La Salle University) is an Associate in the Amsterdam office of DLA Piper and prior to that was an Associate in the Washington D.C. office of DLA Piper. He has represented and advised clients on the False Claims Act, the Foreign Corrupt Practices Act and other white collar matters, the Foreign Sovereign Immunities Act, commercial speech issues, and commercial contractual disputes.

James E. Signorovitch (PhD, Biostatics, Harvard University) is a Vice President at Analysis Group, Inc. He specializes in health economics and outcomes research, including real-world evidence generation and evidence synthesis methods development.

Sean Tiggelaar (MSc, Health Technology Assessment, and BSc, Health Sciences, University of Alberta and Brock University) is an Economist at Analysis Group, Inc. He specializes in health outcomes research and health economic modeling.

Arjun Vasan (MPP, Harvard Kennedy School) is Senior International Health Economist at the US Department of Treasury. He was Senior Program Officer at Results for Development (R4D), working across projects in the Health Analytics team. He specializes in assessing donor policies for health in low and middle income countries.

Francis Vekeman (MA, Economics, Laval University) was employed at Analysis Group, Inc. at the time of this interview. He specializes in health economics and health policy, as well as antitrust and product liability matters.

Richard J. Willke (PhD, Economics, Johns Hopkins University) is Chief Science Officer at the International Society for Pharmacoeconomics and Outcomes Research. He has worked extensively in economic analysis of clinical trials and comparative-effectiveness analysis using real-world data.

Jipan Xie (PhD, University of North Carolina, Chapel Hill; MD, Peking Union Medical College) is Vice President at Analysis Group, Inc. She has extensive expertise in health economics and outcomes research as well as economic evaluations and health technology assessment in the USA and abroad.

Jie Zhang (PhD, Statistics, North Carolina State University) is an Executive Director at Novartis. He has led health economics and outcomes research and global HTA submissions for multiple disease areas, including oncology, rare diseases, and respiratory medicine.

Part I
Introduction

Chapter 1
Introducing *Decision Making in a World of Comparative Effectiveness Research*

Howard G. Birnbaum and Paul E. Greenberg

Abstract Although comparative effectiveness research (CER) is now a widely-established discipline, less attention has been paid to how it has been used in practice by decision makers in the life sciences industry and the extent to which it has satisfied the needs of stakeholders. While CER often provides important insights, valuable information sought by the ultimate "consumers" of this research often is not available to support the decisions they must make. This *Introduction* sets the context for the remaining chapters in *Decision Making in a World of Comparative Effectiveness*. Contributions by senior industry executives, key opinion leaders, accomplished researchers, and leading attorneys involved in dispute resolution consider issues important to life sciences industry decision makers both in the US and globally.

In the past decade, there has been a worldwide evolution in health outcomes research, most prominently in the role that health reform in the United States has provided to comparative effectiveness research (CER). Demands for insights from new types of data sources and analytically rigorous methodologies have moved beyond those that were common even a few years ago and that we considered several years back in a special issue of *PharmacoEconomics* [1]. Consequently, there is a burgeoning, rapidly evolving literature focused on research findings and methods to compare the effects of one medical treatment (e.g., prescription drug, device) versus another. While CER has become a driver of studies primarily using observational data in real-world settings, clinical trials have evolved, too, as in using pragmatic trial methods. Such research and methodological discussions consider the direct evidence of intervention A vs. B using a wide variety of existing data sources (e.g., retrospective data from third-party insurance claims and electronic medical records, patient surveys, and clinical-trial data). In addition, a significant methodological advance now allows for indirect comparison of intervention A and B where each was itself compared to a placebo in a clinical trial. The extent to which these

H.G. Birnbaum • P.E. Greenberg (✉)
Analysis Group, Inc., Boston, MA, USA
e-mail: howard.birnbaum@analysisgroup.com; paul.greenberg@analysisgroup.com

© Springer Nature Singapore Pte Ltd. 2017
H.G. Birnbaum, P.E. Greenberg (eds.), *Decision Making in a World of Comparative Effectiveness Research*, DOI 10.1007/978-981-10-3262-2_1

different approaches and the availability of "big data" satisfy the evolving needs of stakeholders is the focus of the chapters in this book.

As long-time observers and contributors to this literature covering a broad cross section of the health care sector, we have had numerous discussions with the sponsors of CER about how best to structure research in ways that would be most valuable to them. We have been impressed by the frequent observation that the CER that life sciences sponsors have supported often has not fully met their underlying needs. While often providing some insights, valuable information sought by the ultimate "consumers" of this research often is not widely available to support the decisions they must make. Such topics could include: the contribution of an innovative therapy; the value of a new intervention versus existing products; the extent to which new types of data are appropriate sources of information; how the marketplace might react in terms of formulary acceptance or pricing; resolving litigation disputes; and regulatory considerations in the life sciences industry.

From the inception of our effort to assemble material for this book, we envisioned that this volume of contributed chapters would provide a practical guide to the analysis and interpretation of CER that is aimed at "users" and "decision makers" rather than "doers" (i.e., researchers). Our goal throughout has been to assemble material that, while useful to the hands-on research community, would focus on bigger-picture issues rather than technical considerations. We invited a wide range of industry decision makers and other thought leaders to make contributions to this book. We sought out thought leaders in the pharmaceutical and biotechnology industry as well as device manufacturers and also attorneys involved in life sciences litigation. We included authors from both the United States and other countries that commission or conduct CER. Their participation, we hope, has resulted in a book that is relevant to a wide range of life sciences users and decision makers, worldwide. As is made clear in the chapters that follow, stakeholders and researchers offer a divergent range of views on approaches to decision making in the early twenty-first century.

A constant theme in these chapters is the search to optimize the use of information. Questions considered cover a wide range of topics including:

- How does one use information effectively?
- How does one balance the costs of gathering data and evidence before, as well as after, approval to enter the market?
- Where there is less than optimal data available, what research would convince regulators and payers to move forward with approval and coverage?

These are among the topics that are considered in this book. Addressing these and the other relevant questions is a dynamic process involving a continuum of knowledge and interactions among stakeholders and researchers. We hope that this book contributes to this ongoing and evolving process.

Reference

1. Birnbaum Howard G, Greenberg PE (2010) Comparative effectiveness special issue. Pharmacoeconomics 28(10):789–798

Chapter 2
Perspectives of Comparative Effectiveness Research from the World of Decision Making

Dave Nellesen, Howard G. Birnbaum, and Paul E. Greenberg

Abstract Interest in comparative effectiveness research (CER) and reliance on CER for treatment selection and resource allocation decisions continues to grow. However, the goals of CER (i.e., to inform decision making by providing evidence on the effectiveness and consequences of treatment options) are often difficult to attain, and the pace of change in conducting and disseminating the results of CER has been frustratingly slow. This introduction to *Decision Making in a World of Comparative Effectiveness Research* surveys the wide range of viewpoints that reflect on various challenges facing both researchers conducting CER and healthcare decision makers. As a whole, the vision for CER expressed in this book is one that remains optimistic in achieving its goals, where real-world studies and big data analyses are applied to meet pragmatic needs of clinicians, patients, and policymakers.

2.1 Initial Thoughts on the Evolving Conception of CER

Interest in comparative effectiveness research (CER) and reliance on CER for treatment selection and resource allocation decisions continues to grow. However, the goals of CER (i.e., to inform decision making by providing evidence on the effectiveness and consequences of alternative treatment options) are often difficult to attain, and the pace of change in conducting and disseminating the results of CER has been frustratingly slow. This book provides a range of viewpoints that reflect upon various challenges facing both researchers conducting CER and healthcare decision makers. Looking back to the ideas expressed in a special issue of *Pharmacoeconomics* published in 2010, a vision for CER expressed in this book is distinct from CER in its original formulations over the past decade, one that remains optimistic in achieving its goals, where real-world studies and big data analyses are applied to meet pragmatic needs of clinicians, patients, and policymakers.

D. Nellesen (✉)
Analysis Group, Inc., Menlo Park, CA, USA
e-mail: Dave.Nellesen@analysisgroup.com

H.G. Birnbaum • P.E. Greenberg
Analysis Group, Inc., Boston, MA, USA

© Springer Nature Singapore Pte Ltd. 2017
H.G. Birnbaum, P.E. Greenberg (eds.), *Decision Making in a World of Comparative Effectiveness Research*, DOI 10.1007/978-981-10-3262-2_2

Early definitions of CER, many associated with heightened attention and funding from the American Recovery and Reinvestment Act (ARRA) of 2009 and the establishment of the Patient-Centered Outcomes Research Institute (PCORI), were generally optimistic and suggestive of large prospective studies that would definitively establish the best course of treatment [1–3]. Recent definitions are more operational than theoretical, describing CER as a tool for a specific purpose, namely, to provide the specific evidence needed by clinicians, patients, policymakers, health plans, and other payers to make specific treatment and resource allocation decisions [4, 5].

In its original conception, CER would provide evidence to evaluate the comparative clinical and cost-effectiveness of alternative interventions, which would include long-term follow-up assessing patient-centered outcomes, and (eventually) would assess whether or not this evidence actually changes clinical practice. One example that arguably meets this very high standard is the widely disseminated study of the comparative effectiveness of percutaneous coronary intervention (PCI) versus drug therapy alone in patients with stable angina in the COURAGE study [6, 7]. In this study, a modest symptom benefit of PCI relative to drug therapy was observed, but no evidence was found for improved survival or reduction in the rate of myocardial infarction, an absence of effect that persisted over 15 years of follow-up [8]. The observed clinical benefit meant that PCI demonstrated comparative clinical effectiveness, while the small magnitude of the benefit and its incremental cost meant that it was not cost effective at typically accepted thresholds [9]. Moreover, the effect of this new evidence was shown to have a real impact on clinical practice [10]. The full potential of CER is exemplified by studies such as COURAGE (and its successor, ISCHEMIA) [11] and by several others [12–14]; however, the number of clear-cut success stories for large centralized comparative clinical trials has arguably been modest, especially considering the vast number of treatment and resource allocation decisions where evidence is sorely needed.

An emerging conception of CER is quite distinct from the approach taken in these large, broadly inclusive clinical trials. Real-world, observational data often provide evidence of comparative effectiveness, potentially from a range of sources designed to meet the needs of different decision makers, and thus can provide incremental evidence, often supportive or correlative rather than independently definitive. Widely accepted definitions of real-world evidence are broader than the original scope of CER, including electronic health records and medical chart reviews, administrative data, and surveys [15]. While prospective randomized controlled trials are undoubtedly the gold standard for establishing safety and efficacy, observational data sources have the potential to meet pressing needs of decision makers: outcomes observed in real-world populations are typically broadly generalizable, may capture relevant outcomes (both clinical and economic), and, perhaps most importantly, observational data are widely accessible, such that relevant evidence can be generated on a time scale of months instead of years. The need for better generalizability of evidence reflects the current state of evidence for many interventions, particularly for drug treatments. It is widely recognized that participants in clinical trials submitted to the Food and Drug Administration (FDA) for regulatory approval are younger and have fewer comorbid conditions than patients in the general population [16–18]. Although there are widely understood limitations to observational studies including selection bias and unobserved confounders, oftentimes real-world

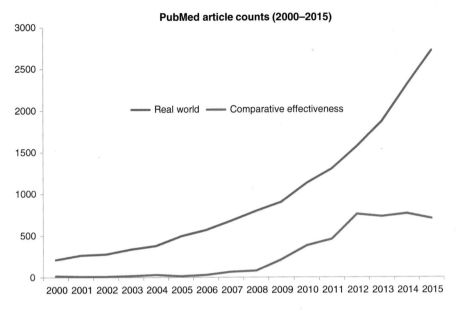

Fig. 2.1 PubMed citation analysis: comparative effectiveness research versus real-world studies (Source: Analysis Group research. The NCBI PubMed database was searched using the terms "real world" and "comparative effectiveness" in title and abstract fields on 9/16/2016)

observational studies are the most appropriate means to validate and expand upon the results of registration clinical trials, in particular to assess whether benefits observed in registration studies are generalizable to real-world populations.

Publication analysis provides some support for such an evolving conception of CER. PubMed citation analysis shows a leveling off in the number of publications that mention CER following the ARRA (after about 2012), while interest in "real-world" studies as measured by publication counts continues to accelerate (Fig. 2.1).

The number of comparative effectiveness clinical trials has also peaked (Fig. 2.2), with roughly the same number of registered CER studies initiated in 2015 as earlier in the decade. One interpretation of this data is that the demand for evidence provided by CER continues to grow, while the capacity to execute prospective clinical trials that assess comparative effectiveness remains relatively limited.

The implication for researchers is that there is a growing reliance on real-world observational studies to supply evidence of comparative effectiveness to meet the needs of decision makers. Powerful, traditional CER study designs (large, broadly inclusive prospective interventional studies comparing two alternative interventions) may find use primarily for "big picture" questions, typically highly prevalent conditions that attract academic interest and substantial government funding. While prospective interventional studies undoubtedly provide valuable data with superior internal validity, the future of CER will reflect the accessibility of observational study designs and the growing power of large databases used for observational studies (aka, big data). In the words of one contributor to this book, the marriage of big data and CER represents the potential to revolutionize decision making and realize the goal of what the Institute of Medicine (IOM) describes as a "learning healthcare system" [19] (Chap. 8 by Berger).

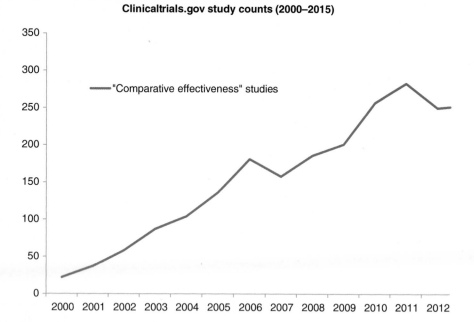

Fig. 2.2 Clinical trial analysis of comparative effectiveness research studies (Source: Analysis Group research. Clinicaltrials.gov was searched using the term "comparative effectiveness" on 9/16/2016)

2.2 A Survey of Key Themes Covered in this Book

Following is an outline of some of the main questions addressed by multiple authors in this book.

2.2.1 How Are Demands for Evidence of Comparative Effectiveness Changing? What Strategies Are Available to Manufacturers?

A consistent theme across several chapters of this book is the growing demand for evidence of both clinical and cost-effectiveness from payers. Conducting studies in payers' own plan populations is one response to challenges that evidence from registrational clinical studies may not be generalizable. In one example, discussed in an interview with Mike Bonney [20] (Chap. 3 by Bonney), a manufacturer provided study drug to support a CER study in several integrated health systems. A second example, cited by Cohen et al. [21] (Chap. 6 by Cohen et al.), was the well-publicized collaboration between AstraZeneca and WellPoint to conduct CER in plan populations. While such collaborations offer the potential to confirm the value of new technologies, the authors note that they also come with some risk, not only in terms of the cost to conduct this research but also the uncertainty of findings. Another approach for manufacturers to engage more directly with payers is to apply the findings from CER in pricing and contracting, described by Frois et al. [22] (Chap. 7 by Frois, Grueger). Although CER evidence has not always been available as an input to pricing negotiations, this has the potential to change, and the authors discuss the use of CER in pricing and highlight several novel strategies for application of CER evidence, in particular using indirect treatment comparison (ITC) analysis prior to approval or for flexible pricing approaches including as personalized reimbursement models and managed-entry agreements, all of which may rely on evidence of comparative effectiveness rather than simply upon evidence of efficacy from pivotal studies.

2.2.2 How Have Methods for Conducting CER Changed? What Is the Emerging Picture?

While the availability and quality of real-world data for conducting CER studies is increasing, several authors argue that clear quality standards and transparency in the analysis of data are still lacking [19] (Chap. 8 by Berger), [23] (Chap. 9 by Willke). Standards for conducting and reporting CER and real-world studies are available, but may not be widely observed. Authors cite several standards for observational studies, including research guidelines by the Agency for Healthcare Research and Quality (AHRQ) and the International Society for Pharmacoeconomics and Outcomes Research (ISPOR), standards for reporting observational studies from the Strengthening the Reporting of Observational Studies in Epidemiology (STROBE) initiative, and for evaluating CER and observational studies from the Good Research for Comparative Effectiveness (GRACE) initiative, all of which lend greater credibility and usefulness in decision making [24] (Chap. 5 by Piech); [23] (Chap. 9 by Willke). Greater acceptance of standards for conducting real-world studies was noted as one means to address skepticism of CER data by payers in the future [20] (Chap. 3 by Bonney).

A wealth of methods and a wide range of study designs are available for conducting CER, some relatively novel and likely to gain prominence in the future with the anticipated growth in availability and interconnectedness of big data resources. Several emerging methods are discussed, including ITC analysis, n-of-1 trials, meta-analyses of real-world studies, and various approaches for analysis of real-world observational data [23] (Chap. 9 by Willke). In particular, methods for conducting ITC have become more sophisticated, with statistical advances that have improved accuracy and expanded application of this method, including matching-adjusted indirect comparisons and simulated treatment comparisons that adjust individual patient data from one trial to match patients from a published comparator trial [25] (Chap. 20 by Signorovitch). New approaches to data analysis also may help to address the heterogeneity of treatment effect (HTE), a key limitation of CER wherein a well-designed study may find no benefit on average, while a specific subset of patients might still gain from the treatment. Onukwhuga et al. [26] (Chap. 19 by Onukwhuga et al.) discuss the value of big data to address limitations of HTE, specifically the technical approach for marginal analysis of patient subgroups, a method with considerable power in the context of larger sample sizes and longer follow-up often characteristic of larger more interconnected data sources. Finally, the advent of big data analyses in CER may open the door to algorithmic or machine learning approaches to decision making. Pinheiro et al. [27] (Chap. 21 by Pinheiro et al.) discuss the potential for application of machine learning to assess comparative effectiveness, especially in areas with large data sets and high complexity, such as analysis of genomic data and biomedical images.

2.2.3   What Has Been the Impact of CER on Public Payers in the United States (US) Compared with Other National Health Systems? How Is this Changing?

Globally, growing interest and reliance on CER is due in part to pressures from publicly funded health systems, namely, to contain costs and maximize value for money spent. Rossiter et al. [28] (Chap. 10 by Rossiter) discuss the use of CER by several public payer decision-making bodies in the USA, including Medicare/ MEDCAC, state Medicaid, and the Veterans Administration. CER has the potential to shape coverage and payment policies for each of these public payers, yet each entity uses CER evidence in different ways subject to varying legislative authority, oftentimes limited resources to carry out decision making, and, in some cases, political mandates that influence their deliberations. Decision-making bodies from national health systems across the world all have varying authority over use and implementation of decisions based on CER. Several authors in this book review the use of CER by national health systems, with detailed discussion of the importance of CER in health technology assessment (HTA) in various systems including the United Kingdom's (UK) National Institute for Health and Care Excellence (NICE), the Swedish Council on Technology Assessment in Health Care (SBU), France's National Authority for Health (HAS), Germany's Institute for Quality and Efficiency in Health Care/Federal Joint Committee (IQWiG/G-BA), and the Canadian Agency for Drugs and Technologies in Health (CADTH). Xie et al. [29] (Chap. 12 by Xie) present case studies from England (NICE), Japan ("Chu-i-kyo"), and the Global Fund (an international organization funding treatment in developing countries), providing examples of the use of CER in the context of HTA for decision-making bodies with different levels of sophistication and decision-making authority. Lefebvre et al. [30] (Chap. 13 by Lefebvre) provide an overview of the process conducted by CADTH for review of evidence of clinical and cost-effectiveness, a process that includes multiple decision makers at national and regional levels. Each of these cases provides examples of a slow but steady trend toward more sophisticated evidence review, in many cases, with growing transparency and centralization of decision making.

2.2.4 What Are Current Constraints or Barriers to the Use of CER, and How Could this Change?

Authors have identified a number of barriers that could stand in the way of wider adoption of CER. A fundamental limitation of CER pointed out by several authors is HTE, the principle that while CER is useful in measuring average effects, findings for an aggregate study population may not actually apply to any one patient, and thus application of CER findings by payers in the context of coverage decisions can lead to suboptimal treatment. Piech et al. [24] (Chap. 5 by Piech) discuss HTE and the potential application (or misapplication) of CER in the context of Medicare Part D protected drug classes, arguing that CER is most useful at a population level for determining policy, while at the individual level differences in response and individual patient preferences should factor into treatment decisions. Better alignment with patient preferences is important to test hypotheses that truly reflect what is most meaningful to patients. In their chapter on the design of pragmatic clinical trials, Hung et al. [31] (Chap. 11 by Hung) point out the relatively limited inclusion of patient perspective in designing CER studies, arguing for greater engagement with patients throughout the entire process of CER, from planning and conducting a study to disseminating results.

Several authors have pointed out barriers in the dissemination of the results of CER studies. Gately et al. [32] (Chap. 15 by Gately) and Kalb [33] (Chap. 16 by Kalb) both discuss limitations on communication of CER findings that relate to uses of a drug product outside of the approved labeling. In particular, pharmaceutical manufacturers have been reluctant to disseminate CER health economic findings due to perceived legal risk. Statutory limitations, in particular Section 114 of the Food and Drug Administration Modernization Act of 1997 (FDAMA 114), that pertain to communication of healthcare economic information suffer from a lack of official guidance regarding the scope of the statute, but are clearly less restrictive than FDA standards for dissemination of scientific information. However, based on recent developments in the law, restrictions on affirmative communication of this evidence may diminish, such that manufacturers may have greater freedom to communicate CER findings outside the product label, with perceived risk alleviated either through proposed amendments to FDAMA 114 or judicial challenges to limitations on commercial speech directed at payers and other sophisticated and experienced consumers of CER findings. Finally, Asche [34] (Chap. 17, Asche) points out that even when CER findings are disseminated through peer-reviewed publication, this information is not consistently or systematically integrated into clinical practice guidelines, an issue compounded by a lack of adherence to the many (thousands) of clinical guidelines of varying quality and often conflicting recommendations.

2.3 Closing Thoughts

The 21st Century Cures Act was approved with bipartisan support in late 2016. This legislation includes substantial changes to regulatory processes for drugs and medical devices, including provisions for the use of real-world observational data to support initial FDA approval and to satisfy post-marketing commitments. Clearly, there is a movement toward a more pragmatic application of real-world data to meet the needs of both regulators and of payers. CER, as it was conceived nearly a decade ago, was framed in a context where the availability of real-world data and methods for analysis were relatively limited, and thus the place of prospective clinical trials to assess comparative effectiveness was central. In the next decade, as the availability and interconnectedness of real-world data increases, comparisons of clinical and cost-effectiveness using real-world data are likely to become the standard, with growing influence on the commercial success of new health technologies.

References

1. CBO (2007) Research on the comparative effectiveness of medical treatments: issues and options for an expanded federal role (Rep. No. 2975). Congressional Budget Office, Washington https://www.cbo.gov/publication/41655. Accessed 11/13/2016
2. HHS/PCORI (2009) Federal coordinating council for comparative effectiveness research. Department of Health and Human Services, Washington, DC
3. IOM. (2009) Initial National priorities for comparative effectiveness research. http://www.nationalacademies.org/hmd/reports/2009/comparativeeffectivenessresearchpriorities.aspx. Accessed 11/13/2016
4. AHRQ. (2015) What is Comparative Effectiveness Research. US Department of Health & Human Services. Agency for Healthcare Research and Quality. http://effectivehealthcare.ahrq.gov/index.cfm/what-is-comparative-effectiveness-research1/. Accessed 11/13/2016
5. GAO. (2015) Comparative Effectiveness Research: HHS Needs to Strengthen Dissemination and Data-Capacity-Building Efforts. United States Government Accountability Office. http://www.gao.gov/assets/670/668804.pdf. Accessed 11/13/2016
6. Boden WE, O'Rourke RA, Teo KK, Hartigan PM, Maron DJ, Kostuk WJ, Knudtson M, Dada M, Casperson P, Harris CL, Chaitman BR, Shaw L, Gosselin G, Nawaz S, Title LM, Gau G, Blaustein AS, Booth DC, Bates ER, Spertus JA, Berman DS, Mancini GB, Weintraub WS (2007) Optimal medical therapy with or without PCI for stable coronary disease. N Engl J Med 356(15):1503–1516. doi:10.1056/NEJMoa070829
7. Weintraub WS, Boden WE, Zhang Z, Kolm P, Zhang Z, Spertus JA, Hartigan P, Veledar E, Jurkovitz C, Bowen J, Maron DJ, O'Rourke R, Dada M, Teo KK, Goeree R, Barnett PG (2008) Cost-effectiveness of percutaneous coronary intervention in optimally treated stable coronary patients. Circ Cardiovasc Qual Outcomes 1(1):12–20. doi:10.1161/circoutcomes.108.798462

8. Sedlis SP, Hartigan PM, Teo KK, Maron DJ, Spertus JA, Mancini GB, Kostuk W, Chaitman BR, Berman D, Lorin JD, Dada M, Weintraub WS, Boden WE (2015) Effect of PCI on long-term survival in patients with stable ischemic heart disease. N Engl J Med 373(20):1937–1946. doi:10.1056/NEJMoa1505532

9. Weintraub WS, Spertus JA, Kolm P, Maron DJ, Zhang Z, Jurkovitz C, Zhang W, Hartigan PM, Lewis C, Veledar E, Bowen J, Dunbar SB, Deaton C, Kaufman S, O'Rourke RA, Goeree R, Barnett PG, Teo KK, Boden WE, Mancini GB (2008) Effect of PCI on quality of life in patients with stable coronary disease. N Engl J Med 359(7):677–687. doi:10.1056/NEJMoa072771

10. Howard DH, Shen YC (2014) Trends in PCI volume after negative results from the COURAGE trial. Health Serv Res 49(1):153–170. doi:10.1111/1475-6773.12082

11. ISCHEMIA Trial (2016) Clinicaltrials.gov Database. International Study of Comparative Health Effectiveness With Medical and Invasive Approaches (ISCHEMIA) (2016). https://clinicaltrials.gov/ct2/show/NCT01471522. Accessed 11/13/2016

12. Rossouw JE, Anderson GL, Prentice RL, LaCroix AZ, Kooperberg C, Stefanick ML, Jackson RD, Beresford SA, Howard BV, Johnson KC, Kotchen JM, Ockene J (2002) Risks and benefits of estrogen plus progestin in healthy postmenopausal women: principal results From the Women's Health Initiative randomized controlled trial. JAMA 288(3):321–333

13. Hersh AL, Stefanick ML, Stafford RS (2004) National use of postmenopausal hormone therapy: annual trends and response to recent evidence. JAMA 291(1):47–53. doi:10.1001/jama.291.1.47

14. Chlebowski RT, Anderson GL, Sarto GE, Haque R, Runowicz CD, Aragaki AK, Thomson CA, Howard BV, Wactawski-Wende J, Chen C, Rohan TE, Simon MS, Reed SD, Manson JE (2016) Continuous combined estrogen plus progestin and endometrial cancer: the women's health initiative randomized trial. J Natl Cancer Inst 108(3). doi:10.1093/jnci/djv350

15. Garrison LP Jr, Neumann PJ, Erickson P, Marshall D, Mullins CD (2007) Using real-world data for coverage and payment decisions: the ISPOR Real-World Data Task Force report. Value Health (The Journal of the International Society for Pharmacoeconomics and Outcomes Research) 10(5):326–335. doi:10.1111/j.1524-4733.2007.00186.x

16. Hutchins LF, Unger JM, Crowley JJ, Coltman CAJ, Albain KS (1999) Underrepresentation of patients 65 years of age or older in cancer-treatment trials. N Engl J Med 341(27):2061–2067. doi:10.1056/NEJM199912303412706 http://www.nejm.org/doi/full/10.1056/NEJM199912303412706

17. Wasilewski J, Polonski L, Lekston A, Osadnik T, Regula R, Bujak K, Kurek A (2015) Who is eligible for randomized trials? A comparison between the exclusion criteria defined by the ISCHEMIA trial and 3102 real-world patients with stable coronary artery disease undergoing stent implantation in a single cardiology center. Trials 16:411. doi:10.1186/s13063-015-0934-4

18. Downing NS, Shah ND, Neiman JH, Aminawung JA, Krumholz HM, Ross JS (2016) Participation of the elderly, women, and minorities in pivotal trials supporting 2011–2013 U.S. Food and Drug Administration approvals. Trials 17:199. doi:10.1186/s13063-016-1322-4

19. Berger ML, Harnett J (2017) Are real-world data and evidence good enough to inform health care and health policy decision making? In: Birnbaum HG, Greenberg PE (eds) Decision making in a world of comparative effectiveness research. Springer, Singapore

20. Bonney M, Birnbaum HG (2017) Perspectives on the use of comparative effectiveness research by life sciences executives: an interview with Mike Bonney. In: Birnbaum HG, Greenberg PE (eds) Decision making in a world of comparative effectiveness research. Springer, Singapore

21. Cohen J, DiMasi J, Kaitin K (2017) Impact of comparative effectiveness research on drug development strategy and innovation. In: Birnbaum HG, Greenberg PE (eds) Decision making in a world of comparative effectiveness research. Springer, Singapore

22. Frois C, Gueger J (2017) Pricing of pharmaceuticals: current trends and outlook, and the role of comparative effectiveness research. In: Birnbaum HG, Greenberg PE (eds) Decision making in a world of comparative effectiveness research. Springer, Singapore

23. Willke RJ (2017) Translating CER evidence to real-world decision-making: some practical considerations. In: Birnbaum HG, Greenberg PE (eds) Decision making in a world of comparative effectiveness research. Springer, Singapore

24. Piech CT, Lefebvre P, Pike CT (2017) Comparative effectiveness research: a pharmaceutical industry perspective on outlook, dilemmas, and controversies. In: Birnbaum HG, Greenberg PE (eds) Decision making in a world of comparative effectiveness research. Springer, Singapore
25. Signorovitch J, Zhang J (2017) Indirect comparisons: a brief history and a practical look forward. In: Birnbaum HG, Greenberg PE (eds) Decision making in a world of comparative effectiveness research. Springer, Singapore
26. Onukwugha E, Jain R, Albarmawi H (2017) Evidence generation using big data: challenges and opportunities. In: Birnbaum HG, Greenberg PE (eds) Decision making in a world of comparative effectiveness research. Springer, Singapore
27. Dadson N, Pinheiro L, Royer J (2017) Decision-making with machine learning in our modern, Data-Rich Healthcare Industry. In: Birnbaum HG, Greenberg PE (eds) Decision making in a world of comparative effectiveness research. Springer, Singapore
28. Rossiter LF (2017) Decision making by public payers. In: Birnbaum HG, Greenberg PE (eds) Decision making in a world of comparative effectiveness research. Springer, Singapore
29. Xie J, Chalkidou K, Kamae I, Dittrich RE, Mahbub R, Vasan A, Metallo C (2017) Policy Considerations: Ex-U.S. Payers and Regulators. In: Birnbaum HG, Greenberg PE (eds) Decision making in a world of comparative effectiveness research. Springer, Singapore
30. Lefebvre P, Lafeuille MH, Tiggelaar S (2017) Perspectives on the Common Drug Review (CDR) Process at the Canadian Agency for Drugs and Technologies in Health (CADTH). In: Birnbaum HG, Greenberg PE (eds) Decision making in a world of comparative effectiveness. Springer, Singapore
31. Hung A, Baas C, Bekelman J, Fitz-Randolph M, Mullins CD (2017) Patient and stakeholder engagement in designing pragmatic clinical trials. In: Birnbaum HG, Greenberg PE (eds) Decision making in a world of comparative effectiveness research. Springer, Singapore
32. Gately ME, Schmitt PD (2017) Challenges and opportunities in the dissemination of comparative effective-ness research information to Physicians and payors: a legal perspective. In: Birnbaum HG, Greenberg PE (eds) Decision making in a world of comparative effectiveness research. Springer, Singapore
33. Kalb P, Greenberg PE, Pike C (2017) Legal considerations in a world of comparative effectiveness research. In: Birnbaum HG, Greenberg PE (eds) Decision making in a world of comparative effectiveness research. Springer, Singapore
34. Asche CV, Hippler S, Eurich D (2017) Application of comparative effectiveness research to promote adherence to clinical practice guidelines. In: Birnbaum HG, Greenberg PE (eds) Decision making in a world of comparative effectiveness research. Springer, Singapore

Part II
The Future of Comparative Effectiveness Research for Evidence Developers: Perspectives from Pharmaceutical Decision Makers

Chapter 3
Perspectives on the Use of Comparative Effectiveness Research by Life Sciences Executives: An Interview with Mike Bonney

Michael W. Bonney and Howard G. Birnbaum

Abstract Mr. Michael W. Bonney, B.A., Partner at Third Rock Ventures, LLC and former CEO of Cubist Pharmaceuticals Inc., discusses his personal experience with using comparative effectiveness research in decision making while presiding over Cubist. He also considers how the introduction of the Affordable Care Act and increasing interest in the budget impact of the pricing of recent pharmaceuticals are changing health care decision making and impacting manufacturer decision making.

Howard Birnbaum In the context of your several decades of leadership in the pharmacological/biotechnology space, how do you relate to comparative effectiveness research (CER) and how have you used it in your work?

Mike Bonney When I hear comparative effectiveness research, I think about the *Affordable Care Act* [1] and PCORI (the Patient-Centered Outcomes Research Institute), the independent nonprofit, nongovernmental institute that was set up though authorization from Congress in 2010—though in the United States (USA) currently there's no legislation that is intended to actually do comparative effectiveness research (which is comparing one drug to another drug in certain clinical situations, etc.).

At Cubist we did use economic data to help pharmacists and clinicians understand who the best patient for our drug might be. This is more about real-world use of drugs and how comparative effectiveness influences usage in the real-world drug market.

M.W. Bonney (✉)
Third Rock Ventures, Boston, MA, USA
e-mail: mwbonney@outlook.com

H.G. Birnbaum
Analysis Group, Inc., Boston, MA, USA
e-mail: howard.birnbaum@analysisgroup.com

© Springer Nature Singapore Pte Ltd. 2017
H.G. Birnbaum, P.E. Greenberg (eds.), *Decision Making in a World of Comparative Effectiveness Research*, DOI 10.1007/978-981-10-3262-2_3

Birnbaum Please tell us a little bit about your personal experience with CER. How have you used CER or seen it used in your various roles over the last several decades?

Bonney Given that our business at Cubist was focused primarily on antibiotics that were administered by IV, a critical component of the information required by hospital decision makers was the economic impact of the product in their own patient populations. In this context we often partnered with hospitals to develop data about length of stay in ICUs, total days in hospitals, and likelihood of readmission within 30 days of discharge to assess the economic impact of use of our products.

Birnbaum I understand that, as a decision maker, the academic approach taken by PCORI and other policy and research centers established under the *Affordable Care Act*, you feel that it is not directly relatable to your work, particularly because it's not real world and it's got a variety of methodological baggage to it. But you do use real-world CER analyses in your work; I'd like to hear more about how you use CER in that context, if you could elaborate on that.

Bonney Personally, I think that we have a very fractionated US health care system, where risk is allocated to different provider groups in almost arbitrary fashion, and so in our experience at Cubist, we chose to focus on the hospital environment, where you have in some respect the advantage of a self-contained decision-making body and reimbursement system, whether it's funded through public or private insurance.

If you could simply demonstrate to the providers who take risk in the form of capitated payment structures (which in our case would be the hospital) that, in the appropriate patient, a new drug that would be inherently more expensive than a generic drug but could confer economic benefits to the system—and you can identify patients where it's appropriate to use the drug—then you have just used CER with a real-life application.

What are the challenges? I think that the whole system in the USA faces, with respect to real-world data, the fractionation of this risk—and we see that with the Sovaldi pricing battle with the big mail-order pharmacies, Medco and Express Scripts. There's little doubt in my mind that Sovaldi is going to save money in the system over the long term, because there'll be fewer liver transplants and there'll be fewer liver cancers that will need to be treated—both of which are vastly more expensive than even the $80,000 cost of 12 weeks of Sovaldi.

The problem is that the risk for the cost-benefit of the medicine has been taken largely by the pharmacy benefit managers (PBMs) and they get no benefits from a reduction in hospitalization costs from reduced transplants and the reduced oncology cost 10 years down the line in a patient's life. That all accrues to the insurer who's hired the PBM to administer the pharmacy benefit. So, we've got misaligned incentives even though I think you could prove that Sovaldi is going to be good for the health care system economically over the long term. The fractionation of the system causes dissonance in the adoption of these medicines, because the people who are taking the risk to pay for these medicines, i.e., the PBMs, don't get any of the benefits, because the benefit is going to show up in lower hospitalization costs, etc., and that accrues to the insurer who's hired the PBM to administer the pharmacy benefit. Lower hospitalization costs include such things as the cost of occupying a bed per day times the number of days, or the cost differential of a bed per day in the ICU versus the cost per day of a bed in a general ward. It also includes, in the US the costs of caring for patients who are readmitted to the hospital within thirty days of discharge with the same diagnosis.

Birnbaum   So, as you think about that analogy, can you apply that lesson to past decision making while at Cubist?

Bonney   No. We had the advantage that, because the vast majority of hospital care in the USA is paid for under a capitated system—for example, Harvard Pilgrim would contract with a group of hospitals to provide emergency services to the Harvard Payer Group (known as the insured) and that would be done based on a diagnostic-related grouping on a capitated basis, we did not have to worry about multi-payer reimbursement issues. So, in that capitated system, if you have primary *S. aureus* bacteremia you are going to get referred to hospital X. That system drives the hospital to look for efficiency but, frankly, it also stifles innovation.

What we were able to prove was that when you find the right patient—and those patients generally had comorbid conditions that included some level of immuno-compromise, or elderly, diabetic, renal failure, etc.—and they have *S. aureus* bacteremia, if you used our drug, you could cut ICU days, hospital days, and readmissions within 30 days of discharge and that would confer an economic benefit to the hospital. So, it's a pretty straightforward analysis because there was a self-contained unit. They bore all the risks for delivering emergent care.

Birnbaum   So, how did you prove that? Did you actually do an economic analysis using real-world data?

Bonney   We contracted with a couple of integrated health systems (IHS) with sophisticated data collection capabilities and we funded and set up several studies. We provided the drug and some support for the analysis, and they did their own work and concluded that these patient types do better on Cubicin than they do on the generic vancomycin.

And so every time an IHS patient has this type of infection with this set of background information, they were going to use this drug, even though it cost a lot more to acquire the drug than other competitors', because we more than make it up in reduced ICU days, reduced full days in the hospital and fewer readmissions within 30 days of diagnosis (which are not paid for by Medicare and other payers).

Birnbaum   I can understand how they would have therefore been the early adopters of your drug; how did you then translate that to other payers and how did you handle their situations and possibly their skepticism?

Bonney   There was some skepticism, though the results of our collaborative studies were published in prominent peer-reviewed health economics journals. One of the big challenges that we faced was that many systems—whether it'd be a payer, an

ACO, or a hospital—would discount the data on the basis of their patient population, saying "it's just different from our patient population."

I think this is one of the big challenges with real-world data: many of the payers in the USA want you to prove it in their population. This is just not economically feasible. Instead of $200 a vial [of our medication] now, we would have to charge $400 a vial to pay for multiple studies for the many [thousand] different payers in the USA.

So, that's one of the big challenges and I think there is a need to develop some kind of standard. The challenge is that nobody has been able to do so over the 30 plus years I've been in the industry. No one has been able to come up with an executable standard that is generally accepted as valid by the payers.

Birnbaum So, it sounds that, in some ways, the challenges remained pretty much what they were 20 years ago. Is that correct?

Bonney I think that we're improving, but very gradually, and the improvement is more in terms of the availability of data. Twenty years ago, we couldn't go to a payer group and expect them to have the sophisticated data systems that allow us to pull data on the outcomes. We can now do that with a variety of payer groups, ACO groups, etc. So, we at least have access to the data. But the methodological challenges that you have with real-world data allow nonparticipating entities to criticize the data in ways that often result in no movement.

We even had one circumstance where a major academic medical center that had proven in their patient population real savings of over $22,000 per appropriate patient with the use of our drug came under economic pressure, hired consultants (who did not have a medical background), and limited the use of the drug based just on acquisition costs.

Birnbaum You mentioned improvements in the availability of data over the past 20 years. Have there been any changes in the methodologies used or in the research questions asked?

Bonney Again, this depends on the setting. Greater availability of data and better, though far from perfect, cost data in hospitals allows better, more accurate assessments within those closed systems. Given my focus on the hospital environment over the last 13 years while at Cubist, I'm not really qualified to discuss methodological improvements for outpatient therapies, which often rely on calculations of cost per quality-adjusted life years (QALYs) to demonstrate value.

Birnbaum It sounds as if the fractured health care system that you described earlier is a system of silos within silos. You seem to indicate that even within a health care system that uses CER to determine best use of resources, there are these silos. Is that true also? I wonder if within the pharmaceutical industry itself, within Cubist for example, did you have people pushing you in different directions in the context of this real-world data effort?

Bonney Yes, there was some trepidation. For example, when we started to expand internationally, we hired a series of health economists to help us frame the issue for new drug development—what the economic argument would be for reimbursement in European Union countries. And what we found very quickly was that the requirements vary country by country, that you couldn't really develop a single set of data, do one study or go into one group of hospitals, collect data and then cut it in ways that each country wanted. They wanted fundamentally different data. So, we had to operate at very high level and then delegate down at the country level to obtain that collection of data for that country, because each country, just like each payer in the USA, wanted data from their own systems, not from somebody else's system. So, we had to develop a central group overseeing all data management and then we had local groups for country-specific reimbursement issues. To me, it was more a reflection of what a customer world was like than any inherent silo type of organization.

Birnbaum Sometimes when I present comparative effectiveness research to my Health Economics and Outcomes Research (HEOR) clients, they tell me that that's all well and good but that within their own organization, when they communicate with their pricing and access group, sometimes there are marketing considerations that govern price-setting for a drug. Do you have any observations to that effect?

Bonney We at Cubist did not have that particular set of challenges, but I am very familiar with the issues that exist within particularly very large companies. The other thing to keep in mind is that the Cubist business model was somewhat different: we were dealing largely with hospital drugs, which, again, are a little bit easier to collect valid data on than it is in the outpatient world.

And in the outpatient world, one of the other considerations is "what's the value of a QALY?" Nobody seems to be able to agree. The British economists and policy makers have done as good a job of standardizing as anyone, but most drugs don't meet what they set as an accessible cost per QALY. And so, if you're a marketing person and you invest the energy to collect this sort of data and then your customers reject it, I think the rational response would be to question why you would spend the money collecting that data, since this will be rejected anyway. Why not just argue the point that the patient should have access to innovation, instead. I think it's a little bit of an ostrich issue in some of these companies. They have to deal with this because that's how virtually every government in the world—except for the USA— is set up to reimburse drugs, as the primary payer, and they're the ones who make the rules.

The problem is that it's very hard to integrate economic considerations early in clinical development. So, even if you can, and you structure data selection in your phase 3 clinical trials, most governments or payers will reject phase 3 data because it's not indicative of practice in the real world. So now, you're trying to launch a drug, get reimbursement, but you've got 3 more years of data collection to satisfy these reimbursement requirements (cost for QALY requirements) and you chewed up 30% of your remaining patent life before you even know what the answer is. So, it's a bit of a conundrum from that standpoint, too. The fallback position is to mobilize the patient, so if I can't collect the data in a timely fashion, then I must convince the government that this drug is worth X dollars. That means I'm going to need to mobilize the patients to put pressure on the government to make the new drug available to them because it could save or change their life to improve the patient's ability to engage in activities of daily life, become more independent etc. So, historically, I would say that pharma companies have done a better job of the latter than the former. But I think they're going to have to get better entrenched in the former going forward.

Birnbaum So you've mentioned QALYs several times and that certainly is relevant in the European context. Our experience has been that a lot of the US payers have not been responsive to that metric. Do you think that that consideration should not receive increased priority in the future?

Bonney I think that there could be movement in that direction in the USA, but, historically, payers have not cared about it, because most insurers prior to the *Affordable Care Act* had full membership turnover about every 2–3 years. And the QALY calculation is a whole life calculation. And so, it was easy for them to argue that the long-term benefit is going to accrue to somebody else (another insurer). It's not going to accrue to us. So we have no incentive for the patient to get this initially more expensive drug.

With the *Affordable Care Act* and the mandate that everyone should have insurance and take it with them, payers can no longer hide behind the fact that the benefit is going to improve beyond the tenure of their membership. So they're beginning to say, "Okay. We understand that if this is truly an advance and has a meaningful impact on the quality of life and though the cost of quality would be a little bit higher with this drug, we will pay for it."

And I think that in the orphan-drug world, people are paying $300,000 or $400,000 or more a year for orphan drugs. The concern they have is that there are 25 million Americans with orphan diseases. And if all of them are taking one or two drugs at $300,000 or $400,000 a year, that's unsustainable; that's going to blow up this system economically, as well.

Birnbaum So, as you think about other situations, what type of advice would you give other decision makers in the pharma industry about this situation going forward over the next few years?

Bonney I think that, for any disease that you're developing a drug for, you have to understand the total economic cost and per-activity cost early in clinical development, so that you're making rational decisions about which product you're advancing. And if your hypothesis for a product is that this will improve a patient's life but it's not going to save any money in the system, it's unlikely that you're going to get the levels of reimbursement that you will need to justify the expense of proceeding with that development—unless one is dealing with a very small and/or very vocal patient population.

So you've got to start earlier—and what we were trying to do at Cubist was to at least develop an equation based on phase 2 clinical trial data that we could then test in phase 3 to see if we were offsetting enough of the other costs in order to be able to price the drug in a way that would generate an acceptable return on investment. For example, concurrent with phase 2 clinical trials we would dive into the costs of care of the patient populations we were testing the drug in. That would allow us to see where the big cost drivers were. Then, based on the results we saw in the phase 2, we could estimate whether we would be able to offset enough of the other costs of that patient population, in order to be able to define a place for the new drug where the price point necessary to continue finding and developing new drugs could be achieved. We were just integrating this into our decision making when Cubist was acquired.

If you're a payer, I think the fundamental question is that you need to get together and develop standards, just like we have standards for safety and efficacy. I don't see a better solution in the near term other than establishing an agreed set of international standards that articulates the worth of cost for QALY. Then you actually have a target you can choose drugs for and you can start talking data and assessing what are your interventions that can get over that hurdle or not!

Take the example of the PCSK9s: the FDA approved two PCSK9s this summer. And just this morning a report from the Boston-based Institute for Clinical and Economic Review has said that they're priced too high—they're sold at $14,000 a piece a year—and that their economic value is really only around $4,000 a year [2]. So the question is how they've arrived at that conclusion, what assumptions are they making? What data are they using to arrive at $4,000 versus the $14,000 that the two companies priced the drug at? I haven't been able to find that information this morning. And the company has been asked to comment, and their reaction is exactly what I just said, which is "we have no idea how they came up with this number. If they're willing to open the kimono and show us, maybe we agree and maybe we don't, but at least we'll have a place to start the conversation. We can't start the conversation at, 'It's $4,000.' 'No, it's $14,000.'"

Part of the rationale that's articulated here is that, at the cost of $14,000 per year, the patient we should give this drug to can't get it because of the ultimate impact on the budget. That's a legitimate concern, but that's also a Sovaldi legacy, in my opinion. The companies, I think, have the obligation then to figure out not just the cost-effectiveness of a drug but also its budget impact.

Birnbaum I have one more question. I wonder if you could comment on the following common scenario: sometimes when we present our budget impact model or comparative effectiveness research and a pharmacological company (the study sponsor) presents this to payers, the payer says, "Well, you know, you have conflicting interests when you do that analysis. I don't know if I trust you." What's the proper response to that? How do you defend against this and say, "Look. We hired these independent people. They did a fair job." And then the payer says, "Yes. But I've never seen an analysis that doesn't show that a drug is cost-effective."

Bonney Even the Congressional Budget Office has determined in the 8 years following the introduction of Medicare part D that improved drug access, and adherence lowers overall health care costs and they've changed the scoring on legislation, but they haven't made this public. I think there are a couple of responses. One is you have to be transparent with the methodology—what data did you collect, how did you assemble that data, what assumptions did you make, and how did you run your statistical programs? I think the second thing is, as I said earlier, a standard, an international standard, or even a US standard that is done with the appropriate methodology and so forth would also be helpful.

Most drugs actually are pretty cost-effective. While this is not true for *all* of them, many of them are. But that's a very bitter pill to swallow for a payer whose margins are getting crushed. And it's fair criticism to say that, basically, nobody ever produces negative data. However, if the customer doesn't view HEOR data as valid, why would biopharma companies spend the money to create the data?

References

1. Patient Protection and Affordable Care Act (2010) HR 3590. 111th Congress, vol 2. USA
2. Institute for Clinical and Economic Review (ICER) (2015) PCSK9 inhibitor therapies for high cholesterol: effectiveness, value, and value-based price benchmarks

Chapter 4
Perspectives on the Use of Comparative Effectiveness Research by Life Sciences Executives: An Interview with a Senior Executive at an International Life Science Company

Senior Life Sciences Executive, Howard G. Birnbaum, and Francis Vekeman

Abstract A Senior Executive at a major pharmaceutical company discusses his experience in managing a health economics and outcomes global department. He considers factors which have impacted his decision-making, how comparative effectiveness research evidence use and methodology has evolved over time, and how such evidence is strategically used within a multinational company and in interactions with diverse payers/payer systems and regulatory agencies across the world.

Howard Birnbaum You have been involved in health economics and outcome research (HEOR) using real-world data for decades. How have the types of questions and the uses of this type of research evolved?

Senior Executive Yes, I have been involved in this area for over 20 years. I joined my firm as the head of the US HEOR group, which gives me responsibilities for all of the health economics research, including comparative effectiveness research (CER).

Over the years, we've certainly seen changes in terms of the evolution of CER, particularly regarding the methodologies. It's now more oriented toward the standard use of electronic medical records (EMRs), which is where it needed to be. When CER first started out, it was not quite as rigorous, and there were a lot of shortcuts. There's much greater transparency to it now. I think the studies that are being done now are of a significantly higher caliber than what was done before.

Senior Life Sciences Executive
Anonymous Identity, Employed at Multinational Pharmaceutical Company,
Boston, MA, USA

H.G. Birnbaum (✉) • F. Vekeman
Analysis Group, Inc., Boston, MA, USA
e-mail: howard.birnbaum@analysisgroup.com

© Springer Nature Singapore Pte Ltd. 2017
H.G. Birnbaum, P.E. Greenberg (eds.), *Decision Making in a World of
Comparative Effectiveness Research*, DOI 10.1007/978-981-10-3262-2_4

There are still things that have yet to be changed, however. CER, at its heart, is a *comparative* science. Payers routinely demand a comparison of our drug versus the current market comparator. However, you still do not see a great deal of that, particularly as companies balance what the demands are in the marketplace versus those required by regulatory agencies. So, while we have grown the field in terms of methodology, we still have not gotten to where we want or need to be, as both the science and marketplace demands grow and change.

Birnbaum What were the big steps in the evolution of comparative effectiveness in terms of data and methodology in general?

Senior Executive Regarding the evolution of the tools and methodology in CER over time, now there are far more data providers than there were before. In addition to the traditional data vendors, there are a growing number of provider groups and other consulting organizations that you wouldn't normally think of in this space, so there are a growing number of sources of data. And, of course, the expansion into EMRs is adding to that even more. This leads to the questions that people are facing now: how do I get my arms around all of these different data sources that are out there? What's the best group to partner with? What's the best source of data? We all know that there is not one magic source of data that will be able to answer all research questions; it doesn't exist. The better approach is in making sure that you have partnered with the right data providers or customers in order to be able to get the data you need for the questions that are being asked. And that's a process that's getting more and more complicated.

And then, on the other side, is the question of which analytic tools to choose. So people nowadays are struggling, first on the data side—what data sets do we need?—but also on the analytic side, what analytic tools do I bring in to be able to appropriately conduct these analyses? And, which ones will give me the greatest range of use? Because probably in most large pharmaceutical companies research of this nature is not conducted in one group. So it's a huge explosion—not only of data that's available and the choices that need to be made but also in terms of the analytical tools and the methods behind them such that it makes it very complicated in terms of trying to get to the bottom of this. One of the key questions that I assembled for an upcoming advisory board meeting is what are other people doing in other companies, to the degree that they're willing to share, relative to those two issues—in terms of choosing data sets and in terms of choosing analytic tools.

Birnbaum So the methodologies have improved, but have improved methodologies led to improved decisions? Has it changed anything from a decision maker's perspective?

Senior Executive There have been instances of documented improvements in decisions, and we have had our successes. At my company, we had a drug that we were putting in front of a major HMO, and they said, "We would like to do some research on the economic value of this drug versus the alternatives that we're currently using."

We jointly conducted that study with them. The findings were presented at the latest scientific congress and in a published manuscript, and the HMO actually said,

"The findings of this are great, and we actually believe in these findings. Because of that, we are going to implement changes to our policy and mandate that this product is used earlier in the treatment paradigm, because we see how cost effective it is." So we do see examples of clear wins. One of the downsides is that when you have one of those clear wins in the company, every other brand manager says, "I want the same thing for my product."

And that's one of the shortcomings that persists despite all the years of health economics and CER and all the education on this topic that we've tried to do. There still seems to be a shortfall in the understanding from the commercial side. They feel that we can generate a value story for our product, irrespective of the product and its attributes and also of the other players on the market. However, we know there are only two ways to make something more cost-effective: (1) you do another clinical trial or a real-world study and get better data or (2) you adjust your price. So there's definitely room for growth in terms of the understanding within the company of the role CER plays. CER is not a marketing tool. It has rigor and science behind it. The results can be used to support marketing messages, but they're fundamentally based on a scientific process.

Birnbaum As the head of HEOR, you are both a consumer and a producer of CER. You commissioned these studies and so you use the information they provide, and then you also are producing these results for other folks both inside your company, on the commercial side, and also externally. How do you think about CER studies as a decision maker, when you actually go about commissioning a study? How far down the road can you extrapolate?

Senior Executive A lot of people start early on by asking, "What are we going to get within the clinical trial program?" I'll give you an example.

We should set up a routine process now where we conduct advisory boards on compounds that are in phase 2 clinical trials and for which the phase 3 program has not been finalized. We should bring in the clinical leader for the compound, present to them the preliminary results of ongoing phase 2 studies and the phase 3 plan, and ask, "What do you feel our needs are going to be in terms of value messages or data—both from a clinical and an economic standpoint—when this compound hits the market? And are the plans that are in place for phase 3 going to hit the mark?" Based on this informed conversation, we should then go back to the R&D team to say, "You need to change some of the things that you are doing." For example, we had an advisory board meeting on a product, and we had major phase 3 studies that were going to be done—one on the use of the drug to prevent a chronic condition, the other one on the use of the drug for when patients had the chronic condition. All of the advisers said, "Currently, we don't treat these patients in the pre-condition stage. However, if you could show data on the value of treating patients prior to the onset of the condition, that would be fantastic. A new market would open up for you. You need to focus on a value message there, and you could have an opportunity to capture some of that market." That information should then be pushed back to the clinical research leaders, to take back to the R&D team, and to, hopefully, make positive decisions.

The other issue that is sometimes seen in this business is that a clinical lead in a company may throw up the target product profile (TPP) and say that the goal is to make superior claims that a drug is better than a competitor's drugs. They will show the phase 3 program, and the advisors will sometimes put their hands up and say, "You have a TPP in which you want to claim superiority, yet your entire phase 3 program has a non-inferiority design. How are you going to accomplish this?" And the answer sometimes is, "Well, we're just hoping it will come out." There is still a lot of educational outreach that needs to be done. Currently, we try to push back into the earlier stages to say, "You need to make changes back here in the R&D process, so that we have data available at time of launch, but also so we'll get a really good sense of the target profile." Because if we're going to do a real-world study, we want to look to the clinical experts to ask what is the probability of success. My company, like probably many of its peers (other life sciences companies), has mandated complete transparency—not only on the clinical trial side but also on the health economics side. So a registry of all studies that are being done has been implemented, containing all comparative sector research studies that are being done within clinical development and real-world and any other aspects of drug development.

Birnbaum So that's a big change?

Senior Executive Yes, it's a big change. The company now mandates that all studies, including retrospective studies, must be posted to the FDA.gov or similar sites—not only the study protocol but also the findings.

Birnbaum Do you think that this mandate has changed how you do business? How you commission studies?

Senior Executive I don't think so, but it might in the future. It makes people think a little more carefully about what they're doing. They can't say, "Well let me try something and see if it works, and if it doesn't, go on to something else." For the most part, other companies are similar, in that our end goal with respect to the studies is publication. We want to publish these results, and there's an obligation to publish these results, which was a policy in place prior to the implementation of this company mandate. In that sense, transparency is pretty much woven into the fabric of how we do business right now, so I don't see a great deal of change left to do.

Birnbaum So what do you do when you get a non-result?

Senior Executive That's published as well. Now we sit down with the commercial sector folks more than we used to in the past, in order to let them know that, if we proceed with a study and we get negative results, we will still be obligated to publish this. So when we go forward, if we get a negative finding and we publish the findings, they don't feel left out of the conversation.

Birnbaum I would have to think that from a payer perspective, for example, this approach would improve their reception. If payers see that you are publishing both positive and negative results, letting the chips fall where they may, this gives you more credibility. Is that part of it?

Senior Executive I think the intent behind this new transparency initiative is to give us more credibility, but there's still a healthy skepticism from our customers around the studies we do. We have to do a combination of both internal and external partner studies—both from a credibility standpoint and because sometimes we can't access the appropriate data sets that we need. For example, if you're working with the SEER database for oncology, you have to work through an academic center, because pharmaceutical companies cannot be granted access.

Francis Vekeman You raised the issue that different groups within a pharmaceutical company may be involved in the CER process: how do you coordinate all of that? And is there a risk that one group does one methodology and then comes up with result A, while another group does methodology B, and comes up with a different result, and then you have a problem? Does that create an issue for decision-making?

Senior Executive Yes. A big discussion that's going on within our global real-life evidence framework is regarding that very issue—different results coming from work that's done internally within a company versus something that's done externally by an academic center, by another company, etc. but there's even the factor of internal differences. To manage this issue, we have mandated a centralized internal global review process. So if you're going to do any effectiveness research, any real-world study, you submit a concept into our trials storage system. The concept is reviewed by the group and then you submit a full protocol, which looks very much like a clinical trial protocol. Then the concept goes up for approval and then the study itself goes to a review committee. Observational studies in our company are defined pretty much how we defined comparative effectiveness research at the beginning of our discussion here—everything real-world and of a retrospective/prospective nature.

So here there is a process to look at what's being done and hopefully catch some of the similarities, but more importantly to look for potential discrepancies between similar research questions. We have had discussions within the global real-life evidence team where we said, "This group will get a study from the US that says we're going to look at the benefits of drug X's use in treating a particular condition and the reduction in a condition-related issue." And that study will go through the review committee and get approved. Well, somebody in another country may submit a similar study, and it will be reviewed and approved. Now, reviewers will look at these two protocols and say, "Are these studies in any way going to conflict with each other?" That's the piece that needs to be looked at. And one of the primary discussions being had is "If we're looking at analytic tools, can we get to a point where the base data can be utilized for all research done in the company?"

For example, we're now looking at patients in one of our target disease states. We will conduct an analysis to understand the demographics of these patients, current treatment patterns, resource use and costs, prevalence and incidence, natural history of the disease, current treatment pathways, background rates of adverse events in the population, etc. Then all of that data will be shared with all of the research groups across the company. So we will all be working off of the basic foundation of the same data and the same results, in order to try to bring some level of cohesion across the different internal research groups.

Vekeman So with the availability of new data and new tools, have you seen a shift in the type of questions that are answered or are the same type of questions being answered, but in a better way, with better data?

Senior Executive No. I think that in many ways the questions are the same. Payers are still asking "What is the incremental value that your product brings and what am I paying for it?" I think that payers, in general, have moved from this earlier under-standing in the field that drugs had to demonstrate cost savings to one of incremental value. However, I think that there is far greater awareness of the fact that real-world data is not just a market access issue but that it spans work in epidemiology or drug safety. Companies recognize that compounds have a risk/benefit balance and that it is important to track and generate data on both sides of that equation. Payers realize that it's about an incremental benefit and their question now is "what is the cost of that incremental benefit? What am I getting for that?" Then the questions they're asking are "I need to have that data and I need to have it on a comparative basis, so I need to know what your drug is giving me that I'm not getting from the current drug on the market."

We actually had an advisory board recently for one of our recently launched compounds. When we had the advisory board, the advisers clearly said, "When you launch this drug on the market, you have an existing compound that is probably the key compound on the market that is used for treatment of the condition in question." And the script data show that to be true. It's about 60% of the market. They said, "You need to do a real-world study that compares your drug to this market leader and that shows that it has a better safety profile in the real world—similarly to what you saw in clinical studies. That will give you huge ground in terms of arguing why this product should be preferred over the market leader if the plans are going to consider a tiering approach." The marketing folks actually came back and said, "We need to do the study next year. We want to give you money for it. We need to do the study because the advisers said we needed to do it." And I was impressed with this, because there are very few marketing groups that want to do that type of a comparative study.

So, to conclude, I think the questions are the same. We're still asking the same questions, we're asking for the same data, and we're now in a market where we're forced in many instances to pay for a lot of these new compounds. We need to make decisions in terms of what's the value that these products bring, so that we can compare them, as well as knowing what's coming in the future—what data are you going to have on future compounds to document their benefit, where are they going to sit in the treatment paradigm, what population are they going to be used in, how often, and how much is it going to cost us—so that they don't have a Sovaldi-like drug that hits their system when they weren't anticipating it.

Birnbaum Since you mentioned Sovaldi, how has the reaction in the marketplace impacted the type of research that you end up doing?

Senior Executive It is actually an interesting subject. A lot of the payers that we talked to said, "If you look at the data, Sovaldi is a truly cost-effective drug. It really is. The problem is the budget impact. The pure budget impact associated with the product if we were to give it to everybody that required it. We simply can't handle the budget impact associated with that compound. We are not arguing whether it's cost effective or not, we already bought into that concept. We're just saying, on the budget side it has a significant impact."

Birnbaum So, in some sense, it is not a CER problem—that part has been answered—it's really "How do you pay for this?" And, "Who pays for it and when do you pay for it?" In other words, it's a payment question.

Senior Executive Right, which is not what we tend to address. If we have a Sovaldi-like compound that is clearly cost-effective, our job is done. Plans might think "It's great, but, you know, the price is prohibitive." Well, that's a separate discussion you need to have on the negotiation side. We demonstrated that the product deserves to be considered, now the issue the plans need to consider is "At what price? What can your budget handle?"

Birnbaum Do your colleagues within the life sciences community then take that change in the main question and turn that into a research request for you? For example, I can imagine this whole notion of the payers sharing the risk of the drug not being effective with the seller, in order to be able to financially afford this drug... Does that then filter back to you in terms of a request for some type of research?

Senior Executive Yes. And that's where you're seeing some of the new things into play in the industry such as risk-sharing agreements, where we're asked to document the value on the back end of it. Recently, a large pharmaceutical company competitor announced that they were going to introduce value-based pricing. If the drug works, they'll charge X. If the drug doesn't work, they will charge X minus Y. And that's going to catch on. One of the things that we have in this whole area that does comparative effectiveness research is the question of the value of a drug/therapy. Now, several initiatives have been launched to determine a fair price for a drug. And these value guidelines are now going to affect how a company sets the price of a drug. So the issue of the growing cost of therapies is being seriously debated, and companies can't just continue to keep charging what they want to charge without a documented cost-benefit (and risk-benefit) argument. We're seeing that there is an end to the clock here, and it may be closer than we think.

Vekeman We talked about changes in the type of data, in the methodologies. Do you feel pressure from the payers or regulators to move forward and advance in the methodologies and the data or it's just really the other way around? You've mentioned that the FDA is becoming progressively more interested in looking at the retrospective data from a safety perspective. So do you feel this kind of pressure from payers and regulators?

Senior Executive I think there is growing pressure in the sense that the payers now want to see these data to help them in their decision-making process. And, since they are skeptical about pharma-produced studies, we have looked at external partnerships with payers, where it makes sense to partner with an organization in order to be able to say, "We did the study with your patients. These are your findings. It's very hard to argue with them because there they are. If you're saying 'I don't believe in the study's results,' then you're saying you don't trust yourself." So we're always looking to increase credibility, because the payers are pushing us on credibility and transparency.

Here, we have a standing expert panel of health economists that we meet with quarterly to get their input on some of the major research projects that we're doing. When we come out with the publication of one of these research projects, we can say that this study was done and the methodologies were reviewed and confirmed by the following expert panel. It's like our "Good Housekeeping" seal of approval.

Payers are also asking for this information earlier: "We want to know what these data look like when the product hits the market." The Institute for Clinical and Economic Review (ICER) is another example—the recent report they did on the PCSK9 new compounds [1]. ICER is expected to produce more of those, and, when we discussed that with the payers, they said, "We're going to watch for that type of information." We're going to take a look at those reports because we know that payers are going to use these to gauge a compound's cost-effectiveness. So we're seeing mounting pressure because of the fact that payers are screaming about the price of drugs and the price increases after the first year of use. And that gets us to an interesting side issue, which is that if you do a comparative effectiveness study for the drug at the time of the launch, and you say, "My drug is cost effective," this calculation is very different from what you would see if you look 5 years down the road, when you have taken annual price increases of X and the pricing of your drug may now be double or triple the initial price. And that's where we see a lot of pushback from the customers. They've said "Since you launched your drug, you have tripled the price of it, and what new value have you given us along with that price?" When you have a new iPhone that comes out and it's more expensive, they say it's more expensive because it has these new features. But, for a drug, payers are pushing back and saying, "You increased the price of your drug, but the last time I checked the clinical data are exactly the same. There are no new data, so where's the value story in terms of that increase in price?" And that's something I think we're going to be pushed to answer more and more as an industry, to be able to justify some of these price increases.

Birnbaum So how does the CER process look from a global perspective? Most of the answers that you've given us so far are in the context of a US market. Do you find that globally things are different in terms of the questions asked? In terms of CER usage or otherwise, how different are they?

Senior Executive The questions asked are similar, but there's much more of a formal and rigorous demand for this information outside of the USA and a demand for cost-effectiveness for cost per quality-adjusted life year (QALY) that we certainly don't see in the USA. And that's where we see differences in terms of the needs of the European countries, for example, Germany, France, etc. They have developed systems to support their health technology assessment (HTA) submissions, which are very formal. So at The National Institute for Health and Care Excellence (NICE) in the United Kingdom (UK), it's a very formal submission, it's a very formal review process, and it's a very formal finding. This can be good or bad. And if you look at the UK marketplace and the success in terms of drugs listings, it's completely different from the US market. It's almost a worst-case scenario there; it's an even tougher environment from the European perspective. Tougher also in the sense that there are less real-world data.

We've looked at expanding part of the global real-life evidence initiative: what sources of data are there, ex-US? It's a very limited environment: the ability to in-license is virtually nonexistent, and partnerships can be very difficult. For example, the German government has their own data, and they look at them, but pharma cannot partner with them, so it's a closed system. So it makes it more difficult to do CER studies in Europe. The demands there are much more formal and rigorous, with the HTA bodies and the HTA review process, but the data to be able to do some of these real-world analyses are far more restricted than in the USA. So it creates a much more challenging situation for a company outside the USA.

Vekeman   Do you think that the more formal process ex-US is due to having more rigorous HTA processes or would you think it's simply because the payer structure basically is pretty much that of a single payer (the government), which allows for a centralized assessment?

Senior Executive   Yes. I think that's a big part of it. For example, Canada has a centralized review body, as do Australia, the UK, etc. And that's where you're seeing more of the restrictive, formalized requests and approach to reviews, and so on. In the USA, there's been ongoing debate of whether there will ever be a centralized review process. And some initially thought that the Patient-Centered Outcomes Research Institute (PCORI) was going to be that, and, while they will come out with some clinical findings, I don't think they're going to meet all the needs of the payers in that regard. Some people are now looking at ICER and saying they might be the US version of NICE. However, I don't know that we'll ever have a centralized process… I don't know that the payers will ever agree to one centralized body for a review process that they will look at and agree to undergo in a formal fashion. They may look at it and say, "We'll take the data under advisement," but I think they are very entrenched—particularly the big players (United, Kaiser)—in terms of conducting their own analysis and their ability to do so. These payers are now sitting on a wealth of data. They have the analytic capabilities to look at it. They have the in-house expertise to conduct research. They come back and say, "We can do our own analysis on compounds and determine whether they're cost effective or not. We're happy to partner with the industry, but we don't need to partner with you, and we don't need an external body to give us findings. We can look at it for ourselves and our own population."

So I don't think that the current multiple-payer system is going to change anytime soon. These other groups will just be another value-add—so if ICER comes out with a positive report on a product, payers will probably say, "Yes, but I'm going to ignore that." If they come up with a very negative report, saying the evaluated drugs are not worth the price, then that would be something payers might pick up and say, "I'll use this to augment my own decision and justify putting restrictions in place."

And some payers are actually doing that now. Florida Blue Cross mandated a new medical policy: for any new product that is going on the formulary, a mandatory step therapy is added. It doesn't matter what the product looks like. It doesn't matter what's its value. It's a mandatory step therapy until they've had a chance to assess it over a year and determine for themselves the value of the product. Therefore, this is mandatory step therapy for all new drugs, and that's that. No exceptions.

Birnbaum So evidence doesn't make any difference?

Senior Executive Evidence will always matter, but payers are arguing on the strength of the evidence relative to other agents that they are looking at. They're using it as a delay process: "We're going to put this in place, because then we can delay it for a year. At the end of the year, we'll see if the added benefit can justify the price for new compounds."

And that opens a separate issue—the issue of late-stage treatments for end-stage conditions. The late stages where you get significantly lower survival benefits, whether that's overall survival or progression-free survival, your cost-effectiveness story is a lot harder to tell. However, if your drug is aimed at the early-stage disease, that's a very different scenario than when you're treating end-stage disease. This quandary of looking at later-stage treatments underlines the need to have a different benchmark of what cost-effective is for these products. You are not going to have a cost per life-year gain of $30,000. It's simply not feasible for that stage of the disease. So how do you wrestle with this issue of these boundaries that have been in place for a number of years? The thresholds of $30–40,000, $50–60,000, and over $100,000?

Birnbaum In some sense, the marketplace, the pricing, and the dynamics with payers and the constraints that payers face are changing the questions that you have to ask, because if these new drugs that are now coming on to the marketplace are so expensive—especially in late-stage—that's in a way reframing what you research…. Maybe it's not changing the question, but it has to change the methodology and the types of approaches that you bring to them.

Senior Executive Yes, how do we look at new ways of doing CER? Do we look at total costs of care across the continuum? So that the late-stage treatment is put into the perspective of the overall cost of the patient, which means it doesn't stand out quite as much. You could argue that you're playing numbers games, but you can also argue that you're putting it into the perspective of natural incidence/prevalence…. There's an ethical debate around the question "Is it worth paying thousands of dollars for this person, so they can gain a limited amount of additional life?" Payers are looking at this question, because they're saying, "Do we even pay for this?" Do we now, as a society, say, "Late stage, forget (treating) it." The debate around that is going to be very significant.

Birnbaum It goes beyond economics?

Senior Executive Oh, yes. We've all had the discussions about costs per life-year gain, the value of a life here. Is the cost per life-year gain of greater value in a 15-year-old than it is in an 80-year-old retired person? We can sit here and argue that, but we all stayed completely away from it. We've always said that a life-year gain is a life-year gain.

Vekeman I was wondering: Where does personalized medicine fit into all of this? Do payers often ask if there is a specific subset of patients for whom this treatment makes the most sense? Is that something that has been coming up as well in recent times?

Senior Executive Yes, it has. It's something that's coming up and it's something that I've discussed internally in our company and I imagine my colleagues have had other similar discussions, such that when they say, "I don't like our cost-effectiveness number that we have for drug x, for example—a limited survival gain across all the patients in the trial. We could pull out that subset of patients that was on the upper end of our survival curve, and we could truly show that it was very cost effective in these people." There's no question that in that group it wouldn't be truly cost-effective, but at this point, we are talking outside of the CER and clinical science, and this is more around the economic aspect and marketing strategy of the company, where the company is saying "I'm niching myself in terms of that subpopulation," and you get to the whole bit of sales and guarantees to the market and analysts and all of that kind of stuff, which is outside of our remit. But yes, there are definitely certain subpopulations where you could say this is where the drug is truly cost-effective.

I'll give you one example—we were trying to work with our drug safety folks on a product where they said, "Basically, there's a certain subpopulation for which we've identified that if a patient has a failure with it, when they get a second product, they're 50% more likely to get a failure again." And we said, "Wouldn't it make sense to be able to identify those people that have this huge likelihood of a second failure, and go out and recommend that these people probably should not get the product?" Then, overall, you're making your compound more cost-effective. This whole issue of identifying patients who are going to be responsive is a big area where payers have growing interest. "If we can get down to a subpopulation where your drug really works," they say, "Your drug will truly be cost effective." And there's no question they will be.

Vekeman Yes, there's a tension between identifying the ideal population versus having a broader target population.

Senior Executive Yes. We could do that work now. We could go back to our clinical trials, and there's enough data in most of our clinical trials to be able to pull out a subpopulation and say, "Here's where the drug is truly cost effective." We can find where it really gave a huge survival advantage and it's truly cost-effective. When it gets to this issue of niching, the payers are continuing to push for it. We are in favor of personalized medicine when it comes to identifying patients that will truly be responsive to a product or, more importantly, those that won't be. And it gets to recent announcements on value-based pricing. We're not going to identify that subpopulation, but what we're going to do is say "When you hit patients in that subpopulation, we're going to charge you this much for the drug. When you hit patients outside of that target population, who don't do so well, then we're going to charge you this much (a lower amount)." And that essentially is that we're charging you more for those patients that truly get the value.

Vekeman Where do you see the next big step in terms of data and what is the piece that you find is missing currently in what's available out there for CER?

Senior Executive I don't know that there's anything missing. If you search, you can find the data you need to answer a question one way or another. There will be more and more data providers that will come out, particularly in the USA. I don't think you're ever going to see a consolidation. You are never going to see one centralized data repository in the USA like the more centralized data sources in Europe; that's not feasible. But the groups are getting bigger in terms of the number of patients they have in their databases. We are never going to have a magic database that will answer all our questions for us. We're always going to have to look into several different sources.

I'd like to see greater collaboration. The payers will also say this. If you get payers in a room, most of them will say, "We have the same goals as you do. Our end goal is to do what's best for the patient. Looking at that subpopulation that's going to get the greatest benefit is obviously one that is of greater economic value to us, but it's also of the greatest value to the patients. Our goals are not disjointed from what you, as an industry, have as goals. We're both for-profit organizations, so within the making of profit, we're both trying to target what is the best for the patients. And we should come together collaboratively more often to look at this issue."

That means both sides have to open up. There has been greater transparency, but there needs to be more. Both sides have to open up and be willing to accept some wins and some losses. The industry has to accept that if we partner with payers and do research with them, if we find that there are suitable subgroups in which a product is truly cost-effective, then we have to be willing to accept that finding, which may mean a smaller market, but in that market we would offer more of an advantage in terms of the value to the patient and, therefore, the value of the drug. Payers have to be willing to accept that if you truly show value in those subpopulations, then they should put those drugs in a first-line position and use them without any restrictions, and they should not base their decisions on the fact that the price of that drug is more than that of a generic. There is a lot of hesitation on both sides around not wanting to open up completely, but both sides have said that that is where we need to get to. We have to come together to do that. There is a whole host of issues in the way that probably will prevent that from ever happening, so these two groups will probably never completely come together. But that would be the ideal.

Reference

1. Institute for Clinical and Economic Review (ICER) (2015) PCSK9 inhibitor therapies for high cholesterol: effectiveness, value, and value-based price benchmarks. Available from: http://cepac.icer-review.org/wp-content/uploads/2015/04/PCSK9_Draft_Report_0908152.pdf. Accessed 23 Nov 2015

Chapter 5
Comparative Effectiveness Research: A Pharmaceutical Industry Perspective on Outlook, Dilemmas, and Controversies

Catherine Tak Piech, Patrick Lefebvre, and Crystal T. Pike

Abstract Comparative effectiveness research (CER) is the process of examining the performance of various medical treatments against each other. Many factors can affect how CER is performed and what conclusions are derived from any given study. Though the gold standard in health-care research remains the randomized controlled trial, CER studies that utilize real-world evidence can be valuable, as long as such studies are fit for purpose, have a rigorous methodology, and are easily interpretable and sources of bias have been minimized and disclosed. Though technological advances make it increasingly easier to perform sophisticated comparative analyses using ever-larger and increasingly detailed datasets, adoption and dissemination of CER by the US pharmaceutical industry has been surprisingly slow. This may be due to myriad of factors including regulatory and legal pitfalls, economic incentives, cultural influences, and public perceptions. Nevertheless, in an environment where both technological and economic pressures require smarter, less-resource intense ways of understanding the value and benefit of all existing treatments, CER has tremendous potential to improve decision making through its evidence-based approach to treatment choices. Alternative ways of conducting, interpreting, and disseminating CER should be a priority for the industry.

C.T. Piech (✉)
Janssen Scientific Affairs, LLC, A Subsidiary of Johnson & Johnson,
Titusville, NJ, USA
e-mail: cpiech@its.jnj.com

P. Lefebvre
Analysis Group, Inc., Montreal, QC, Canada

C.T. Pike
Analysis Group, Inc., Boston, MA, USA

© Springer Nature Singapore Pte Ltd. 2017
H.G. Birnbaum, P.E. Greenberg (eds.), *Decision Making in a World of Comparative Effectiveness Research*, DOI 10.1007/978-981-10-3262-2_5

5.1   Introduction

Knowing how the products or services we buy are compared in terms of characteristics, price, and performance is a fundamental step in making a purchasing choice
in any market. It is expected that buyers will research a product or service in order
to make comparisons among factors such as features, performance, price, terms,
and availability before coming to a final decision, especially for an important or
costly purchase. When a health issue arises and there is a need for treatment, it is a
reasonable expectation that one would ask for and consider information about the
comparative effects of the relevant treatment choices. The availability of comparative effectiveness research (CER) on treatment choices is a natural expectation that
has strong, immediate, and universal appeal for its simplicity. However, experience
working within the drug industry reveals insight into just how challenging this reasonable expectation is and the complexity of the issues that surround CER. By recognizing and understanding these obstacles, stakeholders (i.e., patients, clinicians,
payers, manufacturers, and the greater public) can enter discussions informed, pav-
ing the way for an effective partnership that works toward providing the types of
CER that decision makers expect.

5.2   What Is CER in the Pharmaceutical Context?

CER is defined as "the conduct and synthesis of systematic research to compare the
benefits and harms of existing health care interventions and strategies to prevent,
diagnose, treat, and monitor health conditions" [1, 2]. In the pharmaceutical context, CER can take many forms. It could be a comparison of the ability of two treatments to improve a disease ("efficacy") or a comparison of the side effects of two
medications ("safety") or both. It could also focus on the comparison of costs associated with using one medication over the other. In short, CER for pharmaceuticals
is any *research* study that *compares* how *effective* different treatments are on a given
dimension (or dimensions). The greater goal of CER is to improve decision making,
in an evidence-based medicine approach to personalized treatment choices using
the best and most current information available.

Examples of CER include a clinical trial assessing the difference in survival time
between two different cancer treatments, a study of patient medical charts comparing the rates of side effects between two diabetes treatments, or a survey collecting
and comparing patient-reported outcome measures at various time intervals in
arthritis patients with differing treatments. Comparative effectiveness studies can
also include aspects of cost-effectiveness, which is often of particular interest when
the frequency, type, and costs of medical care are included in the study.

In order to conduct CER, several steps must be undertaken. First, the type of
study needs to be determined. For example, will the study be a clinical trial conducted in tightly controlled conditions or a real-world study? Next, the specific

comparators and dimensions on which effectiveness will be measured need to be chosen. After this is done and the data are obtained, the results need to be disseminated in a manner that meets various government and private entity policies. Each step in this process needs to be managed carefully in order for useful reliable results to be generated. As will be discussed below, there are numerous complexities that arise throughout this process, making CER within the pharmaceutical industry challenging.

5.3   What Are the Current Efforts Toward Using CER?

Currently, the use of CER is most prominent outside the United States (USA) due to the fact that there are centralized payers controlling access and payments for treatments, which are controlled by the government, and the public accepts this. For example, in the United Kingdom (UK), the National Institute for Health and Care Excellence (NICE) actively uses CER. NICE was established in 1999 to determine the most clinically and cost-effective drugs and treatments to be adopted in the National Health Service (NHS) of England and Wales [3]. NICE is generally recognized for its ability to produce authoritative, evidence-based recommendations on technology adoption and health-care guidelines [3], and it was among the founding members of current worldwide health technology assessment (HTA) societies [4, 5]. NICE has both regulatory and advisory authorities; that is, NICE decisions on valuable treatments are the regulatory authority by which England's NHS must comply, and even outside of England's NHS, its clinical practice guidelines are often taken under advisement by other agencies (e.g., All Wales Medicines Strategy) and are often adopted.

The advisory and regulatory authority NICE holds is rare when compared to the international context. In Sweden, the Swedish Council on Technology Assessment in Health Care was established in 1987 to combat rising health-care costs and improve the diffusion of innovative cost-effective technologies. This national agency had no regulatory function, and it, like many other HTA agencies, was simply tasked with conducting assessments to inform providers and policy makers as they determined priority setting and technological disinvestment [6]. France's HTA agency, the National Authority for Health, is an independent scientific authority advising health insurers on health service coverage and reimbursement, and while it also lacks authority to control decisions on pricing and coverage [7], in practice its advice is highly influential. Germany's primary HTA agency, the Institute for Quality and Efficiency in Health Care (IQWiG), has a unique relationship with the country's supreme decision-making body, the Federal Joint Committee (G-BA). Although IQWiG has no decision-making power itself, it acts together with the G-BA, where IQWiG conducts assessments on behalf of the G-BA which are then later appraised by the G-BA [8].

In the USA, there currently does not exist a single systematic process for CER and value assessment unlike in other countries with socialized medical systems, primarily because our health-care system is fragmented and decentralized. Instead, multiple organizations, including health plans, pharmacy benefit managers (PBMs), integrated delivery networks, nongovernmental organizations, consulting organizations, as well as government-sponsored academic research, produce CER and, increasingly, value assessments based on it. This approach can be beneficial, as it promotes a market actively at work driving relative value, and it represents the pluralism, individualism, and independence that define American exceptionalism. Others might characterize it as inefficient, duplicative, and wasteful [9] and may recommend more centralized decision making and control. For instance, in the course of treating patients, the average practicing physician has to deal with dozens of insurance plans, each of which makes independent CER decisions and has different reimbursement requirements. While a physician's concern is primarily the safety and efficacy of the prescribed treatment, the complexity and variability that comes with differences in insurance coverage and utilization rules can be overwhelming. In an ideal state, the CER would help provide information on the comparative effectiveness of the safety/efficacy dimensions, which would then line up with the pricing and financial incentives of payers. However, as will be discussed further below, currently this is not the case.

Still, CER, and the need for and expectation of it, has been growing steadily over time. A review on comparative cost–utility analyses (CUA) over the past three decades showed the proportion of CUAs that focused on pharmaceuticals increased from 34% in 1990–1995 to 47% in 2001–2005 [10]. This steady growth has been in part driven by insurers paying more attention to the comparative costs between the overall efficacy and safety of different treatments, public pressure to contain health-care costs and move toward value-based pricing/care, an aging population and expanded time in which people are living with chronic conditions, and the increased investment by the industry in treatments for serious diseases that have a more transformational impact on patients' health (as compared with developing "me-too" products with little differentiation). As industry turns its focus to transformational impact, often some treatment for a serious disease already exists, and it is unacceptable to not actively treat patients during a clinical trial. Therefore, regulators and insurers increasingly require active controls in randomized controlled trials (RCTs), and as a result, we see product launches that include comparative effectiveness data much more frequently now. Examples include novel oral anticoagulants vs. warfarin, biologics for psoriasis, and sodium–glucose cotransporter type 2 agents vs. metformin or dipeptidyl peptidase-4 inhibitors.

Escalating health-care costs and the recent implementation of health-care reform in the USA, with its new incentives for value-based care, have prompted a growing interest in the relative value of treatments within this new environment focused on maximal efficiency in resource use and budgetary restraint. In addition, technological and scientific breakthroughs have accelerated the availability of data and tools to interpret them, as well as the development of many new treatments. However, while multiple treatments are frequently available in the clinic for a specific condition, there is often a lack of hierarchical recommendations for their use. CER is invaluable in such a scenario, helping determine the best use of finite resources. As the quality of evidence presented and the quality of decision making in health care improves, CER can help improve overall health outcomes and health status in the USA as well as elsewhere in the world.

Currently, several initiatives are ongoing and are expected to impact the face of future treatment valuation. At the governmental level, reports on the comparative effectiveness of treatments are provided by the National Institutes of Health (NIH), the Agency for Healthcare Research and Quality (AHRQ), and the Patient-Centered Outcomes Research Institute (PCORI) [11], the latter of which was formed by the Patient Protection and Affordable Care Act (PPACA or, more simply, ACA) specifically to fund and (methodologically) improve CER in the USA as well as build an infrastructure to enable it. However, unlike its European counterparts, the US government agency efforts can only make suggestions regarding the recommended treatment pathway in a certain disease domain and do not play a formal role in reimbursement decisions made by private payers. The approach of using costs per quality-adjusted life year (QALY) in particular was prohibited in the PPACA, reflecting the political sensitivities and trade-offs needed to pass this landmark legislation.

Governmental input is currently constrained by Medicare regulations on covered vs. noncovered services, provided by the US Department of Health and Human Services (HHS) at the federal level [12]. These regulations are applicable for all parties that would like to be reimbursed by the government, but actual implementation of Medicare regulations can differ at the state or regional level. For drugs covered under Medicare Part D plans, the insurers offering those plans—and not the federal government—make the value-based formulary decisions and negotiate prices with manufacturers.

However, Medicaid program administrators felt a keen need to cite an independent technical/clinical review for their decisions on state Medicaid formularies. The Oregon Health Sciences University has been running its Drug Effectiveness Review Program (DERP) for many years now, and its focus is on clinical CER, not on cost-effectiveness or value assessment. Formulary decisions based on such technical/clinical reviews, and the accountability for them, remain the purview of the state Medicaid administrators, who are the customers buying the DERP-independent evaluations. This approach enables the state administrators to tailor decisions to their unique patient populations, community health needs, and fiscal constraints [13].

Given the constraints on governmental input, a second layer of value review and recommendation comes from several independent nonprofit, nongovernmental organizations that have been established to evaluate the value of competing treatments, such as the Institute for Clinical and Economic Review (ICER) [14]. ICER developed from an academic initiative begun at Harvard and is still located in Boston. After comprehensive discussions with multiple stakeholders (insurers, PBMs, manufacturers, patients), ICER researchers have concluded that value assessment should follow the World Health Organization's (WHO) cost per QALY approach, with evaluated thresholds for incremental cost-effectiveness ratios set at $50,000/$100,000/$150,000 per QALY [15] for US decision makers. In addition,

ICER's position is that the potential budget impact of a drug should also be considered in terms of evaluating the contribution of expected drug costs to health-care spending relative to anticipated growth in the national GDP +1% [11] (based on a cost-cutting model adopted by the state of Massachusetts in 2012—*Chapter 224 of the Acts of 2012* [16]).

In addition, several professional medical and research societies (e.g., the National Comprehensive Cancer Network, the American Society of Clinical Oncology, researchers at the Memorial Sloan Kettering Cancer Center) have developed detailed clinical pathways and their own value assessment frameworks. The impetus for this development was the high cost of oncology drugs, coupled with pressures from the insurance industry to produce clinical pathways that are more prescriptive (considering cost-effectiveness, as opposed to only clinical effectiveness) and present treatment choices according to a given hierarchical preference. Various payers now either endorse these professional society-designed pathways or develop their own recommended clinical pathway. In order to maintain perceived neutrality, some clinical societies have opted out of the value assessment debate, but their members will most likely still be affected by the now rapidly increasing number of clinical pathways and value assessment frameworks. The availability of multiple competing pathway builders is now becoming a new issue for physicians and prescribers: what pathway is the patient's insurance operating under? If a single pathway is preferable, who should develop it? The clinicians who treat the patients with the disease, or the variety of unique insurers who pay for that treatment, or some third party, which is still likely to introduce bias? Recent advocacy efforts are now suggesting that professional medical societies should design a common clinical pathway (focusing on value) which insurers should cover. By adopting this movement, physicians will be able to treat patients more consistently.

5.4   What Are the Current Challenges and Barriers for CER Research?

Despite the push from physicians, payers, and international HTA agencies to further expand CER's utilization, not all industry players are actively promoting its use. Most of the largest pharmaceutical players (Johnson & Johnson, Pfizer, Sanofi, Merck, GlaxoSmithKline, AstraZeneca, Eli Lilly, Amgen, Roche) have officially stated their position regarding the beneficial role of CER in improving health-care decision making, promoting awareness of the impact of health-care interventions on individual and population health, and more efficiently allocating limited resources; but not all drug manufacturers have followed suit. At first this decision is puzzling, particularly given the large demand for this type of data from the payer side [17], but regulatory, technical, and financial issues may prevent manufacturers from being more supportive of undertaking CER or outspoken about the need for disseminating CER findings. There are several possible reasons and obstacles to wider generation and adoption of CER, including (1) the challenges and costs associated with designing and running CER, (2) the natural tension in the motivation underlying the uses of CER by various health-care stakeholders, (3) the barriers associated with regulatory restrictions and other factors associated with disseminating CER, (4) the unique aspects of the US pharmaceutical pricing and reimbursement environments, (5) the fact that even the best CER cannot account for the fact that individuals respond differently to treatment, and (6) the lack of societal consensus on how value should be assessed. This section will discuss each of these challenges.

5.4.1 Challenges and Costs Associated with Designing the Study

5.4.1.1 Randomized Clinical Trials vs. Observational Studies

The first step in conducting CER is to specify and design the study, and although this seems fairly routine, there are numerous challenges that arise. An ideal study would be performed in a non-biased (randomized, controlled design), easy-to-interpret manner (clearly defined and agreed-upon patient-centric outcomes of interest, intent-to-treat assessment), would allow researchers to compare all existing treatments recommended for a specific condition (multiple comparators), would yield results in a timely manner, would provide reasonable information on long-term outcomes, would involve a large diverse population that best represents the actual population of patients suffering from the specific disease/condition (highly generalizable), and would have reasonable costs of execution. In real life, such an ideal design, one that produces perfect information for decision makers, does not exist, and trade-offs will be necessary. The choices are often limited to RCTs that formed the basis for regulatory approval versus observational studies [18, 19]. Both options have their supporters and detractors, but neither can offer every desired quality.

The RCT has long been the gold standard for eliminating bias in health-care research. RCTs offer the benefits of high internal validity (minimizing selection bias), systemized data collection and processing (therefore good data quality), and strict protocols (minimizing variability of some factors that can also impact effectiveness) [20]. RCTs are true experiments, seeking to uncover the effects of a chemical or biologic treatment on a group of humans affected by a specific condition. However, both practical considerations (e.g., financial and time constraints) and methodological limitations necessitate considering other study types. Some factors, such as patient adherence or real-life usage patterns, cannot be adequately studied in the artificially designed RCT environment and instead require a "real-world" setting. By contrast, observational studies have a high external validity, represent real-world situations (not artificial designs), are typically completed in a short time frame (particularly if retrospective), are more representative of the actual treatment population, are inexpensive relative to the large amounts of data that can be collected (e.g., claims data studies), and can more accurately represent system-wide impacts of treatment choice (not just a localized effect) [18, 19, 21].

RCT proponents claim that RCT data are harder to manipulate, in the sense that these trials have pre-analysis plans that are publicly pre-registered, whereas observational studies are inherently biased and have noisy data and dubious methodology. Comparisons of findings of observational vs. RCT data found that, in general, there is agreement between the two types of studies [22–25], but observational studies often report weaker correlations among the studied treatments and their outcomes [26], and the credibility of these weaker associations is often questionable [27]. Furthermore, there are high-profile cases (e.g., the case of hormone therapy in postmenopausal women and its role regarding cardiovascular events [28]) where observational studies misled clinicians and treatment decisions based on them negatively affected the health status of patients. Due to these occurrences, there is a lingering suspicion toward some observational studies, and this apprehension prevents measured, appropriate use of observational data where it does make sense. Finally, RCT proponents underestimate the importance of the external validity of a study. It is the messiness of real-life patterns of behavior that brings value to observations from real-world studies. Imperfect treatment adherence, unexpected patient preference, and unplanned barriers to adoption all play a role in the final effectiveness of a drug in the real-world setting. After all, even a highly efficacious drug will not improve patient outcomes if nobody is taking it, or taking it properly.

Observational study proponents argue that RCTs are expensive to conduct and time-consuming, whereas common data sources for observational studies (e.g., claims data, EMRs, chart reviews) provide fast access at a lower cost as they do not involve treatment administration and monitoring and other administrative costs [18, 19, 21]. Additionally, in real life, patients are exposed to interventions all the time. Observational studies allow for measuring the effectiveness of treatments as used in the populations and manner that will actually occur for most patients. Given that this real-word use is closer to what patients will actually experience and payers will actually pay for, observational studies better represent the ultimate treatment population. Among the reasons why this can be important is that expected treatment compliance may be unrealistic, as patients tend to be more compliant when they go through the consent process in tightly controlled clinical settings, when they know they are being observed (the Hawthorne effect [20]), and when they receive additional attention from health-care professionals in a research setting. Observational studies also generally have larger sample sizes, which helps with the generalizability of the results. Although RCTs can be conducted with larger sample sizes or incorporate diverse settings, such trials would undoubtedly have substantially increased research costs. Lastly, despite the fact that observational studies have some advantages over RCTs, one impediment to these studies is the fact that there are stricter controls on manufacturers regarding who such studies can be disseminated to, and they are typically not accepted by FDA for labeling purposes.

In sum, there are challenges and benefits associated with each type of study design and it is unclear that there is only one correct choice. As a matter of fact, many researchers consider the study types to be complementary, as both RCTs and observational research contribute, each in their own way, to a better understanding of the treatment effects of medicines. Ultimately the choice will depend on the goals of the specific CER study.

5.4.1.2   Determining Comparators

Given the central purpose of CER is to compare across different treatments, a key step in any CER study is to choose the treatments to compare. In a field with an array of therapeutic choices, choosing a comparator involves many considerations, such as focusing on global versus country-specific substitutes, or generic versus brand alternatives, or single vs. multiple comparators, or even different dosing regimens or treatment sequences for the comparators chosen. The choice of the comparators has important temporal and financial consequences. Analyzing more than one comparator would be ideal for decision makers, but it also adds significant patient recruitment, processing time, and financial costs. Additionally, selecting comparators that are no longer relevant in actual practice may limit the external validity or generalizability of the study results. A typical example would be Company A, the developer of Treatment A, doing a comparative clinical trial as part of its registration program against Treatment B, the leading treatment in the disease at the time. However, by the time Treatment A is launched, Treatment C is now the leading treatment, and customers now want a comparative study against it instead. Such a scenario is not unusual in health care. For example, the Clinical Antipsychotic Trials of Intervention Effectiveness (CATIE) study on medications for schizophrenia was groundbreaking at the time it was launched [29, 30]. The study aimed to include the plethora of choices available at the time and definitive recommendations regarding the optimal therapeutic choice. Unfortunately the study was afflicted by high costs, a long study duration, and controversy. By the time the results of CATIE were published, some of the information had become obsolete, other treatments had replaced the standard of care, and the choice of endpoints was heavily criticized. Moreover, its economic comparison component was hindered by lack of real-world cost and outcome data. From a value-of-information perspective, this effort was extremely costly and the study's findings had little impact on clinical practices in schizophrenia.

In other cases, the key concern is the lack of an available comparator and/or the inability to use a placebo to control for nontreatment-related effects. The inability to use a placebo treatment is frequently encountered in oncological treatments and rare disease/orphan drug research (vividly illustrated in the ongoing battle over eteplirsen's approval for Duchenne muscular dystrophy [31]). Within late-stage oncological care, patients have frequently exhausted several lines of treatment and there are few, if any, remaining interventions. There may simply not be an acceptable comparator other than palliative or supportive care, and so the historical outcomes in the absence of treatment become relevant. For rare or high-need diseases (where existing treatments do not adequately alleviate the disease burden), there is often a lack of suitable alternative treatment options and the new medication is the only option available for these patients. In this situation, it can be argued that it would be unethical to give patients a placebo instead of the active, if experimental, treatment, especially if the disease imposes great suffering and if the new treatment has a real chance of improving patient outcomes.

5.4.1.3   Choice of Outcomes

Once a research sponsor has come to a decision regarding the comparators of interest, there is the need to identify the relevant outcomes and patients to compare. However, the outcomes of interest vary tremendously across stakeholder groups (patients, physicians, insurers, manufacturers, society), and it can be difficult to determine how/when effectiveness is being defined and measured. A treatment can have more than one outcome of interest, and the study designer has to decide in advance of the study whether to look at single or multiple effects or whether to consider a primary single or composite endpoint versus co-primary endpoints or several other secondary endpoints. Measures of treatment effects can vary from events that are relatively easy to capture (hospitalization or survival) to those that can be difficult to measure and contain a high degree of variability (patient-reported psychometric evaluations of health status and functioning). The choice of when an effect is measured is another important consideration, as time to treatment response varies. To keep planning complexity and costs realistic, researchers typically choose one or a few time points to measure the outcomes of interest, but in real life, all time points matter to the patient. In addition, a study population can range from one that is narrowly defined through a variety of inclusion and exclusion criteria (including disease stage or severity, comorbidities, prior treatment, age, literacy, etc.) to one that is broad and includes even those with a small likelihood of treatment.

Despite the fact that organizations such as ICER have put forth proposed frameworks that provide recommendations or alerts, consensus is still hard to reach, and there are multiple limitations in using the ICER approach. The ICER approach is heavily focused on outcomes that affect the health-care ecosystem (i.e., health resource utilization or budgetary impact) and less focused on patient-centric needs, expectations, or experiences. ICER does use the QALY approach, which translates measurement of patient-reported quality of life changes into a population-based utility measure to account for treatment impacts on patient experiences, but that approach was specifically disallowed in the ACA for use in research supported by PCORI, as it has many limitations and biases [32]. Patient perspectives are not substantially involved in its value assessments, which rely primarily on network meta-analyses of RCTs and regulatory endpoints. ICER has a strong focus on improving affordability, with a clear goal of pressuring innovative companies to lower prices to what it deems to be a value-based price. Such QALY-derived pricing has been criticized for its potential to stifle innovation [33, 34]. In the long term, over, for instance, a 30-year time horizon, which can include 18–20 years of generic availability and competition after a 10–12-year period of intellectual property or patent protection, pharmaceutical innovation becomes much less costly and its cost-effectiveness is amplified. As much as 85–90% of medicines prescribed today are generic, lending credibility to the value creation of the pharmaceutical innovation cycle.

Another limitation with respect to the ICER approach and value-based care is that it does not reconcile with the part of the PPACA that HHS has recently touted as a broad benefit—that this legislation outlawed lifetime limits on health insurance policies. No lifetime limits on health insurance policies has serious implications for the field of HTA. One could interpret this as saying that essentially any service or treatment at any price—as long as it provides some benefit that exceeds its risk—would be expected to be covered. No lifetime limits on what an insurer is duty-bound to pay is akin to saying there is no financial threshold beyond which care can be limited or refused. Therefore, it implies that it is acceptable to spend hundreds of thousands or even millions of dollars on a single individual. This also potentially creates an issue of having different societal standards for health-care services than for other technologies. It begs to ask the question why it is acceptable to spend $200,000 on a single hospitalization whose benefits in terms of QALYs are unpredictable or nonexistent, versus spending the same on a drug whose benefits can be predicted and measured. Politicians flee from these types of decisions being put directly in the hands of government, as evidenced by the negative reception the PPACA's Independent Payment Advisory Board (IPAB) received from physician and hospital groups, who basically saw the IPAB as a government-led panel to deny resources and reduce access to treatment at the end of life [35].

5.4.1.4 Funding and Perceived Bias

The study design, treatment choices, and outcomes cannot be determined in a vacuum. Funding constraints may impact the choice of study design, including the extent to which multiple outcomes, larger patient samples, or long-term treatment effects can be studied. There are constraints on funding and difficult choices need to be made by government agencies such as NIH as well as by pharmaceutical manufacturers. For manufacturers, if CER studies are too costly or will not significantly impact cost-driven treatment choices, it could lead to decisions to use their limited funding for development of new treatments rather than for CER on existing treatments, or alternatively it might reduce spending on development of new treatments in order to fund CER. Similarly, if the length of time needed to undertake the study is prolonged, the opportunity costs and net present value of information from such studies also become obstacles to their funding.

In addition, the choice of sponsor can affect how the study's findings are perceived. For example, studies sponsored by the industry are often portrayed as potentially biased toward favoring the company's product—due to both reader negative bias [36, 37] and favorable interpretation of results [37–40]. Government-sponsored research can also carry their own biases, as grant applicants sometimes cater to not-so-subtle government objectives, and innovative research is often not supported [41]. In general, the involvement of perceived impartial third parties can alleviate those perceptions. In the USA, the public has been led to believe that the majority of research leading to the market availability of pharmaceutical innovation is sponsored and conducted by the NIH. However, in the USA, the industry has sponsored and conducted the majority (66%) of clinical trials between 2008 and 2012 [42]. In the same time frame, the NIH only sponsored 14% of clinical trials, but has often played a reporting role, indexing study applications and findings [42]. Similarly, most comparative (head-to-head) trials are still being underwritten by industry [38, 43, 44], with nonprofit institutions funding a minority (roughly one third) of such research [43, 44].

Issues with perceived bias have typically been highest among observational studies. The perception has been that the observational nature allows the sponsor to easily manipulate the methods and findings to obtain the result consistent with their goals. There is, however, growing consensus around standards for performing and reporting observational research, in order to increase their rigor, credibility, and usefulness in decision making [45–48]. For example, in the field of health economics and outcomes research (HECOR or HEOR), there are methodological principles on what constitutes a robust analysis and what type of study produces reliable evidence that are put forth by research organizations (e.g., the GRACE principles [45], the ISPOR-AMCP-NPC CER principles [48], and the Methods Guide published by the US AHRQ [46]). Prespecified and transparent analytic plans are necessary to limit the perception that data are perceived as being improperly manipulated in observational studies.

5.4.2 Challenges Stemming from the Potential Uses of CER

5.4.2.1 Multiple Uses Each with Varying Purposes

Ultimately, CER is an art, in the sense of balancing multiple approaches with a rigorous, methodical scientific approach to data collection and analysis. If it were simply used in an evidence-based medicine approach to contribute to the shared decision making between individual patients and their clinicians, the existence of these issues might be manageable, as uncertainty is ubiquitous, and perfect information rarely exists. However, in reality there are a variety of stakeholders, each desiring CER that advances their own goals and challenges. For example, in addition to the clinicians who want CER to improve treatment decisions, payers want CER to help manage costs, and pharmaceutical manufacturers want CER that will differentiate their product and enhance market access or physician use.

When CER is coupled with cost analysis, it can be leveraged as a tool for cost bargaining [49]. For instance, in the absence of CER, payers look to commoditize treatments to drive lower prices, but the absence of CER does not necessarily mean that there are no real differences between treatment options, only that the evidence of differences does not exist. There is a well-worn phrase that captures this: "The absence of evidence of a difference is not equal to evidence of an absence of a difference." Even when a manufacturer makes a significant investment and conducts high-quality CER that scientifically demonstrates the superiority of its treatment, it does not necessarily follow that formulary access or treatment utilization will increase. Some patients may have already tried and either not responded to or tolerated the treatment. Other patients may not prefer it, perhaps because of a different mode of administration or because it recently entered the market. Clinicians are justifiably hesitant to switch stable patients unless the incremental gain from a superior treatment is large enough, as switching takes time and resources, and introduces uncertainty.

In addition, for payers, the outcomes of immediate interest are those that primarily return value in the form of savings to the health-care system, as the financial risk they hold is solely for medical and pharmacy claims. Patients and physicians typically take a broader societal perspective and a longer-term view, with greater concern toward long-term safety and effectiveness, not necessarily cost savings in direct medical costs. Their broader view incorporates the aspects of improved functioning, such as the ability to undertake activities of daily living and reduced symptomatology or suffering. Additionally, their longer view is a lifetime view, not the 1-year view that actuaries typically utilize as part of the US insurance business model. Deciding which outcomes to include in any study, and, most importantly, what perspective to take, can become a tricky proposition given the competing interests, both health and financial, of all involved parties. All stakeholders should be "focusing on the effectiveness and quality of health care rather than simply on cost" [50], and while costs cannot be ignored, there is genuine concern that if it becomes the primary foundation of decision making, patient outcomes will not be optimized. It is critical that a broad view is taken, as without it, CER can be misused, and the well-being of patients and societal benefits on innovation will fall out of what should be a full accounting.

These different uses and agendas impact the risk–reward trade-off for pharmaceutical firms deciding whether to fund CER and complicate the questions of comparators, study design, and outcomes.

5.4.2.2   Challenges Associated with Dissemination of CER

Another important consideration and potential barrier to CER is the regulatory restrictions on dissemination of CER and the lack of clarity in this regard. That is, once research is complete, there are hurdles to proactively share data with customers, prescribers, and payers. A company cannot use a single study with physicians, even if well designed, valid, and robust, to actively promote comparative claims of superiority of treatment A versus B, as the FDA regulations generally require that at least two adequate and well-controlled studies are conducted before it can consider the findings as "substantial evidence" for promotional use [51, 52]. The company can, however, bring data forward in response to an unsolicited request for comparative information [53], and comparative effectiveness is not an uncommon customer question in the current health-care environment. However, there is a lack of clarity on what is permissible and what might be considered off-label and/or misleading communication [54]. Given the large amount of enforcement activity that has existed in the past decade combined with the uncertainty of what regulators consider permissible, dissemination can create risk for the manufacturer and/or the constraints on dissemination necessary to avoid this risk may make the risk–reward trade-off of investing in CER unfavorable.

The strategic choice can boil down to investing in a promotional claim for an existing product vs. investing in the new product pipeline, continuously seeking to bring to market new, transformational treatments. A company can also choose to invest in real-world evidence in order to validate findings from a single RCT and more fully answer unsolicited requests for comparative information from payers. Investing in real-world evidence provides the added benefit of being faster, more relevant, and less expensive. On the other hand, even though real-world evidence is immensely useful, the use of this information could lower barriers for competitors who choose only to invest in real-world evidence and not perform a more rigorous comparative RCT.

Other practical considerations associated with dissemination can affect support for undertaking CER. The publication process is inherently slow when going through peer review, but without credible peer review, the concerns over research quality and inherent sponsor bias are exacerbated. Internal review for compliance and legal purposes can also be protracted, given the variability in interpretation of existing laws and risk adversity, which can also delay prompt dissemination, thereby reducing study value.

5.4.3   Barriers from Current Reimbursement Structures and Pricing Strategies in Pharmaceuticals

CER has limited applicability as a price control/price determination tool. This may seem surprising at first glance: if you have five similar widgets and you can determine which one can give you the best results for the money, shouldn't the law of free market competition apply and make the other four widget makers compete with lower prices? Theoretically yes, but in practice, things are more complicated. Even in industries other than health care, when people compare, for example, home appliances, there is rarely a clear winner. You can find a very good appliance with all the desirable features for a high cost, a very cheap appliance with limited features (and probably lifespan), and an assortment of appliances with various combinations of features of interest (not always comparable) at prices in between. The seasoned consumer, who generally has a budget range and a list of feature priorities, picks the top important features within the budget and compromises on the rest.

However, pharmaceutical pricing does not follow this same pattern, since manufacturers cannot typically design in or design out all the performance features of their products, as those products necessarily interact with the complex biology of human protoplasm. Also, several levels of price setting and negotiation (discounting, rebates, chargebacks, etc.) apply at various steps in the purchase of a drug [55], most of which are not publicly disclosed and are not always taken into account in cost-modeling exercises. Thus, the true relative cost to any payer of two competing products is only known by the payer itself, which inhibits the accuracy of cost comparisons within CER.

In a recent proposed model for value-based pricing of its Part B benefits, HHS suggested the use of indication-specific pricing [56] as a possible approach to achieving a common value-based pricing system. While the idea has some merit in aiming to establish the principle that price and value should be related, it could be difficult to implement because drug pricing is currently based on units (of drug) consumed during treatment and not necessarily outcomes. These and other difficulties with this model have been recently addressed in ICER's evaluation of the implications of designing and implementing the use of indication-specific pricing in drug reimbursement in the USA [57], suggesting a long road before agreement can be reached. Moreover, the pricing assumptions that went into the development decisions for treatments with potential for multiple indications being launched today were made 10+ years ago. Holding these treatments to a different standard when they are ultimately launched could result in lower revenue than expected and increase the financial hurdles for health-care innovation investment.

Another limiting factor associated with the pharmaceutical industry structure is that to protect consumer welfare and encourage both market competition and low prices for consumers, existing legislation mandates that when a generic drug has demonstrated identical effectiveness and bioavailability to a branded drug, the (lower-priced) generic is automatically substituted for the branded drug at the pharmacy (AB-rated substitution) [58]. This impacts the cost–benefit ratio for manufacturers as investments in CER for products facing near-term generic competition are likely to accrue benefit to the generic manufacturers despite the fact that the brand manufacturers bore the costs and risk. In other areas of the pharmaceutical industry, to incentivize investment in areas lacking research, the government has provided patent extensions to manufacturers who undertake costly research. For example, manufacturers conducting studies on pediatric patient populations can be granted extra patent exclusivity. However, such an incentive does not exist for CER.

Similarly, the CER process is, to a great extent, influenced by the reimbursement structure of the market. In most European countries, the use of CER evidence for health-care reimbursement decision making involves a central decision-making process by experts entrusted with a public purse [59]. In the USA, the same process is a hybrid of central decision making by experts entrusted with a multitude of different public or private purses (within legislative and regulatory constraints, such as compulsory coverage for treatments that are deemed "medically necessary" in the case of Medicare Part B Program), combined with a free market process. Decisions are further decentralized to the patients acting in concert with their physician and private insurer (either selected by their employer or selected and paid for directly by the patients) [55]. There is a general reticence in the USA to entrust the government and/or payers as sole decision maker(s)—perhaps given their obvious financial interests—and as a result elaborate appeals processes are part of the US system as well. To present CER evidence in the USA, one must always keep the uniquely

American system and values in mind and also ask "Who will be the decision maker, and what do they require?". Sometimes this may involve multiple entities, with multiple conflicting interests that may change depending on the current business environment (e.g., insurer mergers, evolving business models).

A more consistent perspective on the value of treatments is supposedly provided in Europe by governmental agencies working in concert with clinical experts, typically professional societies that provide clinical guidelines for specific disease areas. In the USA, this perspective is provided by an amalgamation of input and loose oversight from the government, health-care policy research centers, and professional societies. Variability in coverage by payers, in this context, is an indicator of the lack of agreement among payers regarding the quality (value for the money) of the treatment(s) in question. This is generally attributed to lack of consensus from multiple stakeholders on the optimal treatment choice and/or variation in standards and approaches toward evaluating the clinical benefits and cost-effectiveness of drugs.

5.4.4 Challenges Associated with Individual Response to Treatments

CER is useful at a population, policy-determining level, but at the individual level, other considerations—such as individual patient response and preferences—apply. For example, individual patients may be intolerant or have no response to select drugs in a given class, but have an acceptable response to other drugs from the same class. In the absence of this phenomenon of heterogeneity of treatment effect, it would make sense to offer very few choices to drive the lowest price possible. In reality, it is important to offer multiple alternatives so that a patient has access to the best treatment for them. This is vividly illustrated by the protected drug class designation in the Medicare Part D program for six drug classes (antidepressants, antipsychotics, anticonvulsants, antineoplastics, antiretrovirals, and transplant-required immunosuppressants). These drug classes have been deemed vital to Medicare's most vulnerable beneficiaries, and experience has shown that significant heterogeneity of treatment effect exists within them. As such, patients are currently granted immediate access to a broad choice of medicines in those categories. Recent attempts to remove class protection for several drug classes faced widespread criticism, and the federal Centers for Medicare and Medicaid Services (CMS) backed away from pursuing their original cost-cutting proposal [60].

The effect observed *on average* across the entire patient population may differ from that of patient subpopulations of interest (defined by severity or comorbidities or other relevant distinctions). This means that the comparative effectiveness found across the aggregate study population in any given study may not actually apply to any one patient, which in turn can lead to suboptimal treatment decisions based on the CER. Recently, after study completion, it has become increasingly common for payers to be interested in examining whether a particular subpopulation is driving comparative differences. This is done to maximize efficiency from their perspective, to potentially restrict treatment to those patients who are most likely to benefit from the treatment. However, while this approach might make sense from an overall efficiency perspective, it can reduce treatment access to patient subgroups who also respond, just not at the same rate or degree.

5.4.5 *Lack of Societal Consensus Regarding the Measure of Value and Other Ethical Considerations*

The usefulness of CER is reduced because of the underlying lack of consensus on what trade-offs are acceptable in the quest for value (What is value in health care and who decides it? What is the social contract for health care in America? What do we owe each other?). As long as there are significant differences of opinion and a lack of agreement on those larger societal issues, more detail/technical refinement on narrower issues (e.g., methodological aspects) is not sufficient to broach the current gap with regard to the usability of CER and its close cousin, value assessment.

In the USA, how CER is used and by whom has to be aligned with the cultural values of the society. Americans have different expectations from their health-care system than do the citizens of the UK, Canada, or Australia, or other non-English-speaking countries for that matter. What our citizenry values, and how much they value it, will not look the same as in another country with a different set of cultural values. Despite this, there is and always has been a substantial diversity in American values, so identifying values related to health care that are generally agreed upon remains a necessary, if difficult, task. Ignoring those differences will likely lead to failed attempts to institute approaches used elsewhere, which are based on a different set of shared beliefs.

This cultural divide is vividly demonstrated in annual reports on the state of health-care systems around the world, produced by the Organization for Economic Cooperation and Development (OECD). Americans love individualism and choice and are natural optimists—rooting for the underdog, long-shot, or hopeless case—and they love innovation and technology. These values are unmistakably reflected in the allocation of money and resources for health care: a recent report (2013) from the OECD revealed that the USA spends the highest proportion of its GDP on health care, and while Americans pay the highest costs for health care among all analyzed countries, spending on social services is among the lowest [61]. Americans are among the biggest consumers of sophisticated medical technology (imaging tests) and prescription drugs, yet have fewer hospital and physician visits than most developed countries (OECD average) [61]. The USA is a leader in cancer care (and availability of specialists), yet is among the lowest performers among OECD peers regarding population health (chronic conditions such as diabetes, heart disease) [61]. On the one hand, Americans want the benefits of a buoyant health-care market—choice, latest technological breakthroughs, and customization of options—but on the other hand, they are inherently opposed to increasing system efficiency though centralization, reduction of choices to a few limited options, and a "no-frills" approach. In essence, the American health-care system has always suffered from a familiar dichotomy: effective and cutting-edge innovation for a high cost at one end of the spectrum or subpar delivery and inferior outcomes at the low-cost end of the spectrum, with little success at standardization. The same is true in our other public/private markets, like education, legal services, housing, etc.

Finally, even when all parties agree on the need for a comparative assessment, there are still some ethical considerations that make it difficult to agree on a common value framework. Such considerations include deciding how much is too much to spend on a drug (and from whose perspective), should cost-effectiveness or budgetary impact be the deciding factor (as vividly illustrated by the negative publicity around Sovaldi and Harvoni price announcements [62–64]), what role does patient affordability or insurance benefit design play, and is pegging the budget impact metric on national GDP either a relevant or fair measure of value? All of these questions are being actively considered as these are issues that will continue to affect the health-care industry for the foreseeable future.

5.5 What Can Reasonably Be Done in the Near Future in Terms of CER?

While existing CER is inadequate and controversies accompany the many steps in designing, completing, and disseminating the findings of a CER study, the use of CER remains invaluable in the health-care environment. Several possible solutions are emerging to break the logjam. An area that shows particular promise is the implementation of an adaptive pathways approach to trial design [65] and the use of randomized pragmatic trials (RPTs). Adaptive pathways aim to lower the costs of clinical research by shortening trial time with surrogate endpoints, or restricting trial populations to specific selected subsamples, and by allowing collection and use of real-world data [66]. RPTs aim to model real-world effectiveness with the added benefit of reduced sample bias under an initial randomization scheme and limit the use of assumptions due to a prospective design (the effect size is measured in the real-world clinical setting). RPTs, in particular, are an appealing alternative to RCTs from an insurer's perspective. Insurers see this approach as the perfect hybrid of the rigorous scientific method observed in clinical trial design—highly controlled, select population, great internal validity—combined with the real-world grittiness of actual treatment and health system realities. Furthermore, the prospective nature of these trials is also appealing: since endpoints are defined before patients are even assigned to treatment, some of the interpretational bias accusations associated with the use of retrospective data do not apply to prospective RPTs. However, since no solution is perfect, RPTs still present some difficulties due to their use in the clinical practice setting. One such difficulty arises at the recruitment stage, as patients may feel financially unprotected in the event of a harmful side effect [67]. Patients may also not want to participate in what is legitimately described to them as an "experiment" to test the comparative effectiveness of multiple treatments, even if each of those treatments is an approved and acceptable choice. Another criticism of RPTs is that they use "lower standards" of evidence gathering, as intermediate endpoints can be used and observational methodology governs the trial process after the randomization step, and this introduces the "messiness" of real-world data problems, including variability between data systems and missing data.

At a macro level, the learning system approach to health care—creation of a system linked by a common electronic health record and shared databases and implementation of a collaborative approach to data and insight sharing across an entire health-care system [68, 69]—seems the most intuitive way to drive more efficient medical practice and patient care in a world that is increasingly interconnected technologically. The goals of the learning system approach are not only to better integrate available data, but to "ensure innovation, quality, safety, and value in health care" [69], through reforming the existing health-care structure to use new data and approaches to treatment evaluation, to integrate patient input and offer personalized solutions, and to streamline current health-care practices in order to avoid preventable medical errors and inappropriate medical management.

A start toward that road is the PROMIS tools initiative from the NIH [70], which consists of developing, validating, and using a common tracking metric across medical systems and health-care providers. In this specific case, the PROMIS tools assess patient-reported outcomes (clinical information that typically cannot be obtained through objective medical testing, such as patient well-being and quality of life) in the clinical practice setting and are designed to allow comparisons of the test sample with a national database. Several dimensions of the PROMIS tools have been validated against commonly used tools [71–73], though PROMIS has not always had better performance against historic instruments that have been used to assess patient-reported outcomes in both clinical trial and clinical practice [74, 75]. Implementation/adoption challenges clearly remain [76] and the road toward a learning health-care system based on good, shared data has just begun. Another ongoing effort in a similar direction is the establishment of the PCORnet in 2014 [77]—a national network for conducting real-world CER. PCORnet aims to collect and share data from routine patient care (in hospitals, doctors' offices, and community clinics) and from specialty patient networks/registries (rare diseases, common conditions), to provide common data querying and analytic tools, and to support pragmatic clinical trials. PCORnet's ultimate goal is to conduct CER "faster, with more power, and at lower cost" [77]. As with NIH's PROMIS initiative, this effort is still at an initial establishment/early development stage.

Continuing to engage in discussing potential solutions to barriers and reaching consensus on how to best perform CER in a cost-effective manner will no doubt continue to move us forward with proliferation and adoption of such research. There has been a lot of progress over the past decade, and the efforts such as PROMIS and PCORnet will continue to advance the cause.

5.6 Conclusions

Comparative evidence is a necessary component in the decision-making process with regard to any product purchased. It is infinitely more precious when individual and population health is at stake. The current state of CER is rapidly evolving and hotly debated, but, nonetheless, frequently in use and growing in popularity. There simply may not be enough resources or time to develop perfect CER for all decision making (with all the variables considered and uncertainties quantified in a dynamic fashion), so all stakeholders must make difficult trade-offs between the costs and timing and value of this information. This is a developing market that has strong methodological advances on the horizon. Less expensive approaches to develop information whose value in decision making exceeds the costs of its development are getting more attention. RPTs and adaptive pathways show great promise in overcoming some of the obstacles and challenges that prevent more CER from being generated today. Yet CER has still more to offer: there are better, less expensive, and fit for purpose approaches that still add value and do not require the formal resource investment that RCTs do. The greater goal of CER is to improve decision making, in an evidence-based medicine approach to personalized treatment choices using the best and most current information available, without waiting for or demanding perfection.

Regarding the crosswalk between CER and value assessment, the degree to which CER, or the lack of it, is simply used as a blunt instrument to either restrict access to treatments or negotiate lower product costs will signal to treatment developers just how serious policy makers are about individual patient-centered care and maintaining adequate incentives for innovation. Controversy and conflict that stems from treatment access restrictions based on population-level value assessments in a nation focused on optimizing at the individual level will continue. To date there has really been no effort to fully characterize or shape our collective thinking on the American social contract regarding health care in general and access to medicines in particular. Should all aspects of health care, including services, be subject to a cost per QALY assessment? Can that approach be reconciled with "no lifetime limits"? If cost per QALY is unacceptable, what is the alternative? Is relative cost-effectiveness for a specific treatment goal, as in cost per responder or number needed to treat, a more culturally competent approach compared to absolute cost-effectiveness using a dollar value threshold? It may very well be a step in the right direction, as cultural change is possible, although it does take time. For example, many will recall that it took decades for smoking and drunk driving to become uncool and socially unacceptable.

Finally, while we can reasonably predict that CER and its use for resource allocation are expected to grow in the coming decades, given the conflicting interests and the important needed alignment with American values, all viable alternatives and unintended consequences should be carefully considered when thinking about our way forward. All stakeholders in the nation's health system need to step up to resolve these overarching issues and combine the various perspectives and business models into a unified vision for the future.

Acknowledgments The authors wish to acknowledge Mei Sheng Duh, MPH, ScD, of Analysis Group, Inc. for her insights and feedback during manuscript development. Editing assistance was provided by Ana Bozas, PhD, a salaried employee of Analysis Group, Inc.

Disclaimer This research was not sponsored. The views and opinions expressed in this article are those of the author(s) and do not necessarily reflect the official policy or position of (their) current or past employers, or any related entities, or those of scientific collaborations of which they are members.

References

1. The Federal Coordinating Council for Comparative Effectiveness Research (2009) Report to the President and the Congress. U.S. Department of Health & Human Services, Washington, DC, pp 1–77
2. Patient-Centered Outcomes Research Institute (2016) Research we support. www.pcori.org. Patient-Centered Outcomes Research Institute, Washington, DC
3. Drummond M, Sorenson C (2009) Nasty or Nice? A perspective on the use of health technology. Value Health 12:8–13
4. Banta D, Jonsson E, Childs P (2009) History of the international societies in health technology assessment: International Society for Technology Assessment in Health Care and Health Technology Assessment International. Int J Technol Assess Health Care 25(Suppl 1):19–23
5. Hailey D (2009) Development of the International Network of Agencies for Health Technology Assessment. Int J Technol Assess Health Care 25(Suppl 1):24–27
6. Jonsson E (2009) History of health technology assessment in Sweden. Int J Technol Assess Health Care 25:42–52
7. Weill C, Banta D (2009) Development of health technology assessment in France. Int J Technol Assess Health Care 25:108–111
8. Perleth M, Gibis B, Gohlen B (2009) A short history of health technology assessment in Germany. Int J Technol Assess Health Care 25:112–119
9. Institute of Medicine (2013) Best care at lower cost. National Academies Press, Washington, DC
10. Neumann PJ, Fang CH, Cohen JT (2009) 30 years of pharmaceutical cost-utility analyses: growth, diversity and methodological improvement. Pharmacoeconomics 27:861–872
11. Institute for Clinical and Economic Review (ICER) (2015) Evaluating the value of new drugs and devices. ICER, Boston
12. United States Department of Health and Human Services (2015) Better, smarter, healthier: in historic announcement, HHS sets clear goals and timeline for shifting medicare reimbursements from volume to value. www.hhs.gov. United States Department of Health and Human Services, Washington, DC
13. Oregon Health & Science University. Drug Effectiveness Review Project (DERP)|Center for Evidence Based Policy|OHSU [Internet]. Available from: http://www.ohsu.edu/xd/research/centers-institutes/evidence-based-policy-center/evidence/derp/index.cfm
14. National Information Center on Health Services Research and Health Care Technology (NICHSR), National Institutes of Health, Health & Human Services (2016) HSRIC: Comparative Effectiveness Research (CER). Heal Serv Res Public Heal. U.S. National Library of Medicine, Bethesda
15. Institute for Clinical and Economic Review (ICER). Value-Assessment-Framework-One-Pager.pdf, http://www.icer-review.org/wpcontent/uploads/2014/01/Value-Assessment-Framework-One-Pager.pdf
16. Warshaw AL, Sutton JH (2012) Controlling state health care costs: Massachusetts forges ahead. Bull Am Coll Surg. American College of Surgeons, Chicago

17. Moloney R, Mohr P, Hawe E, Shah K, Garau M, Towse A (2015) Payer perspectives on future acceptability of comparative effectiveness and relative effectiveness research. Int J Technol Assess Health Care 31:90–98

18. Armstrong K (2012) Methods in comparative effectiveness research. J Clin Oncol 30:4208–4214

19. Concato J, Peduzzi P, Huang GD, O'Leary TJ, Kupersmith J (2010) Comparative effectiveness research: what kind of studies do we need? J Invest Med 58:764–769

20. Buck S, Mcgee J (2015) Why government needs more randomized controlled trials: refuting the Myths. Laura and John Arnold Foundation, Houston, TX, www.arnoldfoundation.org

21. Motheral BR, Fairman KA (1997) The use of claims databases for outcomes research: rationale, challenges, and strategies. Clin Ther 19:346–366

22. Benson K, Hartz AJ (2000) A comparison of observational studies and randomized, controlled trials. N Engl J Med 342:1878–1886

23. Ioannidis JP, Haidich AB, Pappa M, Pantazis N, Kokori SI, Tektonidou MG et al (2001) Comparison of evidence of treatment effects in randomized and nonrandomized studies. JAMA 286:821–830

24. Kitsios GD, Dahabreh IJ, Callahan S, Paulus JK, Campagna AC, Dargin JM (2015) Can we trust observational studies using propensity scores in the critical care literature? A systematic comparison with randomized clinical trials. Crit Care Med 43:1870–1879

25. Concato J, Shah N, Horwitz RI (2000) Randomized, controlled trials, observational studies, and the hierarchy of research designs. N Engl J Med 342:1887–1892

26. Tai V, Grey A, Bolland MJ (2014) Results of observational studies: analysis of findings from the Nurses' Health Study. PLoS One 9:e110403

27. Grimes DA, Schulz KF (2012) False alarms and pseudo-epidemics: the limitations of observational epidemiology. Obstet Gynecol 120:920–927

28. Goldman N (2010) New evidence rekindles the hormone therapy debate. J Fam Plann Reprod Health Care 36:61–64

29. Manschreck TC, Boshes RA (2007) The CATIE schizophrenia trial: results, impact, controversy. Harv Rev Psychiatry 15:245–258

30. Lewis S, Lieberman J (2008) CATIE and CUtLASS: can we handle the truth? Br J Psychiatry 192:161–163

31. Carroll J (2016) Debate over Duchenne drug hits fever pitch as D-day approaches. Fierce Biotech, Newton

32. Pettit DA, Raza S, Naughton B, Roscoe A, Ramakrishnan A, et al. (2016) The Limitations of QALY: A Literature Review. J Stem Cell Res Ther 6: 334. doi:10.4172/2157-7633.1000334

33. Siegel JP, Rosenthal N, Buto K, Lilienfeld S, Thomas A, Odenthal S (2012) Comparative effectiveness research in the regulatory setting. Pharm Med 26:5–11

34. Stafford RS, Wagner TH, Lavori PW (2009) New, but Not Improved? Incorporating comparative- effectiveness information into FDA labeling. N Engl J Med 361:1230–1233

35. "Health Policy Brief: The Independent Payment Advisory Board," Health Affairs, December 15, 2011.

36. Kesselheim AS, Robertson CT, Myers JA, Rose SL, Gillet V, Ross KM et al (2012) A randomized study of how physicians interpret research funding disclosures. N Engl J Med Mass Med Soc 367:1119–1127

37. Krimsky S (2013) Do financial conflicts of interest bias research?: an inquiry into the "Funding Effect" hypothesis. Sci Technol Human Values 38:566–587

38. Schott G, Pachl H, Limbach U, Gundert-Remy U, Ludwig W-D, Lieb K (2010) The financing of drug trials by pharmaceutical companies and its consequences. Part 1: a qualitative, systematic review of the literature on possible influences on the findings, protocols, and quality of drug trials. Dtsch Ärzteblatt Int 107:279–285

39. Rasmussen N, Lee K, Bero L (2009) Association of trial registration with the results and conclusions of published trials of new oncology drugs. Trials 10:116

40. Bero L, Oostvogel F, Bacchetti P, Lee K (2007) Factors associated with findings of published trials of drug-drug comparisons: why some statins appear more efficacious than others. PLoS Med 4:e184

41. Institute of Medicine (US) Forum on Drug Discovery, Development and T. Challenges in Clinical Research. National Academies Press (US), 2010
42. Anderson ML, Chiswell K, Peterson ED, Tasneem A, Topping J, Califf RM (2015) Compliance with results reporting at ClinicalTrials.gov. N Engl J Med 372:1031–1039
43. Dunn AG, Mandl KD, Coiera E, Bourgeois FT (2013) The effects of industry sponsorship on comparator selection in trial registrations for neuropsychiatric conditions in children. PLoS One. Public Library of Science 8:e84951
44. Flacco ME, Manzoli L, Boccia S, Capasso L, Aleksovska K, Rosso A et al (2015) Head-to-head randomized trials are mostly industry sponsored and almost always favor the industry sponsor. J Clin Epidemiol 68:811–820
45. Dreyer NA, Schneeweiss S, McNeil BJ, Berger ML, Walker AM, Ollendorf DA et al (2010) GRACE principles: recognizing high-quality observational studies of comparative effectiveness. Am J Manag Care 16:467–471
46. Agency for Healthcare Research and Quality (2015) Methods Guide for Effectiveness and Comparative Effectiveness Reviews
47. von Elm E (2007) The Strengthening the Reporting of Observational Studies in Epidemiology (STROBE) statement: guidelines for Reporting Observational Studies. Ann Intern Med Am Coll Physicians 147:573
48. Berger ML, Martin BC, Husereau D, Worley K, Allen JD, Yang W et al (2014) A questionnaire to assess the relevance and credibility of observational studies to inform health care decision making: an ISPOR-AMCP-NPC Good Practice Task Force report. Value Health 17:143–156
49. Jena AB, Philipson T (2007) Cost-effectiveness as a price control. Health Aff (Millwood) 26:696–703
50. Eli Lilly and Company (2015) Comparative Effectiveness Research. www.lilly.com. Eli Lilly and Company, Indianapolis
51. Institute of Medicine (US) Committee on Accelerating Rare Diseases Research and Orphan Product Development. Regulatory Framework for Drugs for Rare Diseases. National Academies Press (US), 2010
52. Chow S-C, Liu J-P (2008) Design and analysis of clinical trials: concepts and methodologies. John Wiley & Sons, Hoboken
53. CDER, FDA (2011) Responding to unsolicited requests for off-label information about prescription drugs and medical devices. Draft Guidance, Report Number: 2402769300, pages 1–15
54. Bi K (2012) What is "False or Misleading" off-label promotion? Univ Chicago Law Rev 82:975–1021
55. Berndt ER, Newhouse JP (2012) Pricing and reimbursement in US pharmaceutical markets. Chapter 8 in Patricia M. Danzon and Sean N. Nicholson, eds., The Oxford Handbook of the Economics of the Biopharmaceutical Industry, New York: Oxford University Press, pp. 201–265
56. Bach PB (2015) Indication-specific pricing for cancer drugs. JAMA 10021:2014–2015
57. Pearson SD, Dreitlein B, Henshall C (2016) Indication-specific pricing of pharmaceuticals in The U.S. health care system: a Report from the 2015 ICER Membership Policy Summit. Institute for Clinical and Economic Review, Boston
58. Cheng J (2008) An antitrust analysis of product hopping in the pharmaceutical industry. Columbia Law Rev 108:1471–1515
59. Sorenson C (2010) Issues in international health policy research in drug coverage and pricing decisions: a six-country comparison. Health Technol Assess (Rockv) 91:1–14
60. AAFP News (2014) CMS Reverses Decision on Medicare Prescription Rules After Outcry From AAFP, Others. http://www.aafp.org/news/government-medicine/20140319partDwithdraw.html
61. Squires D, Anderson C (2015) Issues in international health policy U.S. Health care from a global perspective: spending, use of services, prices, and health in 13 countries exhibits. Commonw Fund 15:1–12
62. Watson J. (2015) Hepatitis C: weighing the price of a cure. Medscape Gastroenterol. WebMD, LLC
63. Hiltzik M. (2015) How a hugely overpriced hepatitis drug helped drive up U.S. health spending. LA Times

64. Rein DB, Wittenborn JS, Smith BD, Liffmann DK, Ward JW (2015) The cost-effectiveness, health benefits, and financial costs of new antiviral treatments for hepatitis C virus. Clin Infect Dis 61:157–168

65. Applied Clinical Trials (2016) O'Donnell P. Adaptive Pathways – Pilot or Plot? http://www.appliedclinicaltrialsonline.com/adaptive-pathways-pilotor-plot

66. Mahajan R, Gupta K (2010) Food and drug administration's critical path initiative and innovations in drug development paradigm: challenges, progress, and controversies. J Pharm Bioallied Sci 2:307–313

67. Health Action International (HAI), The International Society of Drug Bulletins (ISDB), The Medicines in Europe Forum, The Mario Negri Institute for Pharmacological Research, The Nordic Cochrane Centre, WEMOS. "adaptive licensing" or "adaptive pathways": Deregulation under the guise of earlier access. 2015

68. Institute of Medicine (IOM) (2007) The evolving evidence base—methodologic and policy challenges. In: Olsen LA, Aisner D, McGinnis JM (eds) Roundtable evidence-based medicine, National Academies Press (US), p 374

69. Institute of Medicine (IOM) (2007) The learning healthcare system: workshop Summary. In: Olsen LA, Aisner D, McGinnis JM (eds) Roundtable evidence-based medicine, Washington (DC): National Academies Press (US), p 374

70. Broderick JE, DeWitt EM, Rothrock N, Crane PK, Forrest CB (2013) Advances in patient-reported outcomes: the NIH PROMIS(®) measures. EGEMS (Washington, DC) 1:1015

71. Pilkonis PA, Yu L, Dodds NE, Johnston KL, Maihoefer CC, Lawrence SM (2014) Validation of the depression item bank from the Patient-Reported Outcomes Measurement Information System (PROMIS) in a three-month observational study. J Psychiatr Res 56:112–119

72. Schalet BD, Revicki DA, Cook KF, Krishnan E, Fries JF, Cella D (2015) Establishing a common metric for physical function: linking the HAQ-DI and SF-36 PF Subscale to PROMIS(®) physical function. J Gen Intern Med 30:1517–1523

73. Schalet BD, Rothrock NE, Hays RD, Kazis LE, Cook KF, Rutsohn JP et al (2015) Linking physical and mental health summary scores from the Veterans RAND 12-Item Health Survey (VR-12) to the PROMIS(®) Global Health Scale. J Gen Intern Med 30:1524–1530

74. Snyder CF, Herman JM, White SM, Luber BS, Blackford AL, Carducci MA et al (2014) When using patient-reported outcomes in clinical practice, the measure matters: a randomized controlled trial. J Oncol Pract 10:e299–e306

75. Kean J, Monahan PO, Kroenke K, Wu J, Yu Z, Stump TE et al (2016) Comparative responsiveness of the PROMIS Pain Interference Short Forms, Brief Pain Inventory, PEG, and SF-36 Bodily Pain Subscale. Med Care 54:414–421

76. Eisenstein EL, Diener LW, Nahm M, Weinfurt KP (2011) Impact of the Patient-Reported Outcomes Management Information System (PROMIS) upon the design and operation of multi-center clinical trials: a qualitative research study. J Med Syst 35:1521–1530

77. Patient-Centered Outcomes Research Institute (2016) PCORnet: the National Patient-Centered Clinical Research Network. Patient-Centered Outcomes Research Institute, Washington, DC

Chapter 6
Impact of Comparative Effectiveness Research on Drug Development Strategy and Innovation

Joshua P. Cohen, Joseph A. DiMasi, and Kenneth I. Kaitin

Abstract Food and Drug Administration approval of a new drug, or new indication for an existing drug, is a necessary but increasingly insufficient condition for market access in the United States. Payers evaluate a drug's clinical safety and effectiveness profile in relation to existing standards of care. The collected evidence is referred to here as comparative effectiveness research (CER). Augmented with evidence from cost-effectiveness and budget-impact studies, CER elucidates a drug's value, which, in turn, helps inform payer pricing and reimbursement decisions. Such decisions can reach back to impact clinical development programs and hence future innovation, as they indicate key value parameters that payers, health-care providers, and patients are seeking in new pharmaceutical products. In this chapter, we address what CER delivers in terms of information regarding the value of pharmaceutical products, to whom that information is directed, and the impact that CER can have on drug development strategy and biomedical innovation.

6.1 Introduction

Until recently, Food and Drug Administration (FDA) approval of a new drug, or a new indication for an existing drug, implied market access. Patients either paid for drugs out of pocket, or insurers (payers) reimbursed, imposing few limits on coverage. Over the past decade, however, a notable rise in drug prices has made paying the full retail price out of pocket unaffordable for many patients [1]. At the same time, payers increasingly established conditions of reimbursement, including patient cost sharing and other tools of drug utilization management. Given that payers bear most of the cost of new treatments, they play an instrumental role with respect to facilitating market access [2, 3]. As a result, in general, FDA approval of a new drug, or new indication for an existing drug, is a necessary but insufficient condition for market access.

J.P. Cohen (✉) • J.A. DiMasi • K.I. Kaitin
Tufts Center for the Study of Drug Development, Boston, MA, USA
e-mail: joshua.parsons.cohen@gmail.com

© Springer Nature Singapore Pte Ltd. 2017
H.G. Birnbaum, P.E. Greenberg (eds.), *Decision Making in a World of Comparative Effectiveness Research*, DOI 10.1007/978-981-10-3262-2_6

Payers evaluate a newly approved drug's clinical safety and effectiveness profile in relation to existing standards of care. The collected evidence is referred to as comparative effectiveness research (CER). CER weighs the benefits and harms of medical interventions used to prevent, diagnose, treat, and monitor clinical conditions to determine which intervention works best for particular types of patients and in different settings [4]. The emphasis tends to be on investigation of outcomes applicable to real-world clinical practice patterns, as opposed to clinical trial data. This implies the use of post-marketing observational research data in addition to evidence from randomized controlled clinical trials. The latter is often lacking and prohibitively expensive to generate, as well as tending to focus on very short-term impact. Consequently, payers opt for greater use of observational study designs, including pragmatic trials and patient registries. Note, pragmatic clinical trials and patient registries aim to inform post-marketing decision-making by using head-to-head comparisons of alternative treatments, heterogeneous patient populations, and outcomes meaningful to patients, prescribers, and payers.

Augmented with evidence from cost-effectiveness and budget-impact studies, CER elucidates a drug's value to payers, health-care providers, and patients. Here, value is defined broadly as health outcomes achieved per dollar spent. Payer pricing and reimbursement decisions are partly a function of a drug's value, in addition to availability of treatment alternatives and state and federal requirements and regulations. Moreover, such decisions can reach back to impact clinical development programs and hence future innovation, as they indicate key value parameters that payers, health-care providers, and patients are seeking from pharmaceutical products. It is worth noting that clinical development does not end when a drug is approved. As part of post-marketing surveillance, the pharmaceutical sponsor must invest in the generation of real-world data to determine safety, labeling comprehension, and unexpected outcomes as a result of wider usage in a heterogeneous population [5].

In this chapter, we address what type of information CER provides, to whom it is directed, and what the likely impact is on drug development strategy and innovation.

6.2 Pricing and Reimbursement Decisions

The public is not privy to pricing and reimbursement negotiations between the drug industry and payers. Nonetheless, we know the general contours of the pricing and reimbursement decisions, and the role that CER plays in informing that process.

6.2.1 Pricing

Before deciding on a new drug's list price, a pharmaceutical company will often perform a preapproval comparative effectiveness assessment, positioning the drug relative to competitors. This includes an evaluation of the anticipated price sensitivity of payers, patients, and health-care providers. Additionally, if the company believes that the drug provides additional benefits compared to existing treatment, this will be incorporated in the price point offered at launch. Also, if the company anticipates a high willingness to pay on the part of payers as a result of particularly favorable drug properties—e.g., targets a rare disease, has a novel mechanism of action, and addresses an unmet medical need—then that will be reflected in the list price [6]. Finally, when pricing a new drug, the company will take marginal cost of production into account, i.e., the cost of producing each additional unit, such as a pill or vial [7]. The marginal cost of production is substantially higher for large-molecule biologics than for small-molecule pharmaceuticals, because the manufacturing process is more complex.

Once the company sets a price, payers express a willingness to pay at that price or not. A negotiation ensues, which results in a transaction price, which is almost always lower than the list price. In this respect, the mutually agreed-upon transaction price is value based to the extent that it reflects how much purchasers are willing to pay for the drug. However, several key assumptions underlying a perfectly competitive market do not hold, which may cause the price and value of a drug to diverge. Indeed, by design, the pharmaceutical market lacks certain features of a competitive market. First, drugs are patented as single-source (branded) monopoly products for a limited period of time. It is during this period without generic competition that companies can charge a price that is higher than the marginal cost of production. Second, third-party insurance shields end users from the actual cost of the drug, which may lead to overuse in some instances (moral hazard). Third, the existence of asymmetries in information between suppliers and purchasers, as well as physicians and patients (principal-agent problems), can lead to distortions in the market [8].

6.2.2 Reimbursement

For proprietary reasons, payers are usually hesitant to shed light on what they perceive to be a good value proposition for a newly approved drug [9]. Nor have they been forthcoming in spelling out the actual evidence base underlying drug reimbursement decisions [9]. What we do know is that reimbursement decisions are made by pharmacy and therapeutics committees responsible for evaluating newly approved drugs and establishing and prescribing guidelines for payers. In addition to considering company data on a drug's safety and efficacy, payers also assess available CER evidence and cost considerations for prescribing and reimbursement decisions. These considerations include formulary management tools used to steer health-care provider prescribing and patient choices: formulary exclusions, tiered formulary placement, patient cost sharing, prior authorization, step therapy,[1]

---

[1] The practice of first prescribing the most cost-effective drug to a patient (usually this is the least costly) and then prescribing a more costly drug if the first drug does not work.

on- and off-label indication restrictions, and quantity limits.[2] Since 2010, billions of dollars in federal funding of CER have been allocated to academic centers, payers, hospitals, and federal agencies [10]. This has led to an increase in the body of evidence available to gauge the value of approved drugs as well as that of products in development. For example, in January 2011, Medicare carrier Palmetto GBA announced its intent to drop coverage of bevacizumab (Avastin) for metastatic breast cancer [11]. This decision was informed by CER. Armed with this evidence, payers are ratcheting up pressure on the pharmaceutical industry to develop products that deliver value.

6.3   Payer Value-Based Approaches

The buzzword in policy circles is value. To reiterate, our short-hand definition of value is health outcomes achieved per dollar spent. Payers are experimenting with *value-based* approaches to formulary management. These include exclusion lists, value-based insurance design, coverage with evidence development, and risk-sharing agreements.

1. *Exclusion lists*: Several pharmacy benefit managers have removed certain brand products with clinically equivalent alternatives from their formularies entirely [12]. Exclusion lists are purportedly a function of value, with excluded products having less value than recommended drugs. However, a business purpose related to exclusions is extraction of greater rebates. When a pharmacy benefit manager delists a drug in a particular class, it rewards the manufacturers of competing products with an increase in market share in exchange for rebates, which lowers the effective price paid. This has occurred, for example, in the class of hepatitis C drugs. The fact that exclusion lists vary significantly across two major pharmacy benefit managers, with one excluding many products that the other recommends, suggests that rebates may play a greater role than value calculations.

2. *Value-based insurance design (VBID)*: Approximately one-third of US payers have adopted VBID, wherein they have reduced patient cost sharing for a number of "high-value" products in disease categories, such as diabetes, and raised cost sharing for certain "low-value" products. Here, "high-value" indicates a product is relatively cost-effective [13]. Premera Blue Cross, an independent licensee of Blue Cross Blue Shield Association, recently piloted a value-based formulary (VBRx) for the drug benefit offered to its own employees. The Premera VBRx appears to be the broadest application thus far of the VBID concept in the United States. Evidence from the initial rollout of the program suggests that it may lead to modest cost savings for Premera [14]. If applied more systematically across a wide range of payers, VBID could help steer resource

---

[2]The practice of imposing limits on the numbers of prescriptions that can be filled, or dosing per prescription.

allocation toward the development of high-value products that would obtain preferential formulary placement. On the other hand, systematic application of VBID could reduce incentives to invest in the development of innovations with an anticipated low value. This could also imply that there would be less money to invest in orphan disease treatments. This is because the majority of orphan disease treatments are not considered cost-effective, as manufacturers tend to charge high prices to offset relatively small patient markets [15].

3. *Coverage with evidence development (CED)*: CED implies coverage of a drug with the stipulation that payers and manufacturers collect post-marketing data for a specified period of time on the drug's real-world safety and effectiveness. Medicare has instituted CED programs for more than a dozen medical devices and a handful of drugs [16]. To continue to have market access after the evidence-gathering period, manufacturers must demonstrate that their products confer added benefits. In order for beneficiaries to gain access to a drug covered under a CED program, they must consent to being participants in a post-marketing clinical trial to test the safety and effectiveness of the product. The idea of only paying for health technologies that work at least as well in real-world settings as they do in clinical trials is intuitively attractive. Also, with CED, drug manufacturers know that there will at least be some degree of patient access to their new products upon regulatory approval, rather than possible outright denial of coverage by the payer. On the other hand, CED could hinder broader access and, in turn, incentives to innovate, given the burden of an additional post-marketing hurdle. Disappointing enrollment numbers in CED trials to date suggest that CED has had limited success in facilitating access [17].

4. *Risk-sharing arrangements*: Related to CED, risk-sharing arrangements between payers and manufacturers typically involve measurement of the performance of a technology in a defined patient population over a specified period, while tying reimbursement to clinical outcomes. To illustrate, in 2009, Merck and Cigna entered into a successful, long-term risk-sharing agreement—tying rebates for the diabetes medications sitagliptin (Januvia) and sitagliptin/metformin (Janumet) to improved blood glucose management [18]. And recently, Cigna and Aetna struck deals with Novartis, linking the price of the congestive heart failure drug sacubitril/valsartan (Entresto) to a reduction in the proportion of patients admitted to hospital for heart failure [19]. To date, however, there have been only a limited number of risk-sharing agreements in the United States. Challenges to broader implementation include high transaction costs and the absence of suitable data-gathering systems [19].

6.4   Reaching Back to Impact Drug Development

Analysts assert that payers, through their pricing and reimbursement decisions, not only exert influence over the diffusion of new products into the market, but they also impact the direction of future R&D and what technology companies choose to develop [6]. Expansion of data requirements for reimbursement decision-making, both before and after product launch, is a clear signal to drug developers of payers' attempts to better align their coverage and payment decisions with outcomes. It is unclear, however, whether the drug development paradigm is changing in response to increased evidence demands from payers, government policy-makers, and consumers. Anecdotal evidence suggests changes are under way, including the use of active comparators in clinical trials, involvement of stakeholders (e.g., patient representatives, payers, and providers) to help define key Phase IIb and III study design features, incorporation of patient-centered outcome measures, and earlier planning for Phase IV studies.

During clinical development, CER is intended to facilitate communication between regulators, payers, and drug manufacturers by providing feedback on study designs and endpoints needed to justify prescribing and reimbursement decisions. As a result, CER is having an impact on drug development and research strategies in several ways. First, beginning in the late 1980s, drug companies began establishing pharmacoeconomic divisions to conduct and review CER [20]. Second, market access and reimbursement teams are being formed as early as 5 years prior to the projected launch of the product [21]. Third, drug companies are establishing marketing strategies targeting doctors and patients as early as 3 years prior to the anticipated launch [21].

Although publicly available quantitative evidence is scant, there are several examples of companies' terminating or adjusting certain drug development plans on the basis of CER findings. For example, in February 2009, Pfizer announced the termination of two drugs, both of which were in late stages of development, declaring, "we don't believe they provide significant benefit over other therapies." Moreover, highlighting a renewed focus on breakthrough therapies, Pfizer stated its intention to "allocate additional resources to high-potential development programs as part of a continuing effort to deliver greater value to patients and shareholders" [22].

Companies are also establishing new partnerships with payers to analyze post-approval data. For example, AstraZeneca and WellPoint have been collaborating on studies that examine the comparative effectiveness of treatments already on the market, with an emphasis on drugs targeting chronic illnesses [23]. AstraZeneca says it will use the data to inform discussions with WellPoint concerning reimbursement and help with R&D decisions regarding products in the pipeline. In general, CER studies on competing products enable payers to assess the value proposition potential of approved products and compounds in development.

6.5 Discussion and Policy Implications

Different stakeholders may use CER findings in different ways. For the pharmaceutical industry, CER can suggest approaches to coordinate clinical development plans with payers, patients, and health-care providers. This may encourage the development of more innovative products. Value differentiation is crucial to facilitate market uptake in crowded therapeutic classes with high budget impact, such as statins, antidepressants, non-small cell lung cancer, HIV/AIDS, and hepatitis C. For payers, CER provides real-world data to evaluate treatments and manage utilization. For patients, CER is a useful tool for decision support as it provides head-to-head comparisons of safety and effectiveness for therapeutically similar products, which facilitates more informed "treatment shopping." And for physicians, CER offers the basis for evidence-based treatment recommendations. A number of medical professional societies and other organizations have developed and disseminated several high-profile value-based frameworks for assessing medical technologies. Three of these frameworks specifically focus on cancer drugs. These include approaches designed by the American Society of Clinical Oncology (ASCO), the Memorial Sloan Kettering Cancer Center (MSKCC), and the National Comprehensive Cancer Network (NCCN). The fourth—the Institute for Clinical and Economic Review (ICER)—has a broader purview.

In today's cost-conscious climate, reimbursement incentives often influence which treatments health-care providers prescribe and thus to which patients have access. These factors ultimately and directly affect drug manufacturers' revenues. As payers increasingly incorporate value into their product assessments and reimbursement decisions, drug developers are being incentivized to invest in products that they believe payers will view as innovative and therefore have better return on investment. Here, "innovative" refers to meeting significant unmet medical needs and providing greater net health benefits relative to existing treatments. In this context, we characterize a health-care technology as innovative if it (a) addresses a previously unmet health need and (b) offers additional effectiveness in comparison with existing treatment alternatives. There are degrees of innovativeness, along a spectrum from incremental improvements to breakthrough (e.g., new mechanism of action). An example of what would be considered an incremental innovation is the cholesterol-lowering drug rosuvastatin (Crestor), and an example of a breakthrough is sofosbuvir (Sovaldi), which has demonstrated a high cure rate in hepatitis C. As shown in Fig. 6.1, reimbursement incentives can influence pharmaceutical industry R&D strategy and which drug development programs are pursued.

Payer reimbursement and corresponding formulary management policies constitute a key mechanism through which CER and cost evidence may influence physician prescribing practices and patient choices (Fig. 6.1). Here, CER informs reimbursement policies. In turn, these policies affect product developers' return on investment in three ways. The first is by establishing a particular payment level or sales price for a drug approved for marketing. The second is by setting a volume of sales at that payment level, as may occur in the case of competitive bidding. The third is by influencing seller costs associated with development, manufacture, or sale of a drug. In turn, these decisions may have repercussions on drug manufacturer R&D.

The value-based approaches to payer formulary management discussed above beg the question of what constitutes "value" and what evidence best approximates value. Value can correspond to innovation, in that developing and producing a cost-effective technology (i.e., one that improves efficiency) may count as an innovation. However, the relationship between CER, innovation, and value is complicated.

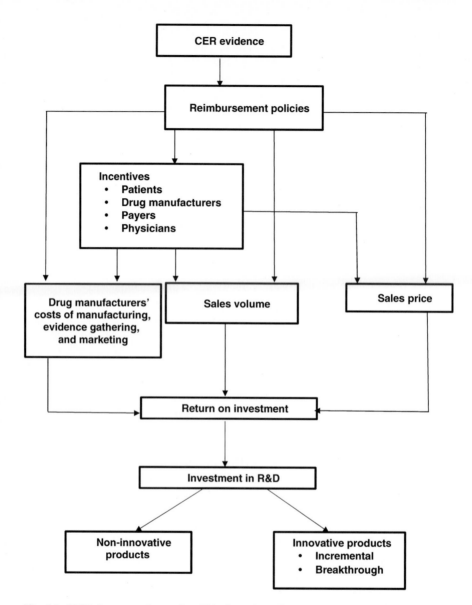

Fig. 6.1 CER's impact on innovation. This figure identifies channels though which reimbursement policies may influence incentives to steer prescribing and reimbursement of pharmaceuticals. It also demonstrates the link between CER and reimbursement and, in turn, investment in R&D

In this respect, it is important not to take too narrow a view of innovation and value and equate the terms with cost-effectiveness. This is because innovative technologies are comparatively effective, relative to existing standards of care, *but not necessarily cost-effective*. To illustrate, the most recently approved orphan drugs would not be considered cost-effective [24]. Each additional unit of benefit (e.g., quality-adjusted life year) derived from a typical orphan drug requires a much larger outlay of resources than the expense needed to obtain the same unit of benefit from a typical non-orphan drug. Yet without considering the cost, orphan drugs are comparatively effective relative to existing standards of care and thus innovative. Therefore, the value concept must incorporate broader terms than cost-effectiveness. Addressing unmet medical need, especially in the absence of treatment alternatives, also adds value.

What all the value-based approaches to formulary management have in common is that they show that value needs to be aligned with price. As previously mentioned, the terms value and price are equal only if certain market assumptions hold. To align with price, a drug's value must be measured in terms of benefits conferred across multiple dimensions [25]. Recently, several methods have been unveiled that seek to determine prices for drugs that commensurate with their value. "Each [method] aims to convert [CER] evidence about the improvement in patient outcomes a drug provides into a price for that drug compared with other treatment options" [26]. Theoretically, there should be a correlation between preferred formulary placement with lower patient cost sharing and greater cost-effectiveness, fewer safety concerns, or greater certainty around evidence. However, at present, it is unclear whether there is an association between these factors and preferred formulary placement [27]. Furthermore, the empirical question of whether patient cost sharing and other conditions of reimbursement imposed on drugs are, on balance, a reflection of evidence concerning their comparative cost-effectiveness profile remains unanswered [28]. This suggests that evidence is not necessarily being applied in such a way that formularies reflect value [29, 30]. Finally, there have been disagreements about how to interpret CER evidence and reach unambiguous prescribing and reimbursement decisions. This is echoed, for example, in a long-standing controversy surrounding the US Preventive Task Force's recommendations concerning breast cancer screening. Among experts, major differences of opinion persist with regard to the interpretation of CER evidence on the efficacy of breast cancer screening [31].

It should also be noted that CER evidence gathering can be perceived as a burden by the pharmaceutical industry, not only in terms of resources spent collecting data but also in the inherent uncertainty attached to CER findings once they are published [5, 32]. Schneeweiss et al. suggest that the increasing demand from payers for CER evidence may slow the process of bringing new drugs to the market, unless manufacturers can successfully incorporate modifications to their drug development programs to anticipate CER's influence on payer, patient, and health-care provider decisions [33]. In light of this, it has been posited that the industry's recent decline in return on investment, as measured in peak sales following product approval, is partly attributable to reimbursement hurdles and other post-approval data requirements [34].

To prevent continued decline in return on investment, it is incumbent on the pharmaceutical industry to develop products that are considered by payers, patients, and health-care providers to add value and that are priced at a level that sustains company profitability and stimulates R&D for future innovation [35]. This implies the need for a close examination of and potential modifications to the clinical development pipeline to better anticipate end user demand.

References

1. Schondelmeyer S, Purvis L (2016) Trends in retail prices of prescription drugs widely used by older Americans, 2006 to 2013 February 2016. http://www.aarp.org/content/dam/aarp/ppi/2016-02/RX-Price-Watch-Trends-in-Retail-Prices-Prescription-Drugs-Widely-Used-by-Older-Americans.pdf
2. Berndt ER, Newhouse JP (2012) Pricing and reimbursement in US pharmaceutical markets. Oxford Handbook of the Economics of the Biopharmaceutical Industry. doi: 10.1093/oxfordhb/9780199742998.013.0008
3. Cohen J (2013) Comparative effectiveness research: does it matter? Clin Ther 35(4):371–379. http://www.sciencedirect.com/science/article/pii/S0149291813000088
4. Garber A, Sox H (2010) The role of costs in comparative effectiveness research. Health Aff 29(10):1805–1811
5. Milne CP, Cohen JP, Felix A, Chakravarthy R (2015) Impact of postapproval evidence generation on the biopharmaceutical industry. Clin Ther 37(8):1852–1858
6. Cohen J (2015) Known knowns and unknowns of U.S. drug pricing. PLoS Med. November 2015. http://blogs.plos.org/yoursay/2015/11/10/known-knowns-and-unknowns-of-u-s-drug-pricing/
7. Howard D, Bach P, Berndt E, Conti R (2015) Pricing in the market for anticancer drugs. J Econ Perspect 29(1):139–162
8. Arrow K (1963) Uncertainty and the welfare economics of medical care. Am Econ Rev 53(5):941–973
9. Schneider E, Timbie J, Fox S, van Busum K, Caloyeras J (2013) Dissemination and adoption of comparative effectiveness research findings when findings challenge current practices. Office of the Assistant Secretary for Planning and Evaluation (ASPE) Report. 2013. https://aspe.hhs.gov/legacy-page/dissemination-and-adoption-comparative-effectiveness-research-findings-when-findings-challenge-current-practices-142166
10. Donnelly J (2010) Health Policy Brief. Health Affairs 1–5. https://www.healthaffairs.org/healthpolicybriefs/brief.php?brief_id=27
11. Pollack A (2011) Medicare coverage for breast cancer drug ends in some states. New York Times, 6 Jan 2011. http://prescriptions.blogs.nytimes.com/2011/01/06/medicare-coverage-for-breast-cancer-drug-ends-in-some-states/
12. Deutsch H (2015) Don't believe what PBMs are saying – the truth behind formulary exclusion lists. https://www.pm360online.com/dont-believe-what-pbms-are-saying-the-truth-behind-formulary-exclusion-lists/ PM360, Dec 21, 2015
13. Bruen B, Docteur E, Lopert R, Cohen JP, DiMasi J, Dor A, Neumann P, Desantis R, Shih C (2015) The impact of reimbursement policies and practices on healthcare technology innovation. Final report for the Impact of Reimbursement Policies and Practices on Healthcare Technology Innovation project. Washington, U.S. Department of Health and Human Services, Office of the Assistant Secretary for Planning and Evaluation
14. Sullivan S, Yeung K, Vogeler C, Ramsey S, Wong E, Murphy C, Danielson D, Veenstra D, Garrison L, Burke W, Watkins J (2015) Design, implementation, and first-year outcomes of a value-based drug formulary. J Manag Care Spec Pharm 21(4):269–275
15. Hyry H, Stern A, Cox T, Roos J (2014) Limits on the use of health economic assessments for rare diseases. Q J Med 107:241–245
16. Guidance for public, industry, and CMS staff, coverage with evidence development. http://www.cms.gov/Medicare/Coverage/Coverage-with-Evidence-Development/index.html Nov 20, 2014
17. Cohen JP, Dong J, Lu CY, Chakravarthy R (2015) Restricting access to Amyvid: medicare coverage criteria inconsistent for drugs and diagnostics. Br Med J. http://dx.doi.org/10.1136/bmj.h3333
18. Garrison L, Towse A, Briggs A, de Pouvourville G, Grueger J, Mohr P, Severens JL, Siviero H, Sleeper M (2013) Performance-based risk-sharing arrangements – Good practices for design, implementation and evaluation: report of the ISPOR good research practices for performance-based risk-sharing arrangements task force. Value Health 16:703–719

19. Miller J (2015) Novartis, Roche find `outcome-based' drug pricing an elusive dream. Reuters. 12 Nov 2015. http://www.reuters.com/article/2015/11/12/us-roche-novartis-drug-pricing-idUSKCN0T11MM20151112
20. DiMasi JA, Caglarcan E, Wood-Armany M (2001) The emerging role of pharmacoeconomics in the R&D decision-making process. PharmacoEconomics 19(7):753–766
21. Ahlawat H, Chierchia G, van Arkel P (2013) Becoming a launch powerhouse. In: Beyond the storm: Launch excellence in the new normal. McKinsey. Insight into Pharmaceuticals and Medical Products series. 2013. http://www.mckinsey.com/~/media/McKinsey/dotcom/client_service/Pharma%20and%20Medical%20Products/PMP%20NEW/PDFs/PMP_Beyond_the_storm_Launch_excellence_in_the_new_normal
22. Chalkidou K (2010) The (Possible) impact of comparative effectiveness research on pharmaceutical industry decision making. Clin Pharmacol Ther 87(3):264–266
23. Davolt S (2011) AstraZeneca, WellPoint Partner to Offset CER Deficit. http://aishealth.com/blog/pharmacy-benefit-management/astrazeneca-wellpoint-partnership-help-lay-foundation-cer AISHealth, Feb 14
24. Largent E, Pearson S (2012) Which orphans will find a home? The rule of rescue in resource allocation for rare diseases. Hastings Cent Rep 42:27–34
25. Bach P, Pearson S (2015) Payer and policy maker steps to support value-based pricing for drugs. JAMA 314(23):2503–2504. doi:10.1001/jama.2015.16843
26. Sorenson C, Drummond M, Burns L (2013) Evolving reimbursement and pricing policies for devices in Europe and the United States should encourage greater value. Health Aff 32(4):788–796
27. Neumann P, Pei-Jung L, Greenberg D et al (2006) Do drug formulary policies reflect evidence of value? Am J Manag Care 12:30–36
28. Dean B, Ko K, Graff J, Localio A, Wade R, Dubois R (2013) Transparency in evidence evaluation and formulary decision-making: from conceptual development to real-world implementation. Pharm Ther 38(8):465–483
29. Neumann P, Weinstein M (2010) Legislating against use of cost-effectiveness information. N Engl J Med 363:1495–1497
30. Neumann P, Greenberg D (2009) Is the United States ready for QALYs? Health Aff 28(5):1366–1371
31. Nelson H, Tyne K, Naik A et al (2009) Screening for breast cancer: systematic evidence review update for the US Preventive Services Task Force. Agency for Healthcare Research and Quality 2009 Evidence Syntheses, No. 74
32. Moloney R, Mohr P, Hawe E, Shah K, Garau M, Towse A (2015) Payer perspectives on future acceptability of comparative effectiveness and relative effectiveness research. Int J Technol Assess Health Care 31(1/2):90–98
33. Schneeweiss S, Gagne J, Glynn R, Ruhl M, Rassen J (2011) Assessing the comparative effectiveness of newly marketed medications: methodological challenges and implications for drug development. J Clin Pharmacol Ther 90:777–790
34. Mullard A (2016) Industry R&D returns slip. Nat Rev Drug Discov 15(7). doi:10.1038/nrd.2015.41
35. Schaeffer S, McCallister E (2015) Paying the piper. BioCentury 2015. Sept 1. http://www.biocentury.com/biotech-pharma-news/coverstory/2014-09-01/22nd-biocentury-back-to-school-issue-time-to-try-new-pricing-schemes-a1

Chapter 7
Pricing of Pharmaceuticals: Current Trends and Outlook and the Role of Comparative Effectiveness Research

Christian Frois and Jens Grueger

Abstract The pricing of biopharmaceuticals has recently received increasing attention, with many calling into question the current model, particularly following the fiscal crunch that occurred in the wake of the 2008 financial crisis. In this article, we review the key elements of a global biopharmaceutical pricing approach – including its specificities compared to other product pricing, the approaches currently used to price biopharmaceuticals around the world, and the role of comparative effectiveness research (CER) – and discuss key trends and the potential future outlook for the pricing of pharmaceuticals. We argue that – despite significant payer pushback, negative press, and other challenges – the current pharmaceutical pricing model has been extremely successful in delivering transformative new medicines and there does not appear to be a credible alternative model. This is no small feat given the considerable uncertainty and complexity involved in biopharmaceutical decision-making for both payers and manufacturers. At the same time, some adjustments to the current pricing model and CER are identified as necessary to address current access hurdles and payer concerns. In particular, we discuss how the European Union payer-directed pricing model needs to evolve, and how innovators must become more adept at evaluating and communicating appropriately the value of their therapy, to ensure that patients worldwide can benefit from drug innovations without undue delay or barriers to access or reimbursement.

C. Frois (✉)
Analysis Group, Inc., Boston, MA, USA
e-mail: Christian.Frois@analysisgroup.com

J. Grueger
F. Hoffmann-LaRoche, Basel, Switzerland

© Springer Nature Singapore Pte Ltd. 2017
H.G. Birnbaum, P.E. Greenberg (eds.), *Decision Making in a World of
Comparative Effectiveness Research*, DOI 10.1007/978-981-10-3262-2_7

7.1 Background

The pricing of biopharmaceuticals has recently received increasing attention. This has been true for a long time outside of the United States (USA) – with the introduction of the National Institute for Health and Care Excellence (NICE) in the United Kingdom (UK) in 1999, the Arzneimittelmarkt-Neuordnungsgesetz (AMNOG) in Germany in 2011, pricing reforms in France and Japan, as well as price-related patent tensions in developing countries (e.g., HIV, vaccines, etc.) [1]. More recently, this has included France's president raising the challenge of high medicine prices for developing countries ahead of the G7 meeting and discussion by top European Medicines Agency (EMA) and national authority representatives of options to reduce medicine prices [2, 3].

However, the situation in the USA is new. In the past few years, the pricing of biopharmaceuticals in the USA has frequently made front-page coverage of major newspapers. This has included articles on large price increases (e.g., for Turing, Horizon, Valeant, or Mylan products), the high price and budget impact of new innovative drugs (e.g., Sovaldi, anti-PCSK9s, etc.), and the increasing focus on rare cancer and orphan drugs which together also lead to an increasing financial burden, etc. [4–6]. Concerns about high prices among US patients have been all the more acute that US payers have been making patients liable for a substantial and increasing percentage of drug costs and particularly for the most expensive therapies (e.g., by moving patient coverage to plans with higher patient coinsurance requirements and simultaneously introducing plans with increasing number of tiers). For example, as of 2016, a majority of Medicare prescription drugs covered by standalone Medicare Part D plans (PDPs) are now subject to coinsurance (58% vs. only 34% in 2015) [7].

This spotlight on drug prices is coming at a time when biopharmaceutical manufacturers have shifted their investments away from traditional disease areas that have seen market genericization to areas of high unmet need where better understanding of the disease biology has led to several new classes of medicines with the potential to transform patient outcomes. This includes curative medicines in immunology as well as cancer immunotherapies. In addition, manufacturers are facing an increasingly heterogeneous payer landscape, with significant payer fiscal constraints, new value frameworks, and market needs.

Many traditional chronic disease areas are now largely genericized by drugs that provide good efficacy and safety when taken regularly/appropriately (e.g., hypertension, dyslipidemia, diabetes, etc.) at low prices, considerably raising the bar for new drug innovation. In the USA, nearly 8 in 10 prescriptions filled are for generic drugs [8]. In the European Union (EU), off-patent medicines now account for 92% of the prescription volume [9], and the prices of these generic drugs are commonly used by health technology assessment (HTA) bodies and payers to benchmark the price of new pharmaceutical innovation seeking reimbursement. The trend can be expected to continue as biosimilar medicines enter the USA and ex-US markets. The US Food and Drug Administration (FDA) has already approved a few biosimilars for commercial sale in the USA: Zarxio (brand reference product Neupogen) in March 2015, Inflectra (brand reference product Remicade) in April 2016, Erelzi (brand reference product Enbrel) in August 2016, and Amjevita (brand reference product Humira) in September 2016. Altogether, there are now over 55 biosimilars in the US FDA Biosimilar Product Development Program referencing over 15 different innovative biologics [10].

This so-called patent cliff has resulted in significant savings for payers. In the USA, we have seen for the first time in recent history a reduction in the overall drug expenditure in 2012, thanks to loss of exclusivity for blockbuster drugs like Lipitor. While the expenditure on specialty care medicines has been increasing significantly recently, the savings generated from primary care medicines has kept the total drug spend relatively stable at 1.2% of gross domestic product (GDP) in Europe and at 2.4% of GDP in the USA over the last 15 years [11].

The payer landscape for biopharmaceuticals is highly heterogeneous. In particular the largest market, the USA, consists of a diverse mix of public and private payers (e.g., with a large number of private health insurance companies and pharmaceutical benefit companies making reimbursement and coverage decisions for commercially, Medicare- and Medicaid-insured patients). Ex-US the landscape is equally challenging, particularly for small companies with innovative new technologies, yet no preexisting global infrastructure. While the EMA provides a common regulatory framework for manufacturers looking to provide their therapies to EU patients, no comparable framework exists on the payer side, and decisions need to be considered country by country. Furthermore, although there can be exceptions, there has been an increasing trend toward devolution of budget responsibility from central fund holders to regional or local payers (e.g., in Spain, Sweden, and the UK). This devolution in turn creates additional barriers to access with different payer requirements and value drivers.

Developing economies represent currently about a third of total pharmaceutical spending, and the percentage is expected to grow in the coming years [12]. The potential for growth from emerging markets is all the more substantial given the disproportionate share of the overall world population in these markets, the increasing convergence in terms of wealth and disease epidemiology compared to the developed world (e.g., in terms of incidence and prevalence of people with hypertension, diabetes, cancer, etc.), and the growing commitment to universal health coverage. The increasing role of emerging markets comes with particularly significant challenges with disparate economic inequalities (and thus ability to pay for new medicines) and limited – but increasing – healthcare infrastructure. Historically, few of these countries have provided significant drug coverage to their citizens – with most paying directly out of pocket for their medicines. However, developing economies have started to introduce national/regional programs to enhance patient access to healthcare and medicines). The first biologic medicines are now included in the World Health Organization's essential drug list (e.g., Avastin, Herceptin, MabThera/Rituximab, Neupogen) [13].

The fiscal crunch that followed the 2008 financial crisis in developed countries, and more recently the slowdown of most developing countries, has aggravated budget constraints and spurred or accelerated healthcare reforms around the world.

Many countries have blocked access to innovative medicines including cancer drug therapies, and requests for price freezes and price reductions are now the norm with many payers (thus creating significant uncertainty on profitability for manufacturers). The crisis has hit some European countries particularly hard (e.g., Greece, Spain, and Portugal) and led to cuts in healthcare investments, with negative growth in health and pharmaceutical expenditure in several countries [14, 15]. In Greece, pharmaceutical prices were reduced by 28% and by more than 50% for certain therapeutic categories during the period January 2012 to August 2013 [16]. Even prosperous Germany introduced a price freeze on reimbursable drugs in August 2010 and imposed a mandatory 16% rebate on non-reference-priced drugs (e.g., excluding generics) [17].

This unprecedented fiscal squeeze on biopharmaceutical spending has increased the focus on relative value demonstration. Most payers in Europe now have well-established processes to assess the value of medicines (both from a clinical and economic standpoint). Value frameworks are now spreading to the USA. This includes the emergence of the Institute for Clinical and Economic Review (ICER) as a public commentator on the appropriateness of new therapies' pricing based on cost-effectiveness assessments – using methods drawn from frameworks used by national HTA authorities such as NICE in England and Wales, though without the formal rules for agency-manufacturer interactions that govern such HTA authorities. Other emerging value frameworks in the USA include those focused on cancer therapies such as the American Society of Clinical Oncology (ASCO)'s Value Framework [18], the Memorial Sloan Kettering Cancer Center's DrugAbacus [19], and the National Comprehensive Cancer Network's evidence blocks [20]. As payers increasingly focus on assessing the relative value of new therapies, this had led to a sharp uptick in requests for head-to-head data to achieve preferential reimbursement and/or price premium (and/or avoid access/price restrictions) vs. standard of care (SoC) comparators.

In part in response to this challenging landscape for the pricing of their biopharmaceutical products, manufacturers have reacted by moving away from their historical primary care blockbuster model to focus on products for higher unmet need, narrower populations, often biomarker-targeted subgroups in specialty care, oncology, rare diseases, etc.

These dramatic changes in the pharmaceutical model and environment could suggest that a revolution in the pricing of biopharmaceuticals is urgently needed. There is certainly a lot of popular support for that view in newspapers [21, 22]. However, while – as we will discuss – important adjustments need to take place in how we price medicines, much of the current approach remains and should remain valid over the coming years. We need to do more and better. A lot of the adjustments we will discuss will aim to better align price, value, and evidence development (with a key potential role for comparative effectiveness research [CER]) – both from an internal and external perspective. These adjustments will help improve pricing practices and outcomes for the biopharmaceutical industry and help further distance itself from non-pharmaceutical finance arbitragers (e.g., Turing, Valeant, etc.). Notably, while the price of new therapies for the treatment of cancer and other drugs

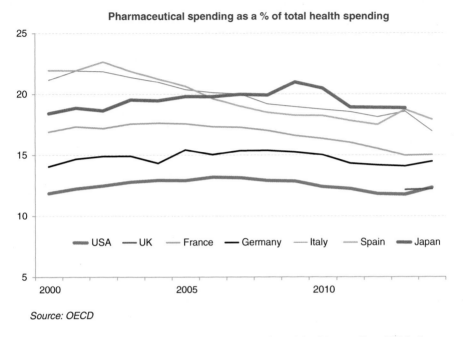

Over the last 15 years, pharmaceutical spending has been flat or declining for the US, Japan and EU-5

Pharmaceutical spending as a % of total health spending

Source: OECD

Fig. 7.1 Pharmaceutical spending as a percentage of total health spending (EU-5, Japan, and USA)

targeting small populations has steadily increased, as Fig. 7.1 shows, overall spending on pharmaceuticals has remained largely flat over the past 15 years across the USA, EU, and Japan (e.g., through large reductions in the costs of treating more prevalent diseases) [23]. In what follows, we start by defining the key elements of a global biopharmaceutical pricing approach, we then review some of the most recent changes that have occurred in biopharmaceutical pricing and its relationship to CER, and discuss the potential future outlook.

7.2   Some Biopharmaceutical Pricing Specificities

Pricing is clearly not a consideration specific to the biopharmaceutical industry. Every industry has a priori some discretion to set the price of its products and adjust it as a function of its costs and external environment including the value of the products perceived by purchasers, competition (potential and existing), bargaining powers of manufacturers and buyers, etc. However, pricing of biopharmaceuticals stands out in a number of ways and in particular for branded biopharmaceuticals.

First, the apparent disconnect between manufacturing/distribution cost and price is particularly stark for biopharmaceuticals. Research and development (R&D) is at the heart of pharmaceutical companies' success, and development failures are a key feature (e.g., with thousands of molecules typically needed to be screened to achieve commercial viability). Due to the high and sunk nature of the cost of R&D, special patent protection has been introduced that creates a temporary monopoly for the inventor/manufacturer and gives companies the power to price at value. Thus – by design – drugs are expected to be priced at a premium to achieve a return on innovation when they provide additional benefits to patients, payers, and societies. This creates challenges for drug manufacturers as it leaves only a short amount of time for a company to capture the rents of innovation, and manufacturers can only recoup these rents on a small subset of products which achieve market approval and success (with only negative returns for drug candidates that fail to achieve marketability).

This can leave drug manufacturers particularly exposed in times of financial pressures on healthcare systems, as medicines are often seen by government healthcare decision makers as the only variable cost (e.g., as opposed to hospital or physician charges). So whenever payers and policy-makers face a financial problem, the focus will typically be on "trimming" drug expenditures first, even if they only represent a small fraction of overall costs (e.g., between 10% and 20% of healthcare spend). Also in an increasingly connected world, it makes manufacturers particularly exposed to criticism of price gouging in the popular media.

Another characteristic of biopharmaceutical pricing is its complexity, as multiple parties may influence the price charged for a patient to use a particular drug, and prices for the same drug may vary substantially across a number of dimensions (e.g., geography, customer, insurer, provider, etc.). While in some (rare) cases the patient pays for the drug fully out of pocket, the primary payer for most drug purchases is typically not the patient, and the patient will only pay a fraction of the overall cost. Furthermore, there can be a web of contractual arrangements between manufacturers and the various actors in the biopharmaceutical supply chain (e.g., providers, wholesalers, pharmacies, public and private insurers, pharmaceutical benefit management companies, patients, etc.) that will affect the actual net price received by manufacturers. Price arrangements may take the form of cash discounts, rebates (i.e., payments made after the fact based on some action by the supply chain actor, e.g., provision of reimbursement, etc.), other forms of conditional pricing (e.g., risk-sharing, outcomes-based contracting, etc.), patient co-payment support programs, etc. For example, manufacturers will frequently provide significant discounts or rebates to these supply chain actors to ensure appropriate reimbursement and/or formulary listing compared to other products. The more competitive a particular market, the more these discounts and rebates may be substantial. Thus, analyzing drug prices will require careful study, and one needs to distinguish public price concepts (e.g., price ex-manufacturer to wholesaler, from wholesaler to pharmacy, from pharmacy to patient) from the net price realized by the manufacturer. Also in many countries, public price points will often consist of price quotes including or excluding value-added tax.

A third aspect that makes pricing for biopharmaceuticals special compared to other industries is that value can be particularly hard to measure and highly heterogeneous across customers/payers. This is clearly the case in therapeutic areas of high unmet medical need where there is little comparative price benchmark. In such therapeutic areas, saving or greatly bettering lives is often part of the value provided by new drugs (e.g., in cancer, many rare diseases, etc.). Value can also be challenging to assess – and vary substantially across stakeholders – in therapeutic areas with well-established SoC. Let's take the example of long-acting injectable antipsychotics. When second-generation injectable antipsychotic agents such as Risperdal Consta and Invega Sustenna were first introduced, some insurers, only considering the impact on their pharmacy budget, saw these drugs having negative value compared to existing alternatives due to their price premium. Fortunately, most insurers also considered the broader benefits that these therapies provided to their companies, for example, in terms of reduced hospitalization costs. However, even this does not reflect the full value provided by these therapies, as these new drugs may also provide benefits beyond health insurers' budgets (e.g., gains in patient productivity, quality of life, etc.). A number of payers and some countries around the world recognized this, which led to a number of innovative pricing arrangements when these drugs were introduced [24–26].

A fourth unusual aspect of pricing for biopharmaceuticals compared to many industries is that drugs – if appropriately used – can act as an insurance to prevent significant adverse outcomes for patients and their payers. Indeed while many drugs are used to respond to an acute condition, the most commonly used drugs are often used preventively to reduce underlying disease risks. This is, for example, the case for statins, antihypertensives, antidiabetes drugs, and many of the other drugs commonly used by patients around the world. Prices for these drugs can be particularly low compared to their value, as payers will often be reluctant to pay for future long-term benefits. While some payers have introduced some frameworks to try and address this (e.g., through long-term health economic modeling), few payers currently have done so. Also, in the USA, the high turnover of commercial insurance membership can act as a disincentive for commercial insurers to consider long-term benefits of therapies (e.g., as early investments to improve future patient health may mostly benefit other payers).

Finally, biopharmaceutical pricing is subject to an unusually high level of regulation around the world. Outside of the USA, there is no "free pricing," and even in the USA, substantial price regulation exists [27]. This is clearly the case for federal or state government program-related purchases (e.g., through price regulations for Medicare, Medicaid, Federal Supply Schedule, Veterans Affairs, etc.). However, private insurer pricing is also impacted. For example, the requirement that discounts provided to the Medicaid program exceed the best private customer discounts or rebates (Medicaid Best Price regulation) frequently leads biopharmaceutical manufacturers to curb net price reduction offers to some of their commercial customers and has been a noted barrier to some biopharmaceutical pricing innovation [28]. Outside the USA, the ability to increase prices once a drug is launched is very limited, and price reductions mandated by governments are common (often more to reflect budget shortfalls than drug/therapeutic area-specific considerations) [29]. Moreover, outside the USA, even the initial pricing of biopharmaceuticals is subject to detailed regulations, price negotiations, and government-imposed pricing approaches.

7.3   Biopharmaceutical Pricing Landscape Overview

A mix of two principal approaches is commonly used today to price biopharmaceuticals around the world; both are value-based pricing (VBP) approaches and try to link the added patient benefit provided by the new products vs. the current SoC to price. The first approach involves an assessment of the added patient benefit followed by price negotiation. Where no added patient benefit can be demonstrated, the price of the new product is automatically referenced against the price of the current SoC (which may include services, devices, or generic medicines). When an added benefit is demonstrated, prices are negotiated between the manufacturer and the national payer, and a more or less significant premium may typically be achieved based on the level of clinical benefit and overall value generated by the innovation. This approach is now commonly used to price products in France and Germany, with the initial patient benefit assessment conducted by the Commission de Transparence and the Gemeinsame Bundesausschuss Federal Joint Committee (G-BA), respectively.

A second approach is that the manufacturer sets an initial proposed price. Payers then review this proposed price on the basis of cost-effectiveness (i.e., the incremental cost of a new therapy compared to its incremental benefit to patients – e.g., measured in terms of quality-adjusted life years as in the UK, Canada, and a number of other countries), budget impact, and/or affordability compared to the clinical and other benefits of the product. Where the price is found not to be appropriate, the manufacturer can reduce the price or accept reimbursement restrictions to a subpopulation in which the product is found to provide an appropriate cost/benefit profile vs. the SoC.

Under both approaches, price negotiations are usually time bound, and prices may be renegotiated by payers when new competitors enter the market, new data become available, or original budget assumptions have not been met. Price reduction will typically take the form of a discount (or rebate) from the original price, and its level may not be publicly disclosed. This is in particular the case for further price reductions or contracting that occurs with subnational level entities (e.g., private and state insurers in the USA or regional/local payers outside of the USA).

Some additional variations exist across countries, e.g., with the use by payers of profit controls (either explicit as in the UK or implicit as in France) or cost-plus pricing (e.g., as in Japan). Also, in addition to price per se, payers may restrict the total cost per patient (episode of care price cap) or per population (budget cap). The methods and metrics to assess the added patient benefit, as well as the choice (or weight) of cost-effectiveness, budget impact, or affordability approach, will also vary by payer or country. In addition, smaller countries may benchmark their prices against those of other countries (international price referencing), and this can alleviate the need for an assessment of added patient benefit or cost-effectiveness. And regional or local purchasers or payers may negotiate lower prices for their patients (e.g., through tenders or other contractual arrangements).

7.4 Biopharmaceutical Pricing Innovation

There have been recurring discussions about the need for innovation in pricing of pharmaceuticals [30–32]. Some of the critiques of current pricing approaches generally include the complexity of current pricing and its limited transparency, the need for a stronger connection between the value of a pharmaceutical innovation and its price, as well as the insufficient affordability of medicines for some patients, payers, or countries.

As discussed above, it can be difficult to measure value objectively to inform pricing decisions, and the value of a new therapy in the real world may differ significantly from what may be assumed based on the results of its pivotal clinical trial results alone. For example, while payers generally recognize the burden of nonadherence or poor adherence to some chronic medications and its impact on health outcomes (e.g., for diabetes or schizophrenia/bipolar disorder), most payers around the world are reluctant to reward innovation that aims to improve nonadherence. Even with life-saving therapies, there is no commonly agreed view across countries and payers of the values of life or an extra amount of life. Furthermore, even for a given payer, the value of such extra life may seem to vary across products/therapeutic areas. As a consequence, significant differences in net prices may exist across countries and also within countries (e.g., across regions).

VBP has emerged as a key area of potential pharmaceutical pricing innovation. In a VBP arrangement, a pharmaceutical manufacturer and payer agree to link payment for a therapy to its value. Other names that have been used to describe such arrangements include (but are not limited to) outcomes-based, risk-sharing, pay-for-performance, and indication-based pricing arrangements. Value has many components. In addition to patient outcomes (mortality, morbidity, and quality of life), this includes improvements in efficiency of healthcare delivery, avoiding unnecessary treatments and procedures, and improving drug administration and compliance in treatment. Notably, when a product has several indications, the value-based price may differ across these indications. In the past, a "weighted average price" approach has typically been used by payers and manufacturers to guide price and access decisions. Previously payers have been reluctant to engage in indication-based pricing due to the complexity of managing an already large set of pharmaceutical products indication by indication, and in some cases, such as in Germany, the pricing framework does not even allow it. However, increasingly payers appear to be keen to move to indication-based pricing, particularly in therapeutic areas such as cancer and rare diseases, where the prices per patient are high and the overall budget impact can be significant. As a consequence, and as we discuss below, a number of countries and payers have started to introduce mechanisms to price-differentiate between indications and patient segments.

VBP also means that prices should vary substantially between and even within countries, with different prices for different value elements, and in particular different ability and willingness to pay (e.g., depending on wealth and healthcare priorities). The need is particularly stark in developing countries where large income/wealth inequalities and limited but growing healthcare infrastructure and public drug reimbursement support more price differentiation than currently exists. Indeed, with only a small fraction of the population of these countries with access to modern healthcare infrastructure and medicines, the prices in these countries have been much more closely pegged to EU and US prices than their relative wealth per capita would suggest. Manufacturers should be willing to work with payers and governments to develop differential pricing or tier-pricing models that can facilitate reimbursement and expand access, along with patient assistance programs to help self-pay patients afford treatment. Such models appear to be much needed to provide a fair and sustainable pricing approach that balances the needs of manufacturers with those of patients and other stakeholders (e.g., payers, providers).

A challenge for VBP, however, is that even when value can be well defined by both manufacturer and payer, it may be difficult for drug manufacturers to appropriately capture the value of their innovation through pricing. For example, devising a pricing model for new therapies that cure or significantly slow disease progression can be particularly challenging for manufacturers. While such therapies are of particularly high value for all healthcare stakeholders, by nature they tend to have a high upfront budget impact on payers, particularly as adoption tends to be quick, while the economic benefits are cumulative over time. Since it can be hard for payers to commit to reimburse appropriately such innovation, this can lead to significant limits on manufacturers' ability to get rewarded for their innovation. Sovaldi is a particular case in point. The therapy offered for the first time the promise of a cure for over 90% of patients with hepatitis C virus (HCV), a condition associated with significant mortality, patient burden, and healthcare costs. Yet, despite launching at a price in line with the current SoC (in part to manage the financial impact for payers) and strong evidence of its cost-effectiveness and positive budget impact over time, the unexpectedly strong uptake of the therapy in its first launch markets led many payers around the world to require significant price reductions and/or limit patient access.

7.5   CER, HTA, and Pricing Today

HTA plays a key role in helping payers and other healthcare stakeholders assess the value of medicines to guide their decisions [33]. HTA can be formally defined as a process of examining and reporting properties of a medical technology used in healthcare, such as safety, efficacy, feasibility, and indications for use, cost, and cost-effectiveness, as well as social, economic, and ethical consequences, whether intended or unintended [34–36]. It is a multidisciplinary field of policy analysis and aims to provide policy-makers with objective information, so they can formulate health policies that are safe, effective, patient-focused, and cost-effective. Many countries and payers have introduced HTA bodies to support their assessment of new drug technologies (e.g., Agency for Healthcare Research and Quality [AHRQ] in the USA, NICE in the UK, la Haute Autorité de la Santé [HAS] in France, G-BA/

Institute for Quality and Efficiency in Health Care [IQWiG] in Germany, etc.). Formal HTA plays today a critical role in the assessment of value and pricing for new drugs for many EU countries, and a neutral or negative assessment by HTA bodies – whether on a clinical or an economic basis – can greatly curtail the pricing prospects for a new drug therapy.

CER has emerged as an important consideration for HTA. CER can be defined as the generation and synthesis of evidence that compares the benefits and harms of alternative methods to prevent, diagnose, treat, and monitor a clinical condition or to improve the delivery of care [37]. The purpose of CER is typically to assist consumers, clinicians, purchasers, and policy-makers to make informed decisions that will improve healthcare at both the individual and population level. In the USA, public CER has recently received a boost in 2009 when the Obama administration – as part of the American Recovery and Reinvestment Act (ARRA) – allocated $1.1 billion for it and in 2010 established the Patient-Centered Outcomes Research Institute (PCORI) to conduct CER, as part of the Affordable Care Act (ACA) [38].

A priori, CER evidence could play a potentially critical role to inform pricing and reimbursement decisions, and the demonstration of an added patient benefit based on CER would be key to achieve premium pricing vs. existing SoC. Evidence is certainly central to manufacturer and payer pricing and reimbursement decisions. Historically, the focus has largely been on pivotal clinical trial data alone as a basis to inform pricing and reimbursement decisions. Recently, the weight and role of real-world evidence has considerably increased to establish the value of a new therapy. Typically this will include evidence on disease burden, as well as evidence on SoC real-world performance, costs and treatment patterns, and related unmet needs.

Yet, despite this central and increasing role of evidence in pricing and reimbursement decision-making, the role of CER has remained limited to date. A key factor in this is that in practice most therapies launch with limited comparative data. There are several reasons for this. First, the costs of running a clinical trial typically limit the number of comparators included in a trial. In this context, and in particular given variations in SoC across countries/payers, it is often difficult for manufacturers to include a sufficient number of SoC comparators to cover all country needs. Second, key regulatory agencies (e.g., FDA, EMA, Japan's Pharmaceuticals and Medical Devices Agency) have historically considered placebo-controlled trial evidence as appropriate for their decision-making, providing limited incentives for manufacturers to pursue more expensive and risky trials vs. active comparators. Another challenge with selecting an appropriate active comparator is that, in addition to variation across countries and payers, the SoC can evolve over time. Successful new competitive entry prior to trial completion or drug reimbursement approval can make past comparators obsolete as the new competitors displace them as the new SoC.

Still the ability to make meaningful comparisons compared to other treatment options remains a key need for payers. As a consequence, a particular type of CER has been flourishing lately: indirect comparisons. With the introduction of more robust methodologies (e.g., network meta-analysis, matching-adjusted based), indirect comparisons are increasingly considered by HTAs and payers and used by manufacturers to support pricing and reimbursement decisions when direct clinical comparison data is missing [39–41]. The impact can be significant, and in some instances, such indirect comparisons have played a key role in demonstrating superior benefits of a particular therapy to health authorities.

7.6   Pricing Uncertainty, Variability, and Risk

In practice, significant price uncertainty and price-related risks exist for manufacturers when they commit resources for the development of new therapies. Key contributing factors include (i) uncertainty about the evidence supporting their product's value and in particular CER evidence (and what may be required by health authorities in this regard), (ii) payer actions (e.g., healthcare reform, price reductions, etc.), and (iii) outside interventions (e.g., negative press coverage about their product, external assessments of the cost-effectiveness or budget impact benefits of the product, etc.).

By nature, pharmaceutical research and development is risky, and clinical trial or regulatory "surprises" can have a dramatic impact on the reimbursement and pricing prospects of a product. For example, the price that new oncology therapies can achieve will be very sensitive to their pivotal trial overall survival results. Germany is a case in point, where proving an incremental overall survival benefit just above 3 months can be the difference between a major and minor therapeutic clinical assessment by the G-BA and thus the ability of achieving substantial premium pricing. In this context, CER plays a critical role in price decision-making. Similarly to Germany, many countries have rules, formal or informal, that can automatically tie the price of new therapies to that of existing therapies if clinical benefits of the new therapy observed in the trials are not seen as meaningful vs. that of current SoC (often even if real-world differences may be expected to exist).

Payer actions also contribute substantially to price uncertainty. A lot of payers will seek gross or net price reductions from manufacturers over time. Often, these price reductions are largely independent of particular therapeutic area considerations and will generally more be a response to a local government fiscal shortfall. These mandated price reductions can considerably increase price uncertainty for manufacturers. First, directly by their cumulative effect over time, they can greatly reduce the actual price achieved by manufacturers. Also, by changing the price benchmark that will be used to evaluate the price of new entrants, these price reductions also create significant uncertainty for new therapies in clinical development.

Population restrictions are another key mechanism by which payers can affect price decisions. Increasingly payers around the world use such restrictions to limit use of new therapies to a subset of the overall patients for whom the therapy has been approved by regulatory health authorities (e.g., FDA, EMA, etc.). Such restrictions tend to encourage manufacturers to seek higher prices, both by shifting the focus to higher unmet need populations and to maintain the viability of their investments.

Outside interventions can also have a dramatic impact on the price of drugs and its perception. It is now common to see news coverage about the high price of pharmaceuticals, and politicians and other stakeholders will frequently raise concerns about the pricing of drugs [42, 43]. The experience of Zaltrap is a particularly stark example of the high impact that such outside interventions can have on drug pricing. In October 2012, two physicians from Memorial Sloan-Kettering Cancer Center wrote in the *New York Times* that the price of Zaltrap, a new therapy for colorectal cancer, was more than twice as high as that of Avastin, another therapy already used for the treatment of colorectal cancer, although they saw Zaltrap as no better clinically than Avastin and that as a consequence patients at their institution would not be prescribed Zaltrap [22]. A month later, the price of Zaltrap was cut in half by its manufacturer.

A high impact is not limited to existing stakeholders, and in an increasingly interconnected world, new actors can also affect pharmaceutical reimbursement and pricing decisions. For example, until recently the ICER was a little known research institute affiliated with Harvard University doing cost-effectiveness research in the USA. In 2015, ICER emerged with a bang, by assessing that an 85% reduction in the list price of anti-PCSK9 medicines would be necessary for them to be cost effective for the treatment of hypercholesterolemia [44]. Although US payers still today seldom use cost-effectiveness to guide their reimbursement decisions, the ICER assessment was used by payers to negotiate significant reductions in net prices with manufacturers of anti-PCSK9s. The anti-PCSK9 case was particularly impactful on the industry as anti-PCSK9s were seen as highly differentiated (with more than 50–60% reduction in LDL cholesterol in patients who previously failed to achieve their therapeutic target on statins and ezetimibe – the existing SoC) and expected then by many analysts as positioned to become the number one drug class given their high efficacy and large target population. Since then, ICER has published several assessments of new therapies being launched in the USA using a similar methodology and summarizing its evaluation with a price reduction assessment.

In theory, risk-sharing and other innovative pricing approaches could help address some of the uncertainty surrounding pharmaceutical prices, particularly as it relates to CER vs. the SoC. Coverage with evidence development was one of the first forays in trying to reduce the pricing uncertainty and risks associated with limitations in the available evidence at the time a drug is launched and priced. Subsequently, a number of other risk-sharing pricing approaches have been introduced and tested, and there is evidence that the appetite from payers and manufacturers is increasing both in the EU and USA [45]. However, the use of such pricing approaches remains limited currently and typically accounts for only a fraction of overall drug entering the market. A key barrier to their broader adoption can be their complexity, for example, as keeping the negotiation and implementation of such arrangements simple can be challenging. Still, recent research suggests that outcomes-based contracts are now better established and are poised – with new frameworks in place – to experience an increase in activity.

7.7 Trends and Future Outlook

The current pharmaceutical business and pricing model – despite significant payer pushback, negative press, and other challenges – has been extremely successful and provided powerful incentives for industry to deliver new medicines with transformative benefits for patients in areas of high unmet need. Thanks to these new medicines, we continue to see dramatic improvements in mortality, morbidity, and quality of life over the last 40 years. Notable areas of achievement include controlling deadly diseases like HIV, cancer, HCV, rare diseases, and significantly reducing mortality from chronic diseases (e.g., cardiovascular diseases). Currently there is a lack of a credible alternative model. The lack of pharmaceutical R&D success in countries that have tried more directed/restrictive price approaches is stark. This is clearly the case in socialist markets such as China, India, Russia, etc., but also in market economies that have tried to put significant price barriers for new pharmaceutical innovation. For example, price barriers following the introduction of NICE and cost-effectiveness in the HTA process in the UK led to suboptimal access to new cancer therapies and forced the UK government to introduce Cancer Drug Funds to palliate for this. As a result, the UK system is widely seen as suboptimal and does not address the needs for better access therapies including outside oncology. There are even some concerns that the UK's approach to drug pricing and reimbursement decision-making may be leading to a de-investment in the UK by pharmaceutical companies and since 2011 there has been a repeated decline in overall pharmaceutical research and development spend [46]. Also, the complexity of the current pricing system largely reflects the ability of the current model to accommodate widely differing needs across stakeholders.

Still, more flexible pricing approaches and CER are needed to address important access hurdles and payer concerns. Clearly, the demand for CER can be expected to continue unabated going forward, with supply constraints being the more limiting factor. On the pricing side, given the dynamics discussed above, the trend toward more differential or tiered pricing can also be expected to continue. We are likely to see more price variations across countries – for example, between the USA and EU, across European countries, between the USA/EU and emerging markets, and across emerging markets – as well as within countries, e.g., between and within public and private payer segments. Although it is difficult to comment too generally at the net level, given the limited discount/rebate transparency, in a number of therapeutic areas (e.g., oncology, rare diseases, etc.), there is an increasing price gap between the USA and EU prices for innovative new medicines. Since the USA and EU are the largest pharmaceutical markets, the trend may have a significant impact on decision makers. Increasingly, the EU market is being perceived as unwilling to pay for innovation. Already some manufacturers are evaluating for some of their assets whether to delay their launch in EU markets to prioritize other markets (e.g., Japan) and avoid price contagion (e.g., due to international price referencing). Such moves could greatly reduce the access of EU patients to pharmaceutical innovation.

Also, there is increasingly a requirement for integrating early on a drug's evidence strategy with the development of pricing assumptions, to establish the value of a therapy and support pricing decisions. This has been essential for rare disease therapies (e.g., gene therapy, rare cancer drugs) and is now a must-do for any therapy expected to have a meaningful impact on payers' budgets. An integrated evidence strategy comprises early consideration of potential clinical trials, real-world data, and modeling (e.g., including economic analyses, indirect treatment comparisons, extrapolations beyond trial duration and patient population, etc.) from a broader perspective than just that of regulators; a payer perspective is clearly essential given the crucial role of market access to a drug's commercial success. Other important perspectives include physician or physician group advocacy as well as patient advocacy. Such an integrated evidence approach should be initiated from the start of the product development's process and be informed by both clinical/scientific and potential pricing/commercial considerations. To help with this, engaging with HTAs and payers early on to review potential evidence needs and align on clinical development plans is key. This may be done through some of the scientific advice initiatives developed by some HTA and regulatory bodies or through direct research with some of these bodies. Companies that ignore this do so at significant risk. For example, we have seen a number of companies who used clinical trials designed primarily to gain prompt access with US regulators, facing significant hurdles to achieve access in Europe as well as in the USA despite significant clinical innovation associated with their drugs (e.g., anti-PCSK9s, cancer therapies).

The potential solutions to provide more flexible pricing approaches going forward include (i) personalized reimbursement model frameworks, (ii) international differential pricing (e.g., to address varying country affordability), as well as (iii) managed-entry agreements (e.g., to address payer needs for evidence on the clinical effectiveness of a medicine or budget/economic impact). Potential personalized reimbursement model frameworks comprise (i) multiple indication pricing models that allow a medicine that is approved in different indications and combinations to be priced according to the benefit it delivers in each indication and combination, (ii) combination pricing models that ensure that the benefits of the combination therapy are reflected while taking into consideration the limits of payers' healthcare budgets, as well as (iii) pay for performance models that base the level of reimbursement on a patient's response to a medicine over a specified time period.

So, how will the pharmaceutical pricing model need to evolve to better foster future innovation and address stakeholder needs? The US pricing model is likely to remain critical for most international biopharmaceutical companies; however, it faces a very uncertain future. On the one hand, the USA remains the largest pharmaceutical market in terms of revenues for the foreseeable future [47]. On the other hand, as happened with Germany with AMNOG, the outlook could considerably evolve and do so rapidly. Only a few years ago, considerations of cost-effectiveness were mostly absent historically in the USA. Now, with the emergence of ICER as a key assessor of how drug prices relate to value based on a cost-effectiveness methodology calculation, cost-effectiveness has increased in relevance for biopharmaceutical companies looking to launch a product in the USA. This has happened despite the fact that most US payers remain skeptical of cost-effectiveness as a tool to guide their formulary decision-making. Similarly, US political campaigns regularly encourage the emergence of new legislative ideas to try and curb biopharmaceutical drug prices. The most recent presidential campaign has been no exception, and both presidential candidates supported enabling Medicare to negotiate drug prices with manufacturers [48]. In 2015, Democratic members of Congress introduced draft legislation, the Medicare Prescription Drug Price Negotiation Act of

2015, to do just that. Although such legislation would be challenging to pass in a divided Congress, it shows how exposed the biopharmaceutical industry could be to significant changes in its economics and pricing model in the USA. As the pharmaceutical industry continues to invest in high unmet need – often rare – disease areas, another important action point to make the US price model more manageable for patients is for US payers to curb the recent trend to shifting a large share of healthcare costs to patients (e.g., through the use of coinsurance and increased plan tiering). As noted in previous research, patients should not have to decide between potentially life-altering treatment and debilitating medical debt, particularly when it affects patients with rare diseases where the financial insurance risk can be pooled across the entire population [49].

The EU experience in this regard can help manufacturers and payers avoid some of the pitfalls of such approaches. All geographies need to participate in rewarding biopharmaceutical innovation. We need greater social solidarity across leading economies and restrict price referencing and trade to countries with similar economic situations. Currently, higher US brand prices appear to subsidize in large part drug innovation for the rest of the world. This is unlikely to be sustainable, and Europe in particular needs to assume a larger role and share. Absent this, we run the risk of significant hurdles for EU patients to access the most promising medicines, as was seen in the UK for cancer patients prior to the introduction of Cancer Drug Funds (and by some benchmarks even since then). By introducing ever more byzantine and disparate frameworks, EU authorities increasingly make it difficult for biopharmaceutical manufacturers to continue to prioritize innovation for EU patients. This is clearly the case for small biopharmaceutical manufacturers with promising new drugs, yet limited ex-US capabilities and experience, but is also increasingly a consideration for established large companies as well.

The US pricing model, with its emphasis on fostering drug innovation, is a more promising template for future medical innovation around the world, and it can be credited for much of the success in redirecting the pharmaceutical industry toward highly innovative therapies for cancer and rare diseases. However, it is key for biopharmaceutical innovators to further distinguish themselves from fringe companies in the business of pharmaceutical pricing arbitrage. Going forward, it is critical that innovators' price decisions are clearly articulated around the value created by the new medicine. Innovators must become more adept at evaluating and communicating appropriately the value of their therapy. All parties would benefit from a stronger alignment between a drug's reimbursement decisions and a scientific assessment of the strength and quality of its clinical, health economics, and humanistic evidence. To that end, substantial changes in the way prices are set may become necessary in the future, for example, with price adjustments (up or down) becoming more of the norm as more information on the value of a therapy (or data on additional indications/uses) becomes available. Innovating pricing approaches such as outcomes-based contracting or including value-added/integrated healthcare services will need to be part of this. However, we see such changes as evolutionary and already largely under way. Both payers/governments and manufacturers have an important role to play, and any change in policy should carefully consider the long-term impact on drug innovation and as much as possible avoid raising the risks for innovators to pursue their R&D.

References

1. Clinton GJ (2016) Clinton Foundation helped 9 million with lower-cost AIDS drugs. Politifact. http://www.politifact.com/global-news/statements/2016/jun/15/hillary-clinton/clinton-clinton-foundation-helped-9-million-lower-/
2. Hollande F (2016) Towards a global agenda on health security. Lancet 387 (10034):2173–2174. doi:10.1016/S0140-6736(16)30393-2. http://dx.doi.org/10.1016/S0140-6736(16)30393-2
3. European Medicines Agency Can regulators influence the affordability of medicines? http://www.ema.europa.eu/ema/index.jsp?curl=pages/news_and_events/news/2016/05/news_detail_002529.jsp&mid=WC0b01ac058004d5c1. Accessed 28 Sept 2016
4. Millman J (2014) The drug that's forcing America's most important – and uncomfortable – health-care debate. The Washington Post. https://www.washingtonpost.com/news/wonk/wp/2014/07/24/the-drug-thats-forcing-americas-most-important-and-uncomfortable-health-care-debate/
5. Dennis B (2016) Rattled by drug price increases, hospitals seek ways to stay on guard. The Washington Post. https://www.washingtonpost.com/national/health-science/rattled-by-drug-price-increases-hospitals-seek-ways-to-stay-on-guard/2016/03/13/1c593dea-c8f3-11e5-88ff-e2d1b4289c2f_story.html
6. Johnson C (2015) A defining moment in modern health care. The Washington Post. https://www.washingtonpost.com/business/economy/a-defining-moment-in-modern-health-care/2015/09/23/d6496468-6229-11e5-b38e-06883aacba64_story.html?tid=a_inl
7. Pearson CF (2016) Majority of drugs now subject to coinsurance in medicare part D olans. Avalere. http://avalere.com/expertise/managed-care/insights/majority-of-drugs-now-subject-to-coinsurance-in-medicare-part-d-plans
8. U.S. Food & Drug Administration Facts about Generic Drugs. Updated June 28, 2016. http://www.fda.gov/drugs/resourcesforyou/consumers/buyingusingmedicinesafely/understand-inggenericdrugs/ucm167991.htm
9. IMS Institute (2015) The role of generic medicines in sustaining healthcare systems: a European Perspective. http://www.imshealth.com/files/web/IMSH%20Institute/Healthcare%20Briefs/IIHI_Generics_Healthcare_Brief.pdf
10. Frois C, Mortimer R, White A (2016) The potential for litigation in new era of biosimilars. Law360. http://www.law360.com/articles/842318/the-potential-for-litigation-in-new-era-of-biosimilars
11. IMS Health (2013) Pharmerging markets. https://www.imshealth.com/files/web/Global/Services/Services%20TL/IMS_Pharmerging_WP.pdf
12. Modern Medicine Network (2013) U.S. spending on medicines declined in 2012, IMS reports. http://drugtopics.modernmedicine.com/drug-topics/content/tags/ims-institute-healthcare-informatics/us-spending-medicines-declined-2012-im
13. World Health Organization (2015) WHO model list of essential medicines, 19th List. http://www.who.int/medicines/publications/essentialmedicines/EML_2015_FINAL_amended_NOV2015.pdf?ua=1
14. European Commission (2015) Study on enhanced cross-country coordination in the area of pharmaceutical product pricing. http://ec.europa.eu/health/systems_performance_assessment/docs/pharmaproductpricing_frep_en.pdf
15. Leopold C, Mantel-Teeuwisse AK, Vogler S, Valkova S, de Joncheere K, Leufkens HG, Wagner AK, Ross-Degnan D, Laing R (2014) Effect of the economic recession on pharmaceutical

policy and medicine sales in eight European countries. Bull World Health Organ 92(9):630–640d. doi:10.2471/blt.13.129114

16. Olga SC, Kaitelidou CD, Panagiota LS et al (2014) Investigating the economic impacts of new public pharmaceutical policies in greece: focusing on price reductions and cost-sharing rates. Value Health Reg Issues 4:107–114

17. Fernando S (2013) German drug rebate future at centre of debate as industry versus payer battle intensifies.FinancialTimes.http://www.ft.com/cms/s/2/6f533284-b100-11e2-80f9-00144feabdc0. html?ft_site=falcon&desktop=true#axzz4NLdUrqab

18. Schnipper LE, Davidson NE, Wollins DS, Tyne C, Blayney DW, Blum D, Dicker AP, Ganz PA, Hoverman JR, Langdon R, Lyman GH, Meropol NJ, Mulvey T, Newcomer L, Peppercorn J, Polite B, Raghavan D, Rossi G, Saltz L, Schrag D, Smith TJ, Yu PP, Hudis CA, Schilsky RL, American Society of Clinical O (2015) American Society of clinical oncology statement: a conceptual framework to assess the value of cancer treatment options. J Clin Oncol 33(23):2563–2577. doi:10.1200/jco.2015.61.6706

19. Memorial Sloan Kettering Cancer Center DrugAbacus: Evidence Driven Drug Pricing Project. http://www.drugabacus.org/

20. National Comprehensive Cancer Network NCCN Clinical Practice Guidelines in Oncology (NCCN Guidelines) with NCCN Evidence Blocks. https://www.nccn.org/evidenceblocks/

21. The New York Times Editorial Board(2016) Another drug pricing ripoff. The New York Times. http://www.nytimes.com/2016/08/25/opinion/another-drug-pricing-ripoff.html?_r=0

22. Bach PB, Saltz LB, Wittes RE (2012) In cancer care, cost matters. The New York Times. http://www.nytimes.com/2012/10/15/opinion/a-hospital-says-no-to-an-11000-a-month-cancer-drug.html

23. Howard DH, Bach PB, Berndt ER, Conti RM (2015) Pricing in the market for anticancer drugs. J Econ Perspect 29(1):139–162. doi:10.1257/jep.29.1.139 http://www.aeaweb.org/articles?id=10.1257/jep.29.1.139

24. Joubert S (2001) Accélération de l'Accés à l'Innovation Pharmaceutique : Etat des lieux et perspectives. [Thesis] University of Angers. http://dune.univ-angers.fr/fichiers/20061382/2015PPHA5045/fichier/5045F.pdf

25. Jaroslawski S, Toumi M (2011) Market access agreements for pharmaceuticals in Europe: diversityofapproachesandunderlyingconcepts.BMCHealthServRes11:259.doi:10.1186/1472-6963-11-259

26. Lu CY, Lupton C, Rakowsky S, Babar ZU, Ross-Degnan D, Wagner AK (2015) Patient access schemes in Asia-pacific markets: current experience and future potential. J Pharm Policy Pract 8(1):6. doi:10.1186/s40545-014-0019-x

27. Vernon JA (2002–2003) Drug research and price controls. Health & Medicine (Winter):22–25. https://object.cato.org/sites/cato.org/files/serials/files/regulation/2002/12/v25n4-7.pdf

28. Garrison LP Jr, Carlson JJ, Bajaj PS, Towse A, Neumann PJ, Sullivan SD, Westrich K, Dubois RW (2015) Private sector risk-sharing agreements in the United States: trends, barriers, and prospects. Am J Manag Care 21(9):632–640

29. Carone G, Schwierz C, Xavier A (2012) Cost-containment policies in public pharmaceutical spending in the EU. European Commission. http://ec.europa.eu/economy_finance/publications/economic_paper/2012/pdf/ecp_461_en.pdf

30. Goldman D, Nussbaum S, Linthicum M (2016) Rapid biomedical innovation calls for similar innovation in pricing and value measurement. Health Affairs Blog. http://healthaffairs.org/blog/2016/09/15/rapid-biomedical-innovation-calls-for-similar-innovation-in-pricing-and-value-measurement/

31. Siddiqui M, Rajkumar SV (2012) The high cost of cancer drugs and what we can do about it. Mayo Clin Proc 87(10):935–943. doi:10.1016/j.mayocp.2012.07.007. http://dx.doi.org/10.1016/j.mayocp.2012.07.007

32. Gilman D, Dowden N (2016) Is value-based drug pricing compatible with pharma innovation? NEJM catalyst. http://catalyst.nejm.org/is-value-based-drug-pricing-compatible-with-pharma-innovation/

33. Jipan X, Kalipso C, Isao K, Rebecca ED, Rifaiyat M, Arjun V, Cinzia M (2017) Policy considerations: ex-U.S. payers and regulators. In: Birnbaum HG, Greenberg PE (eds) Decision making in a world of comparative effectiveness research. Springer, Singapore

34. Haute Autorité de Santé (2007) General method for assessing health technologies. http://www. has-sante.fr/portail/upload/docs/application/pdf/general_method_eval_techno.pdf
35. European Commission Policy. http://ec.europa.eu/health/technology_assessment/policy/index_en.htm
36. Inahta HTA Tools & resources. http://www.inahta.org/hta-tools-resources/
37. Academy of Managed Care Pharmacy (2011) Comparative effectiveness re-search glossary of terms. http://www.amcp.org/CERglossaryofterms/
38. Ali R, Hanger M, Carino T (2011) Comparative effectiveness research in the United States: a catalyst for innovation. Am Health Drug Benefits 4(2):68–72
39. Haute Autorité de Santé (2009) Indirect comparisons methods and validity http://www.has-sante.fr/portail/upload/docs/application/pdf/2011-02/summary_report__indirect_comparisons_methods_and_validity_january_2011_2.pdf
40. National Institute for Health and Care Excellence (2013) Guide to the methods of technology appraisal2013.https://www.nice.org.uk/process/pmg9/resources/guide-to-the-methods-of-technology-appraisal-2013-pdf-2007975843781
41. Signorovitch JE, Sikirica V, Erder MH, Xie J, Lu M, Hodgkins PS, Betts KA, Wu EQ (2012) Matching-adjusted indirect comparisons: a new tool for timely comparative effectiveness research. Value Health J Int Soc Pharmacoeconomics Outcomes Res 15(6):940–947. doi:10.1016/j.jval.2012.05.004
42. Lee TT, Gluck AR, Curfman G (2016) The politics of medicare and drug-price negotiation (Updated).HealthAffairsBlog.http://healthaffairs.org/blog/2016/09/19/the-politics-of-medicare-and-drug-price-negotiation/
43. Lagrange C (2016) France's Hollande to seek regulation of medicine prices at G7, G20 summits. Reuters. http://uk.reuters.com/article/uk-france-health-who-idUKKCN0WP2JZ
44. Nazareth T, Ko JJ, Frois C, Carpenter S, Demean S, Wu EQ, Sasane R, Navarro R (2015) Outcomes-based pricing and reimbursement arrangements for pharmaceutical products in the US and EU-5: payer and manufacturer experience and outlook. Paper presented at the ISPOR 20th Annual International Meeting, Philadelphia, PA
45. Bradshaw J (2016) 'I would not want to be a cancer patient in England' says Pfizer boss. The Telegraph. http://www.telegraph.co.uk/finance/newsbysector/pharmaceuticalsandchemicals/12150564/I-would-not-want-to-be-a-cancer-patient-in-England-says-Pfizer-boss.html
46. Institute for Clinical and Economic Review (2015) PCSK9 inhibitors for treatment of high cholesterol: effectiveness, value, and value based price benchmarks https://icer-review.org/wp-content/uploads/2016/01/Final-Report-for-Posting-11-24-15-1.pdf
47. IMS Health (2015) Global medicines use in 2020: outlook and implications . http://www.imshealth.com:90/en/thought-leadership/quintilesims-institute/reports/global-medicines-use-in-2020
48. Gurnon E (2016) Where trump and clinton stand on health care and medi-care http://www.forbes.com/sites/nextavenue/2016/08/12/where-trump-and-clinton-stand-on-health-care-and-medicare/#540859451eb0
49. Balch A (2015) The need to eliminate barriers to personalized medicine. Am J Modern Care. http://www.ajmc.com/journals/evidence-based-oncology/2015/august-2015/the-need-to-eliminate-barriers-to-personalized-medicine

Part III
Evolving Stakeholder Considerations: Patients, Physicians, Regulators and Payers

Chapter 8
Are Real-World Data and Evidence Good Enough to Inform Health Care and Health Policy Decision-Making?

Marc L. Berger and James Harnett

Abstract We are entering the era of digitized real-world data to inform health-care and health policy decisions—including medical claims, electronic health records, sensor/real-time monitoring data, etc. This will increasingly be merged with novel other data sources—including purchasing preferences self-reporting of experience via social networks, weather, etc. Together this big data has the potential to revolutionize decision-making and help to realize the goal of a "learning health-care system." However, there are concerns that the data are not of sufficient quality—both in terms of accuracy and completeness. Are data good enough today? If not, when will we know when they are good enough? There are multiple ongoing and potential applications of real-world data/big data including understanding the epidemiology of disease and unmet medical need, informing the development of precision medicines, informing health-care benefit design, informing quality improvement efforts, informing health technology assessments regarding access to and pricing of new therapies, assessing the incidence/prevalence of adverse events associated with marketed medications to inform regulatory labelling, informing bedside shared decision-making between patient and provider, and informing regulatory labelling decisions regarding indications, dosing, benefits in subpopulations, etc. The criteria for what are good enough are not the same across these applications. For some uses, the data we have today are already good enough. For other uses, it remains controversial whether the data are good enough. A framework is discussed that allows end users to decide whether data are of sufficient quality to inform decision-making.

We are entering the era of digitized real-world data—including the use of medical claims, electronic health records, sensor/real-time monitoring data, and electronic patient-reported outcome (ePRO) data—to inform health-care and health policy decisions. These data will increasingly be merged with novel other data

M.L. Berger (✉) • J. Harnett
Pfizer Inc., New York, NY, USA
e-mail: marc.berger@pfizer.com

© Springer Nature Singapore Pte Ltd. 2017
H.G. Birnbaum, P.E. Greenberg (eds.), *Decision Making in a World of
Comparative Effectiveness Research*, DOI 10.1007/978-981-10-3262-2_8

sources—including "omics" (e.g., genomics, proteomics, metabolomics, micro-biomics), purchasing preferences, self-reporting of experience via social networks, weather, etc. Together, big data have the potential to revolutionize decision-making and to help realize the goal of a "learning health-care system" [1]. However, there are concerns that the data are not of sufficient quality, both in terms of accuracy and completeness, and that lack of transparency and varying levels of analytic rigor characterize observational research leveraging such data [2]. This raises the question of whether real-world data and the real-world evidence derived from observational research are good enough today. If not, when will we know when they are good enough?

In terms of data quality, significant efforts and progress have been made in the increased adoption and use of the electronic health record (EHR). For its application to research, work is underway to standardize data collected in the EHR and increase completeness through extraction of information from physician notes via automated algorithms, along with a continued focus on interoperability and data linkages across settings and sources, and physician incentives for data collected that will continue to promote the increased quality of real-world data—as discussed elsewhere [2].

There are multiple ongoing and potential applications of real-world/big data to develop evidence that informs understanding the epidemiology of disease and unmet medical need; the development of precision medicines/interventions; health-care benefit design, guideline development, and treatment pathways; quality improvement efforts; health technology assessments regarding access to and pricing of new therapies; the incidence/prevalence of adverse events associated with marketed medications to inform regulatory labelling; bedside shared decision-making between patient and provider; and regulatory labelling decisions regarding indications, dosing, and benefits in subpopulations. The criteria for what are good enough data are not the same across these applications. For some uses, the data and evidence we have today may already be good enough. For other uses, it remains controversial.

A guiding principle of evidence-based medicine is that the treatment providers should use the best available evidence when taking care of an individual patient [3]. Physicians practice medicine based upon both evidence and clinical judgment, since the evidence base supporting specific treatment recommendations is relatively sparse, especially when considering the unique circumstances of an individual patient [4]. As one moves from the individual to population health management, it may be reasonable to expect that the evidence base to support treatment guidelines, for example, should be more robust, especially for prevalent chronic conditions. One might even expect a higher level of evidence to support regulatory decision-making—what claims can be made by a manufacturer in a drug label. However, many experts and health-care stakeholders do not accept such a framework for consideration of observational clinical research and seek to limit the use of such evidence until it has the same quality (i.e., credibility) as randomized controlled clinical trials (RCTs), regardless of its application. This is illustrated by the following story published in the New York Times Magazine of October 3, 2014 [5].

When a helicopter rushed a 13-year-old girl showing symptoms suggestive of kidney failure to Stanford's Packard Children's Hospital, she was quickly diagnosed with lupus, an auto-immune disease. One of the physicians (Dr. Jennifer Frankovich) thought that the patient's particular combination of lupus symptoms—kidney problems, inflamed pancreas and blood vessels—was similar to other lupus patients that she'd seen who developed life-threatening blood clots. However, the clinical team didn't think there was cause to give the girl antico-agulant drugs. She could not find any studies in the scientific literature, so she queried the electronic medical record STRIDE database (Stanford Translational Research Integrated Database Environment) for all the lupus patients the hospital had seen over the previous five years, focusing on those whose symptoms matched her patient's, and ran an analysis to see whether they had developed blood clots. Of the 98 patients in the pediatric lupus cohort, 10 patients developed thrombosis and the prevalence was higher among patients with protein-uria and pancreatitis. This review took less than 4 hours. She brought her analysis to the team and the patient was treated with anticoagulants, without being certain whether it was the right decision, but the attending team felt that it was the best they could do with the limited information they had. After Dr. Frankovich wrote about her experience in *The New England Journal of Medicine* [6], her hospital warned her not to conduct such analyses again until a proper framework for using patient information was in place to ensure that such data was used with appropriate patient consent and because of the concern that it was not a rigorous scientific analysis.

In terms of evidence for population-based recommendations, evidence-grading frameworks that have been developed and leveraged in the practice of evidence-based medicine and health technology assessment reinforce a hierarchy of evidence in which observational research is often rated inferior in terms of quality to RCTs, especially for assessment of treatment benefits. In the United States (USA), the Drug Effectiveness Review Project (DERP)—which provides research to inform drug coverage decisions to 13 US states [7], supports recommendations for patients in Consumer Reports Best Buy Drugs [8], and collaborates with the Canadian Agency for Drugs and Technologies in Health (CADTH)—considers observational research for evaluation of effectiveness only when clinical trials are considered flawed, although comparative cohort and case-control studies are included for all safety evaluations [9]. McDonagh et al. reported that only seven DERP reviews included observational studies for evaluation of benefits, while all reviews included observational studies for safety at the time of their publication [9]. The Grading of Recommendations Assessment, Development and Evaluation (GRADE) framework has been used or endorsed by many organizations, including the World Health Organization (WHO) and the United Kingdom's National Institute for Health and Care Excellence (NICE). In general, GRADE suggests randomized trials as high level, observational studies as low level, and all other evidence as very low level types of evidence [10].

There is no doubt that the RCT is the single most important methodological advance in clinical research in the last 100 years. However, in the spirit of designing a "learning health-care system" [1], the use of real-world evidence to inform clinical decision-making in the absence of, as well as to supplement, good published clinical-trial evidence is where medicine needs to go. One would imagine most patients referred for elective surgery would like to know what were the outcomes of the last 500 patients just like them (e.g., same gender, similar age, similar health status) who were operated on by that surgeon and how they compared with other surgeons; certainly this is better than relying solely on reputation and which hospital they are associated with.

When one moves from the individual patient bedside to considering health policy recommendations for populations of patients, a quick analysis of an EMR database does not suffice. But when is real-world evidence (or evidence from clinical experience) good enough? There is no simple formulaic answer to this question. It requires an overall assessment of what are potential benefits, and most importantly harms, from such policy recommendations—or to put it another way, how much regret would there be if the recommendations were wrong? This requires an assessment of the relevance and credibility of each piece of scientific evidence, whether it is from RCTs or real-world observational studies.

Perhaps the most prominent initiative in this regard has been the standardized approach for Strengthening the Reporting of Observational Studies in Epidemiology (STROBE) [11], which consists of a checklist of 22 items that relate to the title, abstract, results, and discussion sections of articles. Under methods, there are items that address study design, setting, participants, variables, data sources/measurement, bias, study size, quantitative variables, and statistical methods. The latter addresses issues of confounding, subgroups and interactions, missing data, and sensitivity analyses. In May 2010, one of the largest health-care insurers, WellPoint, was the first in the USA to introduce its evaluation criteria for comparative effectiveness and observational research in formulary decision-making, classifying this real-world evidence as useful, possibly useful, or not useful based on 20 criteria that assess scientific credibility and applicability to the WellPoint population [12].

As part of more recent collaboration among the International Society for Pharmacoeconomics and Outcomes Research (ISPOR), the Academy of Managed Care Pharmacy (AMCP), and the National Pharmaceutical Council (NPC), a group of experts developed a questionnaire in 2014 that explicitly assists decision-makers in assessing the relevance and credibility of published observational studies [13]. The questionnaire consists of 33 items, divided into two domains: relevance and credibility. Relevance addresses the extent to which findings, if accurate, apply to the setting of interest for a decision-maker. Credibility addresses the extent to which the study findings accurately answer the study question and is divided into the following sections: design, data, analyses, reporting, and interpretation. The relevance domain consists of four questions addressing the population, interventions, outcomes, and context. The credibility section consists of subsections that address study design, data, analyses, reporting, interpretation, and conflicts of interest. Under the design subsection, the questionnaire asks whether study hypotheses or goals were specified a priori, whether comparison groups were concurrent, whether there existed a formal study protocol including an analysis plan prior to executing the study, sample size, and power calculations, whether design minimized or accounted for confounding, and whether the appropriateness of the follow-up period was considered, as well as the appropriateness of sources/criteria/methods for selecting participants, and the similarity of comparison groups. Under the data subsection, there are questions related to the sufficiency of the data sources, the validity of how exposure was defined and measured, the validity of how primary outcomes were defined and measured, and whether the follow-up across comparison groups was comparable. Under the analysis subsection, there are questions that address potential measured and unmeasured confounders, subgroups and interactions, and sensitivity analyses. Under the report subsection, there are questions related to how the final sample was defined, descriptive statistics of study participants, description of key components of statistical analyses, the reporting of confounder-adjusted estimates of treatment effects, description of statistical uncertainty, missing data, and the reporting of absolute and relative treatment effects. Under the interpretations subsection, there are questions about whether the results are consistent with known prior information, whether results are clinically meaningful, whether the conclusions are supported by the data and analyses, and whether the effect of unmeasured confounding was discussed.

What distinguishes this questionnaire is that it asks for real-world evidence to be generated in a fashion akin to RCTs and that it is part of the requirements for publication in top peer-reviewed medical journals [14]. It is also consistent with the criteria by which the Food and Drug Administration (FDA) [15] decides whether evidence is substantial. Specifically, the questionnaire advocates for a formal protocol, with study goals explicitly specified a priori. Further, decision-makers should expect replication of results, as is the case with RCTs, whenever possible. If multiple observational studies, using different data sources that employ different analytic approaches, all point to similar results, one should find such conclusions as credible as anything considered in evidence-based medicine. Indeed, the FDA has been reviewing the potential for expansion in the leveraging of real-world evidence in regulatory decisions beyond safety analyses and the rare diseases space. In part, this has been prompted by the 21st Century Cures initiative [16]. A white paper from this initiative observed that the "FDA's review of supplemental applications for new uses or changes to a product are governed by pathways established at a time when computers could not identify trends in statistical or clinical data anywhere close to the degree they can today, let alone what they will be capable of doing tomorrow. Considering these ongoing developments, should we be rethinking the supplemental approval processes and how real-world data can be leveraged?" [16]. In fact, the FDA recently announced their intent to leverage the Sentinel Initiative, originally designed for active drug safety surveillance studies, to develop the "Guardian System" for conducting drug effectiveness research [17]. Outside the USA, the Innovative Medicines Initiative is bringing together regulatory and reimbursement agencies, manufacturers, and other health-care stakeholders for the GETREAL project that aims to investigate the use of real-world data for drug development [18].

In sum, the widespread use of real-world data in health care is critical for recognizing the vision of a rapid learning health system. As such, the robustness of real-world data will continue to improve in terms of accuracy and completeness. In addressing whether evidence generated from real-world data is good enough, we propose that one must consider its application first. At the patient level, it is possible that real-world evidence may be sufficient. For population-level decisions, there are tools to evaluate the quality of observational research; however, the key will be consistent adoption of such resources by decision-makers. Further, efforts to promote the transparency of observational studies are critical to acceptance of this data. A pharmaceutical industry member recently argued that observational studies should be preregistered on public websites as clinical trials are preregistered on the US National Institutes of Health's website, www.clinicaltrials.gov [19]. Only by putting real-world evidence on the same footing as RCTs can we expect to see its more widespread adoption in health-care policy decision-making. Lastly, momentum is building in consideration of leveraging real-world data for regulatory purposes beyond safety; this will certainly reinforce the utility of real-world data for population-level decision-making in the future.

References

1. Institute of Medicine (2007) The Learning Healthcare System. Roundtable on Evidence-Based Medicine; Roundtable on Value & Science-Driven Health Care. Olsen L, Aisner D, McGinnis JM, editors. National Academies Press, Washington, D.C. p. 374
2. Miani C, Robin E, Horvath V, Manville C, Cave J, Chataway J (2014) Health and healthcare: assessing the real-world data policy landscape in Europe. RAND Corporation. p. 1–102. Available from: http://www.rand.org/content/dam/rand/pubs/research_reports/RR500/RR544/RAND_RR544.pdf. Accessed 22 Apr 2016
3. Sackett DL (1997) Evidence-based medicine. Semin Perinatol 21(1):3–5
4. Goldman JJ, Shih TL (2011) The limitations of evidence-based medicine—applying population-based recommendations to individual patients. Virtual Mentor 13(1):26–30
5. Greenwood V. Can big data tell us what clinical trials don't? The New York Times. 2014 Oct 3; Available from: http://www.nytimes.com/2014/10/05/magazine/can-big-data-tell-us-what-clinical-trials-dont.html?_r=2. Accessed 15 May 2016.
6. Frankovich J, Longhurst CA, Sutherland SM (2011) Evidence-based medicine in the EMR era. N Engl J Med 365(19):1758–1759
7. OHSU Center for evidence-based policy. Participating Organizations. In: [website]. OHSU Center for evidence-based policy. 2016. Available from: https://www.ohsu.edu/xd/research/centers-institutes/evidence-based-policy-center/evidence/derp/participating-organizations.cfm. Accessed 15 May 2016
8. A look at the evidence. In: Consumer reports magazine. 2011. Available from: http://www.consumerreports.org/cro/magazine-archive/2011/march/health/best-buy-drugs/a-look-at-the-evidence/index.htm. Accessed 15 May 2016
9. McDonagh MS, Jonas DE, Gartlehner G, Little A, Peterson K, Carson S et al (2012) Methods for the drug effectiveness review project. BMC Med Res Methodol 12(1):140
10. Criteria for assigning grade of evidence. 2016. Available from: http://www.gradeworking-group.org/FAQ/evidence_qual.htm. Accessed 15 May 2016
11. von Elm E, Altman DG, Egger M, Pocock SJ, Gøtzsche PC, Vandenbroucke JP (2008) The Strengthening the Reporting of Observational Studies in Epidemiology (STROBE) statement: guidelines for reporting observational studies. J Clin Epidemiol 61(4):344–349
12. WellPoint. Use of comparative effectiveness research (CER) and observational data in formulary decision making evaluation criteria. WellPoint. 2010. pp 1–5. Available from: https://www.pharmamedtechbi.com/~/media/Images/Publications/Archive/ThePink Sheet/72/021/00720210012/20100521_wellpoint.pdf. Accessed 15 May 2016
13. Berger ML, Martin BC, Husereau D, Worley K, Allen JD, Yang W et al (2014) A questionnaire to assess the relevance and credibility of observational studies to inform health care decision making: an ISPOR-AMCP-NPC Good Practice Task Force report. Value Health 17(2):143–156

14. International Committee of Medical Journal Editors [homepage on the Internet]. Recommendations for the Conduct, Reporting, Editing, and Publication of Scholarly Work in Medical Journals. International Committee of Medical Journal Editors (ICMJE). 2016. Available from: http://www.icmje.org/icmje-recommendations.pdf. Accessed 26 Apr 2016
15. U.S. Department of Health and Human Services Food and Drug Administration, Center for Drug Evaluation and Research (CDER), Center for Biologics Evaluation and Research (CBER) (1998) Guidance for industry: providing clinical evidence of effectiveness for human drug and biological products. US Department of Health and Human Services. pp 1–23. Available from: http://www.fda.gov/downloads/Drugs/GuidanceCompliance/RegulatoryInformation/Guidances/UCM078749.pdf. Accessed 15 May 2016
16. Energy and Commerce Committee of the U.S (2012) House of representatives. 21st century cures: a call to action. Energy and Commerce Committee of the U.S. House of Representatives. pp 1–5. Available from: http://energycommerce.house.gov/sites/republicans.energycommerce.house.gov/files/analysis/21stCenturyCures/20140501WhitePaper.pdf. Accessed 26 Apr 2016
17. Wechsler J (2016) FDA sentinel initiative expands to support clinical research. Applied clinical trials; Available from: http://www.appliedclinicaltrialsonline.com/fda-sentinel-initiative-expands-support-clinical-research. Accessed 15 May 2016
18. Innovative Medicines Initiative (IMI). Welcome to GetReal. 2016. Available from: https://www.imi-getreal.eu/. Accessed 15 May 2016
19. Sutter S. Industry Need FDA (2016) "Engaged" before investing in observational studies. In: The pink sheet. Informa Business Intelligence, Inc. p. Article # 00160314004. Available from: https://www.pharmamedtechbi.com/Publications/The-Pink-Sheet/78/11/Industry-Need-FDA-Engaged-Before-Investing-In-Observational-Studies? Accessed 15 May 2016

Chapter 9
Translating Comparative Effectiveness Research Evidence to Real-World Decision Making: Some Practical Considerations

Richard J. Willke

Abstract For quite some time, there has been an understanding by both clinicians and payers that a given treatment may not be appropriate or justifiable for a broad patient population. The recently increased emphasis on comparative effectiveness research (CER) has heightened both the level of evidence and awareness related to the potential for better targeting of treatments. However, a number of practical hurdles to implementation of such targeting by clinical and payer decision makers remain. Such hurdles can be related to the ability to interpret or confirm the evidence relating to differences in patient physiologic responses to treatment, to the practicality of predicting such responses for individual patients in real-world treatment settings, to understanding differences in patient behavior and/or the cost consequences related to treatment that may affect choices, or to ethical considerations, among other things. This chapter will review and discuss these practical considerations, not only reviewing the literature but also providing some specific case examples. The goal of this chapter is to provide some perspective and guidance on the translation and implementation of CER and evidence on heterogeneity of treatment effects to real-world patients and their treatment.

9.1 Introduction

Both comparative effectiveness research (CER) and personalized medicine are terms that have been used extensively in the medical arena in recent years. Although definitions vary, one might think of personalized medicine as the application and implementation of comparative effectiveness (and safety) research that pays close attention to individual patient characteristics, including behavioral considerations,

R.J. Willke, PhD
International Society for Pharmacoeconomics and Outcomes Research,
San Francisco, CA, USA
e-mail: rwillke@ispor.org

in real-world medical care [1]. The path from CER to everyday implementation of personalized medicine, however, can be a long and arduous one.

There has long been broad recognition that not only patients themselves, but also the particular realizations of the health conditions they face, and the effects of specific treatments on them, can vary dramatically. However, there is a strong tendency to standardize treatment approaches, especially in primary care and in formulary management, partly resulting from cost control considerations and partly from greater emphasis on evidence-based medicine to limit discretionary treatment by physicians. This seeming contradiction is not due to a lack of desire to provide the most effective medical care. Rather it is due more to the difficulties inherent in conducting, communicating, and interpreting CER, along with the costs involved in differentially diagnosing patients well enough to utilize the CER evidence that is available. The hurdles involve the ability to generate, interpret, or confirm the evidence relating to differences in patient physiologic responses to treatment, to the practicality of predicting such responses for individual patients in real-world treatment settings, to understanding differences in patient behavior and/or the cost consequences related to treatment that may affect choices, or to ethical considerations. The goal of this chapter is to provide some perspective and guidance on the translation and implementation of CER, particularly with respect to personalized medicine, to real-world patients, and to their treatment.

9.2   CER and Heterogeneity of Treatment Effects

CER essentially seeks to identify differences in outcomes across treatments, either at the population level (i.e., drug A routinely works better than drug B for most people) or differences in outcomes associated with specific patient characteristics. The latter situation is commonly called heterogeneity of treatment effects (HTE) and is a basis for personalized medicine. Kravitz et al. (2004) define HTE as generally being due to one of four factors: responsiveness to treatment, vulnerability to side effects, baseline risk, and patient preferences for outcomes [2]. Among these factors, baseline risk differences among individual patients have been the most commonly studied and used for treatment choices, while the other factors are, for the most part, still emerging areas for systematic study and use.

In many cases, randomized controlled trials (RCTs) have already provided sufficient CER evidence to drive initial treatment recommendations in general populations. A well-known example is the use of statins relative to bile acid sequestrants (a.k.a., resins) for treatment of hyperlipidemia and cardiovascular (CV) risk. Most patients experience greater responsiveness (i.e., reduction) of low-density lipoprotein (LDL) levels to treatment—and hence reduced CV risk—with statins relative to resins, making statins the usual first-line recommendation for treatment [3]. However, some heterogeneity can occur in patients' vulnerability to—and disutility (i.e., reduced well-being) due to—their side effects, such as muscle pain with statins or bloating with resins, which may subsequently affect treatment choices. CER evidence to date has not provided clear predictions of such vulnerability to side effects for these drugs, either within or across these drug classes, although there is always the possibility of finding genetic determinants. Finally, baseline risk—mainly either current LDL or history of CV events—is likely to be a factor in choice of dosages or even of classes, depending on the side effects. Patient preferences for avoiding CV risk could also be a factor in treatment or dosage selection at the patient-physician decision-making level. Overall, in the case of hyperlipidemia, the availability of CER evidence and small cost of LDL testing have led to the standard use of those data in basic treatment choices.

Nevertheless, further research into differences in responsiveness to statin treatment may provide a greater ability to personalize treatment once it has overcome a variety of hurdles to implementation. For example, Dorresteijn et al. (2013) show how a set of 11 individual characteristics can help differentiate the likelihood of treatment benefit with statins [4]. Boekholdt et al. (2014) show how much the observed change in LDL varies across patients assigned to a given fixed-dose statin regime, using data from a number of large RCTs [5]. In general, such results, based on high-quality data and analytic approaches, could help refine how statin treatments are prescribed at an individual level but are not currently being used in most clinical practice situations. While part of the reason for this is a legitimate scientific need for replication and verification of those results, some clinical practices in other areas—such as selection of treatment regimens in oncology—are guided by arguably scantier evidence.

The statin case is only one of many examples where good CER evidence exists, but the importance of its use could be enhanced for more individualized patient care. In many other situations, however, good CER evidence is lacking. Not mentioned above, but also relevant, are the costs of diagnosis and treatment and their affordability at either the patient level or health system level. While costs are not an element of HTE per se, high out-of-pocket costs may affect compliance with treatment by the patient and thus its effectiveness. For both the patient and the health system, costs and affordability may affect the feasibility of implementing CER evidence as personalized medicine in the clinic and hence will be factors also considered here. Another factor that must be taken into consideration in many analyses is the presence—and different likelihoods—of competing risks, which may not affect HTE in a pure sense but can differentially affect the observation of outcomes across individuals [6].

9.3 Generating CER Evidence

Real-world treatment decisions often involve a variety of patient characteristics, treatment options, and pathways. To generate high-quality, prospectively declared, RCT evidence on all possible—or at least all of the clinically relevant—combinations of subpopulations and treatments would require many more RCTs and much larger sample sizes than seems feasible in today's environment. Nevertheless, available RCT evidence, maintaining at least some of the benefits of randomization, can be retrospectively analyzed to investigate more treatment comparisons and subgroup effects. Some methods involve combining data from multiple RCTs, generally referred to as meta-analysis. Within the class of meta-analyses, there are a number of options, including traditional meta-analysis of primary endpoints for a given treatment comparison across a set of trials or comparison of subgroups or secondary endpoints across those same trials, with due consideration for the issues inherent in retrospective subgroup analysis [7–10]. Indirect treatment comparisons, generally of the primary endpoints, can create comparisons across treatments that were not directly compared head-to-head within any given RCT, but at least have common comparators within the set of trials examined; the most common approaches here involve network meta-analysis [11]. More extended meta-analysis can involve regression-type controls for known patient characteristics; meta-regression [12] and model-based meta-analysis [13] are two approaches here that involve relatively familiar regression-type approaches to revealing effects of patient characteristic on outcomes. RCT data can also be interrogated with more advanced methods, such as predictive risk modeling, including classification and regression tree analysis, latent growth and growth mixture models, finite mixture models, quantile regression, and selected nonparametric methods [14]. While less familiar to most, these methods can sometimes identify heterogeneous treatment effects in combinations of characteristics that place patients into subgroups that may be difficult to identify prospectively. These advanced methods are best used when individual patient data from an RCT, or set of RCTs, are available but could also potentially be used if detailed subgroup effects are reported across a set of similar RCTs.

An RCT design that bears mention here is the series of n-of-1 trials approach. Each n-of-1 trial involves a single patient, treated sequentially, in a blinded and randomized fashion, with alternative treatments. Such trials are most appropriate for chronic conditions where treatment effects wear off quickly. For example, an n-of-1 trial in a chronic pain condition may involve randomizing a patient to treatment with drug A or drug B for 3 months, recording the outcome, followed by a week washout, then re-randomizing to treatment A or B for another 3 months, recording the outcome, and repeating this for a total of 3 or 4 treatment periods. When the same n-of-1 trial is repeated across a moderate-sized set of patients (e.g., 50), creating a "series," the combined data can be quite powerful in terms of capturing effects of individual characteristics on outcomes [15].

Real-world observational data with very large sample sizes and increasingly better detail on patient characteristics are also a rich potential source of CER evidence. They can be analyzed with traditional statistical/econometric methods as well as with the predictive analytics methods mentioned above for RCTs. However, the lack of randomization to treatment in observational data creates well-known concerns about selection bias into treatment and unobserved confounders, even with more advanced analytical techniques. While the literature that details and/or utilizes techniques to mitigate such biases is very extensive, a recent set of Good Practices reports from the International Society of Pharmacoeconomics and Outcomes Research provide a useful summary [16–18]. Occasionally "natural experiments" occur in the real world—situations where treatment selection is determined by a clearly independent, non-confounded, factor—and can be especially useful in these analyses [19].

Meta-analysis can be especially useful for detecting HTE. Both RCTs and real-world studies often include a variety of subpopulations, but if those subpopulation results are not consistently reported, meta-analysis becomes difficult. Kent et al. [20] present recommendations for assessing and reporting heterogeneity in treatment effects in clinical trials; a similar approach could be relevant for real-world studies.

Not all data that may be useful to analysts for extended CER studies is universally available. RCT data may be held proprietarily by life sciences companies, and real-world data from administrative or electronic medical record sources may be proprietary, or expensive to obtain, or difficult to combine because of differences in structure. Even government-based data sources may not be available to all researchers in a timely way. However, an increasing variety of collaborative approaches to strengthening data sources for research purposes have been initiated in recent years, such as combining data from RCTs submitted to the Food and Drug Administration for regulatory purposes, creating common data structures for administrative data, or joint research ventures between private sources, such as pharmaceutical companies and health insurance companies [21]. If observational data are going to be both rich enough to allow detailed analysis of important variables and their results transparent and replicable, such collaborations will be crucial to the credibility of CER evidence. In addition, there are differing incentives and returns across stakeholders for generating such evidence; the public value of CER information is likely higher than the private value, meaning that public funding of CER is important to generating all the evidence that is valuable from a societal point of view [22, 23].

9.4   Communication and Interpretation of CER Evidence

The generation of evidence per se, even that of the highest quality, is only one step in deriving useful supportive information for health care decision making. Such evidence must be made available in one way or another, identified by a user, understood by that user, and judged to be credible and applicable to their situation. The substantial evidence standard as recognized by regulatory authorities for product labeling, based on strong, usually replicated, RCT data, makes it most easily communicated and readily available, but is generally expensive and difficult to achieve for CER evidence, and thus relatively rare. Making CER evidence available usually means presentation at a scientific conference and/or publication in a peer-reviewed journal. Such evidence is not promotable by a manufacturer per regulation in the USA but can generally be distributed on request and is discoverable through literature searches or published systematic reviews. There is some debate about the merits of easing communication standards for CER evidence to decision makers that may be relatively more capable of judging the quality of the evidence [24, 25].

US-based payers see real-world CER evidence as useful for formulary and utilization management but cite their ability to interpret real-world evidence as a significant barrier to its use [26, 27]. There is work that is useful as an aid for the decision maker in examining and interpreting the evidence, such as checklists for retrospective database research as well as meta-analyses and indirect comparisons [28–30]. Among published work that must be interpreted by the decision maker, journal quality is often the first indicator of the credibility of the evidence, but there are many more study-specific factors that should be considered. These factors include study design, data or evidence base used, analytic approach, reporting quality and transparency, interpretation of results, and conflict of interest. Given good credibility, relevance to the decision maker can be evaluated via such questions as: was treatment given under similar conditions, to similar patients, with similar outcomes and comparators, in the study as it would be applicable in the decision maker's situation?

Whether RCT or real-world evidence is being considered, replicability and synthesis of results is a key concern, hence the common view that meta-analysis is the highest level of evidence [31]. While meta-analysis is most often used with RCT results, evidence synthesis is also useful with real-world studies. Even though evidence based on observational evidence is generally considered to be of lower quality than RCT-based evidence, if a set of independent observational studies, employing different techniques, yield similar results, and common unobserved confounders across studies appear unlikely, the combined evidence may need to be considered seriously. Certainly, individual real-world study results can vary depending on the data set used, even when using common methods across studies, as Madigan et al. (2013) have demonstrated [32]. However, it has also been shown that real-world studies, when meta-analyzed, can provide average results that are very similar to RCT results [33].

9.5   Other HTE Considerations

In the real world, patient adherence to prescribed treatment regimens, usually less monitored than in a clinical trial, can vary substantially and thus result in actual differences in treatment effects, even when the underlying physiological effects may not be different. Adherence can be affected by side effect tolerance and patient preference for certain outcome factors, as mentioned above [2]. However, adherence is also known to be affected by both psychological and economic considerations. A common psychological aspect is the "healthy adherer" bias [34]. Here, patients' underlying behaviors may lead them both to adhere better to the treatment being studied and to engage in other "healthy" activities; these other healthy activities may affect the observed treatment outcomes. A related consideration is the effect of medication co-pays: lower income or other personal economic factors may affect the patient's ability or willingness to make these co-payments. These same economic limitations may affect the patient's ability to engage in other healthy activities, possibly aligning the economic bias with the "healthy adherer" bias. On the other hand, if the patient is choosing to prioritize paying for other medical treatments over the co-pay for the treatment being studied, the bias may work in the other direction or be indeterminate.

Behavioral and economic factors may come into play on the provider side as well. Quality measures, practice organization (e.g., medical homes), "early adopter" mentality, insurance generosity, etc. can affect the range and severity of patients that physicians choose to treat or how much the patients' adherence is monitored and influenced. Any of these factors can introduce confounders to treatment outcomes that may or may not be well controlled by covariate adjustment or other estimation approaches in a real-world CER study.

Competing risks for individuals—other disease risks that may affect the occurrence of outcomes of the disease in question—can be relevant in certain treatment situations. The most common ones are those where it does not seem appropriate to treat patients for a situation when the outcome risk is mainly long-term and the patient's shorter-term mortality risk is high. Common examples are deciding against treatment of nonaggressive prostate cancer in males over 80 years old or against cholesterol treatment for terminal cancer patients.

Thus, the task of those who seek to use CER evidence for formulary management is complicated by needing to know the behavioral and economic influences present in the CER study and how those factors may translate into outcomes for their own populations and subpopulations. More naturalistic RCTs are useful in that they are likely to relieve biases that result from the stricter monitoring in typical regulatory trials. Actuarial adjustments may be able to predict adherence differences due to co-pay levels [35]. Covariate adjustments in the CER study may allow for extrapolation of outcomes to other population covariate values. However, some confounding influences are difficult to adjust for—if one is even willing or able to do that much work—so the most fundamental consideration usually distills down to whether the CER study population, and the provider setting in which it was conducted, is largely similar to the population or subpopulations thereof, for which the decision is being made. The larger insurers are now likely to have their own data and analytics units and may do their own CER studies on their own insured populations, either ab initio or to check or adapt the results of an external study.

9.6   CER, HTE, and Reimbursement

It is not uncommon for regulatory studies, especially in Europe, to involve active comparators, or to be directed at a subpopulation of a given disease, and for that evidence to be incorporated into product labeling. However, those making coverage decisions may or may not grant reimbursement strictly according to the approved label. For example, O'Neill and Devlin (2010) found that during 2006–2009, the United Kingdom's (UK) National Institute for Health and Care Excellence (NICE) made "mixed" reimbursement decisions over half of the time, i.e., neither "yes" nor "no" for the entire label-indicated population [36]. In the majority of these mixed decisions, coverage was recommended for less than half of the indicated population, and decisions were often made on the basis of cost-effectiveness results, where limited effectiveness in some subgroups did not appear to justify the additional cost of treatment. Partial coverage is not uncommon in other countries, either, though most have not been as systematically studied as the UK.

Both the potential and the risks of using limited evidence from subgroup analysis to target treatment and its reimbursement are illustrated by the recent case of Medicare coverage of implantable cardiac defibrillators (ICDs), as discussed in Mohr and Tunis (2014) [37]. The Multicenter Automatic Defibrillator Implantation Trial (MAD-IT II) in 2001 showed significant survival benefit of ICDs in a broad population of individuals at risk of sudden cardiac death, regardless of previous history of arrhythmia (previously Medicare had only covered ICDs in those with a history of ventricular arrhythmia). Expanding coverage based on these primary results of MAD-IT II had substantial budget implications. However, a post hoc subgroup analysis of the trial suggested that the effect was much stronger in those with a wide QRS interval on EKG (a result consistent with higher risk of arrhythmia), with minimal benefit in those with a normal QRS interval. It was decided to expand benefits only to those with the wider QRS interval, pending the results of another, ongoing trial. When this ongoing trial was completed, however, it did not confirm the MAD-IT II QRS interval subgroup finding, and Medicare subsequently broadened its coverage for ICDs. The broader survival benefit has since been confirmed by the analysis of a national ICD registry [38]. As this example shows, while the intent to manage insurance coverage costs by focusing on those most responsive to treatment is reasonable, it is important to do so based on replicable findings.

Many clinical implementations of known HTE involve a diagnostic test to determine the appropriateness of a given treatment or treatment approach. Frueh (2013) [39] examines a variety of such approaches, divided into three categories: (1) tests developed in association with a drug; (2) tests developed after the drug has reached the market; and (3) tests not directly associated with a particular drug. Examples are most common in oncology, but also include areas such as HIV-1 and cardiovascular products, and are increasing in some rare-disease areas. From a US perspective, the path to regulatory approval, insurance reimbursement, and utilization tends to be most direct for tests developed in association with a drug and is increasingly more difficult the less clearly tied a test is to a treatment and its outcomes. Payne et al. (2013) [40] provide a related European perspective, which also emphasizes the importance of a strong evidence base for health technology assessment purposes, as well as the need for service delivery considerations.

Teagarden (2014) states that "necessity and fairness require that health plans limit the products and services they cover" [41]. Necessity is often based on budget limitations—allowing all patients low-cost access to all available treatments would likely result in unacceptably high premiums. Fairness involves providing access to treatments that provide acceptable outcomes for as broad an array of conditions as possible. With limited information about treatment effect heterogeneity, however, fairness can only be applied bluntly. However, both necessity and fairness constraints could be satisfied, with improved patient outcomes, given better CER evidence about HTE and its implementation in formulary management. Well-targeted treatment can be more cost-effective if ineffective treatment steps, and their associated costs, are skipped. Fairness in the presence of evidentiary uncertainty about HTE tends to rely on additional "filters" for access to services that are not deemed appropriate as first-line therapy based on population averages. While these filters (such as step edits, prior authorization, tiering that results in different patient co-pay levels, or patient access schemes that require significant patient effort to enter) can also mitigate moral hazard concerns (i.e., over-demand for treatment due to low out-of-pocket costs), fairness is obviously better served by allowing less filtered access to treatment for patients who will benefit most from it.

9.7   Closing Thoughts

The amount of CER evidence in the literature—both for general populations and subgroups who may experience differing treatment effects—is growing rapidly and has the potential to inform both clinical and formulary decision making. Much of it is based on real-world data (RWD) analyses, due to the increased availability of such data, the collaborative efforts to enhance them, and the wealth of methods to analyze those data. However, the variety of real-world treatment situations and influences can create biases that make such analyses difficult to interpret confidently. The major challenge for CER is to close the gap between CER evidence generation and its implementation in health care decision making. This chapter has discussed a range of the considerations involved in that gap and in closing it.

For CER/HTE evidence to be deemed credible, especially when not based on primary endpoints from RCTs, it is crucial for analysts to be transparent about the data and methods used and about whether those methods were prospectively declared and followed. Disclosures in publications, and perhaps posting of RWD protocols, would help reassure those who are reviewing and interpreting that evidence. The information available from both RCT and RWD sources is too valuable to be disregarded due to concerns about data mining.

Increasing education of decision makers regarding interpretation of CER studies is also part of the way forward. A recent survey study found that participation in continuing education regarding the use of CER increased pharmacist attendees' ability to use CER research in general [42]. However, the survey also found that a majority of attendees reported that such CER capabilities had not changed formulary decision making in their organization. Whether this limited ability to effect change was due to a lack of relevant and compelling CER evidence, or to insufficient penetration of such CER-related education across the organizations, was not explored. With respect to HTE, a similar survey [43], following an educational program to enhance awareness and understanding of HTE for managed care pharmacists and medical directors, found that the program was successful in those objectives, and that participants thought they were more likely to use HTE considerations for determining necessity and prior authorizations. Such efforts, if continued and expanded, can help enable CER to fulfill its promise for improving the efficiency and equity of health care.

References

1. Epstein RS, Teagarden JR (2010) Comparative effectiveness research and person-alized medicine: catalyzing or colliding? PharmacoEconomics 28(10):905–913. doi:10.2165/11535830-000000000-00000
2. Kravitz RL, Duan N, Braslow J (2004) Evidence-based medicine, heterogeneity of treatment effects, and the trouble with averages. Milbank Q 82(4):661–687. doi:10.1111/j.0887-378X.2004.00327.x
3. Stone NJ, Robinson JG, Lichtenstein AH, Bairey Merz CN, Blum CB, Eckel RH, Goldberg AC, Gordon D, Levy D, Lloyd-Jones DM, McBride P, Schwartz JS, Shero ST, Smith SC Jr, Watson K, Wilson PW, American College of Cardiology/American Heart Association Task Force on Practice Group (2014) 2013 ACC/AHA guideline on the treatment of blood choles-terol to reduce atherosclerotic cardiovascular risk in adults: a report of the American College of Cardiology/American Heart Association Task Force on Practice guidelines. J Am Coll Cardiol 63(25 Pt B):2889–2934. doi:10.1016/j.jacc.2013.11.002
4. Dorresteijn JA, Boekholdt SM, van der Graaf Y, Kastelein JJ, LaRosa JC, Pedersen TR, DeMicco DA, Ridker PM, Cook NR, Visseren FL (2013) High-dose statin therapy in patients with stable coronary artery disease: treating the right patients based on individualized prediction of treat-ment effect. Circulation 127(25):2485–2493. doi:10.1161/CIRCULATIONAHA.112.000712
5. Boekholdt SM, Hovingh GK, Mora S, Arsenault BJ, Amarenco P, Pedersen TR, LaRosa JC, Waters DD, DeMicco DA, Simes RJ, Keech AC, Colquhoun D, Hitman GA, Betteridge DJ, Clearfield MB, Downs JR, Colhoun HM, Gotto AM Jr, Ridker PM, Grundy SM, Kastelein JJ (2014) Very low levels of atherogenic lipoproteins and the risk for cardiovascular events: a meta-analysis of statin trials. J Am Coll Cardiol 64(5):485–494. doi:10.1016/j.jacc.2014.02.615
6. Varadhan R, Weiss CO, Segal JB, Wu AW, Scharfstein D, Boyd C (2010) Evaluating health outcomes in the presence of competing risks: a review of statistical methods and clinical appli-cations. Med Care 48(6 Suppl):S96–105. doi:10.1097/MLR.0b013e3181d99107
7. Assmann SF, Pocock SJ, Enos LE, Kasten LE (2000) Subgroup analysis and other (mis)uses of baseline data in clinical trials. Lancet (London, England) 355(9209):1064–1069. doi:10.1016/S0140-6736(00)02039-0
8. Cook DI, Gebski VJ, Keech AC (2004) Subgroup analysis in clinical trials. Med J Aust 180(6):289–291
9. Pocock SJ, Assmann SE, Enos LE, Kasten LE (2002) Subgroup analysis, covariate adjustment and baseline comparisons in clinical trial reporting: current practice and problems. Stat Med 21(19):2917–2930. doi:10.1002/sim.1296
10. Rothwell PM (2005) Treating individuals 2. Subgroup analysis in randomised controlled trials: importance, indications, and interpretation. Lancet (London, England) 365(9454):176–186. doi:10.1016/S0140-6736(05)17709-5
11. Hoaglin DC, Hawkins N, Jansen JP, Scott DA, Itzler R, Cappelleri JC, Boersma C, Thompson D, Larholt KM, Diaz M, Barrett A (2011) Conducting indirect-treatment-comparison and network-meta-analysis studies: report of the ISPOR Task Force on Indirect Treatment Comparisons Good Research Practices: part 2. Value Health J Int Soc Pharmacoecon Outcomes Res 14(4):429–437. doi:10.1016/j.jval.2011.01.011
12. Baker WL, White CM, Cappelleri JC, Kluger J, Coleman CI, Health Outcomes P, Economics Collaborative G (2009) Understanding heterogeneity in meta-analysis: the role of meta-regression. Int J Clin Pract 63(10):1426–1434. doi:10.1111/j.1742-1241.2009.02168.x

13. Milligan PA, Brown MJ, Marchant B, Martin SW, van der Graaf PH, Benson N, Nucci G, Nichols DJ, Boyd RA, Mandema JW, Krishnaswami S, Zwillich S, Gruben D, Anziano RJ, Stock TC, Lalonde RL (2013) Model-based drug development: a rational approach to efficiently accelerate drug development. Clin Pharmacol Ther 93(6):502–514. doi:10.1038/clpt.2013.54

14. Willke RJ, Zheng Z, Subedi P, Althin R, Mullins CD (2012) From concepts, theory, and evidence of heterogeneity of treatment effects to methodological approaches: a primer. BMC Med Res Methodol 12:185. doi:10.1186/1471-2288-12-185

15. Zucker DR, Schmid CH, McIntosh MW, D'Agostino RB, Selker HP, Lau J (1997) Combining single patient (N-of-1) trials to estimate population treatment effects and to evaluate individual patient responses to treatment. J Clin Epidemiol 50(4):401–410

16. Berger ML, Mamdani M, Atkins D, Johnson ML (2009) Good research practices for comparative effectiveness research: defining, reporting and interpreting nonrandomized studies of treatment effects using secondary data sources: the ISPOR Good Research Practices for Retrospective Database Analysis Task Force Report – Part I. Value Health J Int Soc Pharmacoecon Outcomes Res 12(8):1044–1052. doi:10.1111/j.1524-4733.2009.00600.x

17. Cox E, Martin BC, Van Staa T, Garbe E, Siebert U, Johnson ML (2009) Good research practices for comparative effectiveness research: approaches to mitigate bias and confounding in the design of nonrandomized studies of treatment effects using secondary data sources: the International Society for Pharmacoeconomics and Outcomes Research Good Research Practices for Retrospective Database Analysis Task Force Report – Part II. Value Health J Int Soc Pharmacoecon Outcomes Res 12(8):1053–1061. doi:10.1111/j.1524-4733.2009.00601.x

18. Johnson ML, Crown W, Martin BC, Dormuth CR, Siebert U (2009) Good research practices for comparative effectiveness research: analytic methods to improve causal inference from nonrandomized studies of treatment effects using secondary data sources: the ISPOR Good Research Practices for Retrospective Database Analysis Task Force Report – Part III. Value Health J Int Soc Pharmacoecon Outcomes Res 12(8):1062–1073. doi:10.1111/j.1524-4733.2009.00602.x

19. Rosenzweig MR, Wolpin KI (2000) Natural "natural experiments" in economics. J Econ Lit 38(4):827–874. doi:10.1257/jel.38.4.827

20. Kent DM, Rothwell PM, Ioannidis JP, Altman DG, Hayward RA (2010) Assessing and reporting heterogeneity in treatment effects in clinical trials: a proposal. Trials 11:85. doi:10.1186/1745-6215-11-85

21. Willke RJ, Crown W, Del Aguila M, Cziraky MJ, Khan ZM, Migliori R (2013) Melding regulatory, pharmaceutical industry, and U.S. payer perspectives on improving approaches to heterogeneity of treatment effect in research and practice. Value Health J Int Soc Pharmacoecon Outcomes Res 16(6 Suppl):S10–S15. doi:10.1016/j.jval.2013.06.006

22. Meltzer D, Basu A, Conti R (2010) The economics of comparative effectiveness studies: societal and private perspectives and their implications for prioritizing public investments in comparative effectiveness research. PharmacoEconomics 28(10):843–853. doi:10.2165/11539400-000000000-00000

23. Towse A, Garrison LP Jr (2013) Economic incentives for evidence generation: promoting an efficient path to personalized medicine. Value Health J Int Soc Pharmacoecon Outcomes Res 16(6 Suppl):S39–S43. doi:10.1016/j.jval.2013.06.003

24. Griffin JP, Godfrey BM, Sherman RE (2012) Regulatory requirements of the Food and Drug Administration would preclude product claims based on observational research. Health Affairs (Project Hope) 31(10):2188–2192. doi:10.1377/hlthaff.2012.0958

25. Perfetto EM, Bailey JE Jr, Gans-Brangs KR, Romano SJ, Rosenthal NR, Willke RJ (2012) Communication about results of comparative effectiveness studies: a pharmaceutical industry view. Health Affairs (Project Hope) 31(10):2213–2219. doi:10.1377/hlthaff.2012.0745

26. Cahill J, Learner N (2010) Managed care pharmacy sees potential of comparative effectiveness research to improve patient care and lower costs. PharmacoEconomics 28(10):931–934. doi:10.2165/11535610-000000000-00000

27. Malone D, Avey SG, Mattson C, McKnight J (2016) Use of real-world evidence in payer decision-making: fact or fiction? In: International Society of Pharmacoeconomics and

Outcomes Research 21st annual international meeting: Issue Panel. International Society of Pharmacoeconomics and Outcomes Research

28. Berger ML, Martin BC, Husereau D, Worley K, Allen JD, Yang W, Quon NC, Mullins CD, Kahler KH, Crown W (2014) A questionnaire to assess the relevance and credibility of observational studies to inform health care decision making: an ISPOR-AMCP-NPC Good Practice Task Force report. Value Health J Int Soc Pharmacoecon Outcomes Res 17(2):143–156. doi:10.1016/j.jval.2013.12.011

29. Jansen JP, Fleurence R, Devine B, Itzler R, Barrett A, Hawkins N, Lee K, Boersma C, Annemans L, Cappelleri JC (2011) Interpreting indirect treatment comparisons and network meta-analysis for health-care decision making: report of the ISPOR Task Force on Indirect Treatment Comparisons Good Research Practices: part 1. Value Health J Int Soc Pharmacoecon Outcomes Res 14(4):417–428. doi:10.1016/j.jval.2011.04.002

30. Motheral B, Brooks J, Clark MA, Crown WH, Davey P, Hutchins D, Martin BC, Stang P (2003) A checklist for retrospective database studies – report of the ISPOR Task Force on Retrospective Databases. Value Health J Int Soc Pharmacoecon Outcomes Res 6(2):90–97. doi:10.1046/j.1524-4733.2003.00242.x

31. Guyatt GH, Sackett DL, Sinclair JC, Hayward R, Cook DJ, Cook RJ (1995) Users' guides to the medical literature. IX A method for grading health care recommendations. Evidence-Based Medicine Working Group. JAMA 274(22):1800–1804

32. Madigan D, Ryan PB, Schuemie M, Stang PE, Overhage JM, Hartzema AG, Suchard MA, DuMouchel W, Berlin JA (2013) Evaluating the impact of database heterogeneity on observational study results. Am J Epidemiol 178(4):645–651. doi:10.1093/aje/kwt010

33. Concato J, Feinstein AR (1997) Monte Carlo methods in clinical research: applications in multivariable analysis. J Investig Med Off Publ Am Fed Clin Res 45(6):394–400

34. Petitti DB (1994) Coronary heart disease and estrogen replacement therapy. Can compliance bias explain the results of observational studies? Ann Epidemiol 4(2):115–118

35. Einav L, Finkelstein A, Schrimpf P (2015) The response of drug expenditure to nonlinear contract design: evidence from medicare part D. Quar J Eco 2(130):841–899. doi:10.1093/qje/qjv005

36. O'Neill P, Devlin NJ (2010) An analysis of NICE's 'restricted' (or 'optimized') decisions. PharmacoEconomics 28(11):987–993. doi:10.2165/11536970-000000000-00000

37. Mohr PE, Tunis SR (2014) Medical and pharmacy coverage decision making at the population level. J Manag Care Spec Pharm 20(6):547–554. doi:10.18553/jmcp.2014.20.6.547

38. Al-Khatib SM, Hellkamp A, Bardy GH, Hammill S, Hall WJ, Mark DB, Anstrom KJ, Curtis J, Al-Khalidi H, Curtis LH, Heidenreich P, Peterson ED, Sanders G, Clapp-Channing N, Lee KL, Moss AJ (2013) Survival of patients receiving a primary prevention implantable cardioverter-defibrillator in clinical practice vs clinical trials. JAMA 309(1):55–62. doi:10.1001/jama.2012.157182

39. Frueh FW (2013) Regulation, reimbursement, and the long road of implementation of personalized medicine – a perspective from the United States. Value Health J Int Soc Pharmacoecon Outcomes Res 16(6 Suppl):S27–S31. doi:10.1016/j.jval.2013.06.009

40. Payne K, Annemans L (2013) Reflections on market access for personalized medicine: recommendations for Europe. Value Health J Int Soc Pharmacoecon Outcomes Res 16(6 Suppl):S32–S38. doi:10.1016/j.jval.2013.06.010

41. Teagarden JR (2014) Managing heterogeneity in prescription drug coverage policies. J Manag Care Spec Pharm 20(6):564–565. doi:10.18553/jmcp.2014.20.6.564

42. Augustine J, Warholak TL, Hines LE, Sun D, Brown M, Hurwitz J, Taylor AM, Brixner D, Cobaugh DJ, Schlaifer M, Malone DC (2016) Ability and use of comparative effectiveness research by P&T committee members and support staff: a 1-year follow-up. J Manag Care Spec Pharm. doi:10.18553/jmcp.2016.22.6.618

43. Warholak TL, Hilgaertner JW, Dean JL, Taylor AM, Hines LE, Hurwitz J, Brown M, Malone DC (2014) Evaluation of an educational program on deciphering heterogeneity for medical coverage decisions. J Manag Care Spec Pharm 20(6):566–573. doi:10.18553/jmcp.2014.20.6.566

Chapter 10
Decision-Making by Public Payers

Louis F. Rossiter

Abstract Public payers could be the most important players in a world of comparative effectiveness, because of the large populations covered and the amount of money spent. Moreover, they frequently set the payment rules for other private payers to follow. However, they are also constrained by program-specific legislative authority and regulations. This chapter explains the framework, openness, and use of findings from their own or other research that public payers use to shape determinations about coverage and payment at the state and federal level. Medicare and Veterans Administration are compared to each other as well as three of the largest Medicaid states – California, New York, and Texas. Medicare uses comparative effectiveness findings largely in a bottom-up approach with local coverage decisions informing national coverage decisions sometimes vetted by a very open centralized advisory panel that relies upon research done by the Agency for Healthcare Research and Quality. Veterans Administration is largely internally driven and closed decision-making by the clinicians and staff, but with the help of a program of their own that synthesizes evidence-based research. The Medicaid programs focus on pharmaceuticals and use comparative effectiveness to develop their automated drug utilization review rules and their preferred drug formulary rules. Medicaid is less likely than Medicare or Veterans Administration to use comparative effectiveness decision-making for devices, procedures, or services. Risk-sharing and outcome-based contracting models are increasingly relying on comparative effectiveness findings to drive performance measures that in turn drive payments.

10.1 Background

Public payers could be the most important users of comparative effectiveness when viewed in the context of dollars spent and numbers of people affected. However, public payers' decision-making processes are constrained by program-specific legislative authority and regulations. In addition, managers of public payer programs

L.F. Rossiter
William & Mary, Raymond A. Mason School of Business, Williamsburg, VA, USA
e-mail: lfross@wm.edu

© Springer Nature Singapore Pte Ltd. 2017
H.G. Birnbaum, P.E. Greenberg (eds.), *Decision Making in a World of Comparative Effectiveness Research*, DOI 10.1007/978-981-10-3262-2_10

work under assumptions and resource constraints that differ from those followed by commercial payers. Moreover, because they each have their own process for making decisions about payment policy based on comparative effectiveness analysis, applying a single rating to the use of comparative effectiveness considerations in these programs is a challenge.

The purpose of this chapter is threefold: first, explain the framework under which public payers incorporate comparative effectiveness considerations into their determinations about coverage and payment; second, assess the extent to which public payers are open to input from the public and other stakeholders when making decisions; and, third, determine when and how public payers use findings from research on comparative effectiveness to shape coverage and payment policies at the federal and state levels.

At the federal level, there are nine health coverage plans: Medicare, Medicaid, Children's Health Insurance Program, the Department of Veterans Affairs (VA), marketplace insurance under the Affordable Care Act (ACA), Federal Employees Health Benefits Program, the Department of Defense programs TRICARE and TRICARE for Life, and the Indian Health Service. This chapter reviews the use of comparative effectiveness by the three largest government payers: Medicare, Medicaid, and the VA. Because each state defines and manages its own Medicaid plans, which means there are at least 50 different processes for making decisions, this chapter examines the three largest state Medicaid programs in terms of spending: California, New York, and Texas.

10.2   Medicare, Medicaid, and Veterans Administration

The three largest public payers cover 37.1% of Americans. Based on the total US population, that coverage level comes to more than 51 million Medicare beneficiaries, 62 million Medicaid recipients, and nearly 15 million members covered by military health care [1]. Figure 10.1 puts these figures in perspective. Nearly

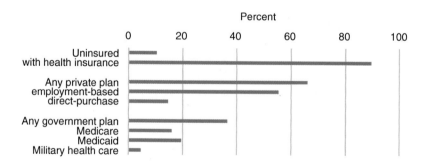

Fig. 10.1 Percentage of people by type of health insurance coverage: 2015 [1]

91% of Americans had health coverage in 2015, leaving just over 9% uninsured. Most coverage, or 67%, is through private plans, either employment based or directly purchased. That leaves 16% covered by Medicare, 19% covered by Medicaid, and just over 4% covered by military health care. However, for the purposes of estimating the population covered by decision-making using comparative effectiveness, these numbers can be deceiving. These three large public payers practice multiple layers of decision-making, depending upon the level and type of managed care in place.

In this context, managed care describes a practice in which government programs contract with commercial insurance companies to provide public benefits through private plans to the people they serve. Under these circumstances, the private plan administers the benefits, subject to applicable law and regulation. The plan then decides how and when its own comparative effectiveness reviews will be applied to coverage decisions regarding medical services, drugs, and devices. The percentage of members covered by private managed care plans indicates the extent to which the covered population is subject to decision-making based on comparative effectiveness. Public payers may include certain regulations and practices in their contracts with private plans that could preempt the regular decision-making routines of private plans. This ability to preempt might occur, for example, when a state Medicaid agency requires a private plan to use its fee-for-service formulary for drugs rather than its own formulary, which is likely to be more restrictive. Otherwise, the private plan is expected to administer benefits with its own comparative effectiveness decisions for the purpose of improving quality, lowering costs, or both. Table 10.1 compares the nature of funding, the level of spending involved, and the percentage of members covered by managed care private plans across Medicare, the VA, and the three largest state Medicaid programs.

Table 10.1 Characteristics of public health coverage plans 2014

Characteristics	Medicare	Medicaid			Veterans Administration
		CA	NY	TX	
Nature of funding	Federal taxes	Federal, state, and local taxes	Federal and state taxes	Federal and state taxes	Federal taxes
Covered persons (millions)	50.5	8.1	4.5	4.2	8.3
Spending dollars (billions)	$618.7	$63.9	$54.2	$32.2	$57.1
Percent managed care	31% medical 100% drug	77.0%	77.8%	88.0%	NA

Source: Centers for Medicare & Medicaid Services, Kaiser Family Foundation 2016
Note: Managed care means private plan administration including formulary and benefit management

10.2.1 Current Medicare Spending and Managed Care

Medicare is funded by federal payroll taxes on individuals and employers along with general federal revenues that come primarily from individual and corporate income taxes and deficit financing. In 2014, Medicare spending was $618.7 billion. Of the over 50 million Medicare beneficiaries in the program, 31% were enrolled in managed care plans called Medicare Advantage. What this means is that about 70% of Medicare beneficiaries were subject to the federal fee-for-service decision-making as applied to the medical services and devices provided by their plan. The remaining 30% experienced decision-making in the context of a private Medicare Advantage plan. Of those enrollees in Medicare Advantage, 72% were served by just six national plans: United Healthcare (20%), Humana (17%), Blue Cross Blue Shield-affiliated plans (17%), Kaiser Permanente (8%), Aetna (7%), and Cigna (3%) [2]. Thus, the decision-making model is quite split between the federal fee-for-service program and a small number of top Medicare Advantage plans.

Medicare expects all its beneficiaries to have private drug plans, regardless of whether they are enrolled in Medicare Advantage plans. These plans – more than 800 in 2016 – are offered nationwide to everyone on Medicare as either a stand-alone plan or as part of a Medicare Advantage plan [2], and these private plans make decisions involving drug coverage based on comparative effectiveness assessments. Put differently, Medicare "outsources" comparative effectiveness decision-making for each beneficiary's prescription drug coverage benefits to private drug plans, their formularies, and their pharmacy benefit management systems.

10.2.2 Current Medicaid Spending and Managed Care Coverage

The total federal and state Medicaid spending was $495.8 billion in 2014, including $63.9 billion for California, $54.2 billion for New York, and $32.2 billion for Texas. Federal general fund revenues cover half the spending, with the other half coming from state taxes. The exact percentage varies by state according to a legislative formula. Going one step further, California has a formula set by state legislation whereby localities share in the cost of Medicaid. The comparative effectiveness decision-making in these three state Medicaid programs reflects differences depending upon state legislation, regulations, contractors used, and the way state Medicaid staff administer the program.

Of those members covered by managed care, Medicaid has the highest percentage. The three largest states have between 77% and 88% of recipients enrolled in private plans – a percentage roughly equal to the 77% average across all states with managed care plans [3]. Following the expansion of insurance coverage through the ACA, Medicaid recipients are mostly women, children, and childless adults. Most states still struggle with enrolling severely disabled or frail elderly recipients in managed care plans, although the trend is toward enrolling everyone in managed care. The six largest national plans named above with Medicare Advantage contracts are also major players in Medicaid managed care. The share of enrollees depends on which state(s) these national plans have signed contracts to provide coverage. This means that the remaining roughly 25% of the fee-for-service Medicaid recipients in California, New York, and Texas, as well as nationwide, are exposed only to the state's Medicaid agency comparative effectiveness decision-making.

Most of this decision-making review revolves around access to prescription drugs. It is less likely that the review is applied to medical services and devices. Comparative effectiveness for medical services is hardly used at all by state Medicaid agencies for several reasons. First, states have difficulty with physician participation rates. Medicaid is almost always the lowest-paying insurer in any state, and states do not want to overburden providers willing to accept Medicaid patients. In addition, states may not have the medical expertise and staff to implement comparative effectiveness reviews for medical services.

The three large states cited in this chapter – California, New York, and Texas – apply comparative effectiveness reviews for drug coverage, and their administrative procedures show how this process works. Drugs are usually dispensed in 30-day supplies. Prior approval is normally required for any drug for the so-called "off-label" use (meaning the provider is free to prescribe the drug, but the Food and Drug Administration has not approved it for that use). Off-label use typically must be reviewed by state Medicaid programs for medical necessity prior to dispensing. Prior authorization is also typically used by the state Medicaid agency in conjunction with the state's preferred drug list. California, New York, and Texas use comparative effectiveness analysis extensively to decide which multiple-source drugs have the most favorable clinical and economic outcomes. They also factor in the price of the drug and use cost-effectiveness reviews to decide which drug(s) to include on the preferred drug list for favorable prior authorization treatment. This process is applied much less often with devices and hardly at all, as mentioned previously, for medical services.

10.2.3   Current Department of Veterans Affairs Spending and Managed Care Coverage

The VA provides health services to 8.3 million veterans at an annual cost of $57 billion. In sharp contrast to Medicare and Medicaid, the VA does not use managed care as the term is defined here to mean coverage through private plans. However, most of the care is provided through 150 hospitals and 1400 clinics operated directly by the VA and its 53,000 employed licensed providers. So while enrollment in private managed care plans is not a factor, care received by veterans is still heavily managed through the VA's central office. As discussed below, this office defines care guidelines and treatment pathways and applies comparative effectiveness to its decision-making [4].

10.3   Public Payer Framework for Decision-Making with Comparative Effectiveness Research (CER)

This section examines how CER affects decision-making. The process involves comparing the level of interest group involvement and the degree to which decision-making is open to the public. It also studies how decisions are made – how topics are selected, cost-effectiveness is used in voting procedures, and appeals are processed. This framework was adapted from a study of the impact CER has on decision-making internationally [5]. The following discussion highlights differences across Medicare, Medicaid, and the VA. The focus throughout is only on the centralized fee-service coverage decisions of Medicare and state Medicaid programs and the central office decisions of the VA. The decisions of private managed care plans are not covered here, although these private plans may defer to fee-for-service coverage decisions or, in some cases, be required to adopt them, in which case they would be based on what is described here.

10.3.1   How Do Public Payers Structure Decision-Making with CER?

Table 10.2 summarizes the way public payers structure decision-making. Medicare has two broad processes for coverage decisions, which incorporate comparative effectiveness decision-making differently: national coverage determinations, which may incorporate external technology assessment or the results of Medicare Evidence Development and Coverage Advisory Committee (MEDCAC) independent expert reviews,[1] and local coverage determinations. Local coverage determinations are made by Medicare Administrative Contractors (MACs) and their medical directors and are used in a particular geographic region to develop claims processing edits that approve, deny, or set a claim aside for a review. National coverage determination reviews focus primarily on medical necessity. A recent study shows that it is becoming increasingly difficult to gain coverage and reimbursement for new medical interventions [6]. An annual report on national coverage determinations has been made to the US Congress each year since 2003. As noted, they may or may not be informed by the results of reviews conducted by MEDCAC, an independent panel of experts. In addition to its role in advising Centers for Medicare & Medicaid Services (CMS) in certain national coverage determinations, CMS also may convene MEDCAC in order to gain advice on important emerging and cross-cutting issues, such as defining the standard of care in wound therapy and identifying an appropriate framework for the evaluation of diagnostic tests.

---

[1] National coverage determinations are driven more by a technology assessment, than by comparative effectiveness concerns, although it is an evidence-based process.

Table 10.2 Framework for decision-making with CER: interest group involvement

Characteristics	Medicare	Medicaid			Veterans Administration
		CA	NY	TX	
Principal CE body (year current process established)	Medicare Evidence Development and Coverage Advisory Committee (MEDCAC) (1998/2015)[a]	Department of Health Care Services (DHCS) (unknown)	New York State Department of Health (1996)	Texas DUR Board (2015)	Pharmacy and Therapeutics Committee (P&T) (mid-1950s) and Evidence-based Synthesis Program (ESP) (2007)
Stated purpose	Advise on whether medical items and services are reasonable and necessary	Ensures that prescriptions are appropriate, medically necessary, and not likely to result in adverse medical results	Establish medical standards for drug review, educational interventions for providers, collaboration with managed care organizations, review of therapeutic classes	Develops prior approval criteria, standards, and the preferred drug formulary for Medicaid	P&T: Improved patient safety Appropriate drug use, improved access to pharmaceuticals Promotion of a uniform pharmacy benefit Reduction in the acquisition cost of drugs ESP: Clinical policies informed by evidence Effective services to improve patient outcomes and to support VA clinical practice guidelines

(continued)

Table 10.2 (continued)

Characteristics	Medicare	Medicaid			Veterans Administration
		CA	NY	TX	
Summary of process	Panels of members review and evaluate medical literature, review technical assessment, and examine data and information on the benefits, harms, and appropriateness of medical items and services that are covered or eligible for coverage under Medicare. No more than 15 members with knowledge specific to the topic in question serve on the panel for each meeting	Drug Use Review (DUR) Board reviews and makes recommendations to DHCS in a collaborative effort with Xerox State Healthcare, LLC, and the University of California, San Francisco. They evaluate drug safety, effectiveness, essential need, potential for misuse, and cost. The Medi-Cal Contract Drug Advisory Committee (MCDAC) recommends additions, deletion, or retention of drugs on the Medi-Cal List of Contract Drugs	New York Health Commissioner appoints 19 members who recommend prior authorization requirements and preferred drugs	Texas Health and Human Services Commission appoints 18 members who recommend preferred drugs, prior authorization requirements, educational interventions for Medicaid providers, and drug utilization review	P&T: The VA National Formulary (VANF) lists products that must be available for prescription nationally. If not on the VANF, a regional formulary committee imposes reasonable clinically driven restrictions that are not based solely on economic considerations. A Medical advisory panel establishes criteria-for-use for VANF drugs and pharmacological management guidelines. ESP: Regional centers with academic partners are funded to prepare reports that are peer reviewed by clinical and policy experts. Reports are disseminated in VA to implement clinical guidelines and impose regional restrictions primarily through electronic health record reminders or decision support tools

Groups with membership on CE body	Consists of 100 appointed members from authorities in clinical and administrative medicine, biologic and physical sciences, public health administration health-care data and information management and analysis, the economics of health care, and medical ethics	DUR Board consists of seven MDs and PharmDs with expertise in critical care, pharmacy, prevention, and rheumatology. MCDAC membership not public	Consists of the Department of Health chairperson, six physicians, six pharmacists, one nurse practitioner, two DUR experts, and three consumers	Membership is made up entirely of 17 physicians and pharmacists from around the state from different specialties, plus two Medicaid managed care representatives, and one consumer advocate	P&T: The regional director in each region assigns a pharmacist executive to manage the regional formulary committees. Groups with membership on the regional formulary committees are not specified. ESP: VA staff and academic ESP center directors. Groups are not specified
Stakeholder involvement	Of the 100 members, six represent patient advocates, and six represent industry interests. Members submit detailed information concerning financial holdings, consultancies, and research grants or contracts to evaluate conflict of interest	DUR Board members are primarily academic schools of medicine and pharmacy. MCDAC membership not public	Consumers must have consumer advocacy involvement affecting Medicaid recipients	None noted	None noted

aFederal Register Vol. 63 No. 239 Monday, December 14, 1998 Amended Federal Register Vol. 80 No. 59 Friday, March 27, 2015

As the names suggest, MEDCAC examines data and information on medical devices and services to determine their benefits, harms, and appropriateness. In doing so, it helps decide whether Medicare should pay for them, consistent with current scientific understanding. While advisory recommendations to the administrator of the CMS or the Secretary of Health and Human Services, the recommendations of this group are usually adopted. For example, in 2013, the MEDCAC determined that the use of positron emission tomography (PET) scanning for imaging of beta-amyloid plaque in the brain is not supported by the current evidence and should not be covered by Medicare.

In terms of state Medicaid programs, California and New York focus decision-making at the agency level, and Texas empowers its Drug Utilization Review[2] (DUR) Board with decision-making. The VA has two comparative effectiveness bodies. One is the Pharmacy and Therapeutics Committee, which was established in the 1950s. The second, initiated in 2007, is the Evidence-based Synthesis Program. These comparative effectiveness-focused entities have a specific purpose: to advise on reasonable and necessary care, establish standards of care, develop prior approval criteria, and reduce acquisition costs.

Medicare has the most flexible process for making decisions regarding comparative effectiveness, and it involves drawing on its large panel of experts to provide the most appropriate expertise for each separate review. If the comparative effectiveness review is about an orthopedic device or procedure, Medicare would select from among orthopedic and internal medicine experts. Meetings for review are limited to 15 voting participants, with a majority determining the decision. There are different experts for each subject-based meeting.

California, New York, and Texas have DUR Boards with fixed membership made up of experts appointed by the relevant cabinet-level agency or commissioner. California is distinguished from New York and Texas in that it has a standing relationship with the University of California at San Francisco for comparative effectiveness expertise. California also has two comparative effectiveness bodies for drugs. The DUR Board advises on utilization review criteria, and the Medi-Cal Contract Drug Advisory Committee advises on the makeup of the preferred drug list. This committee explicitly considers cost-effectiveness in its deliberations, as well as contract price from manufacturers. New York and Texas house the DUR and preferred drug list decisions in the DUR Board.

While the VA has long had a national formulary that must be made available to all veterans across the country, the regional formulary committees can impose restrictions on the use of those drugs, provided the decisions are based on comparative effectiveness and not cost-effectiveness considerations. In other words, decisions must be based on the relative clinical outcomes achieved by different therapies, not on the relative cost of achieving those clinical outcomes. In addition,

---

[2] Drug utilization review is limited to such concerns as appropriate doses, duplication of prescriptions, drug interaction, drug contraindications, and other clinical issues that are more often driven by the Food and Drug Administration's approved labeling than by comparative effectiveness concerns, although it is an evidence-based process.

a medical advisory panel establishes the criteria, reminders, and pop-ups that are embedded in the VA electronic medical record, and an Evidence-based Synthesis Program can conduct reviews and studies to do the same.

Across all the payers discussed, comparative effectiveness bodies consist primarily of physicians and pharmacists. Medicare specifically calls for additional experts in public health administration, health-care data and information management and analysis, economics, and medical ethics. While it does not call for representation from Medicare beneficiaries, it does call for six patient advocates. Stakeholder involvement at Medicare also includes six representatives with industry interests. These representatives are required to submit detailed conflict of interest statements on a regular basis. Texas and New York include consumer representation; California does not. And the membership of its Medi-Cal Contract and Drug Advisory Committee is not public. The VA's Pharmacy and Therapeutics Committee is made up entirely of intramural staff with no representation from veterans and no public disclosure of its membership.

10.3.2   What Is the Process for Public Payer Decision-Making with CER?

Table 10.3 shows the process for public payer decision-making with CER. Medicare and all three state Medicaid programs reviewed give advance notice of their meetings and the agenda. They also explain how the public can attend and/or speak at the meeting. The VA does not. Medicare's MEDCAC meets between four and eight times during the year. The California DUR Board meets once per quarter, and New York meets as necessary. Texas also meets quarterly, and the VA does not publicly disclose how often its internal decision-making body meets. The public is allowed to make comments when the comparative effectiveness bodies at Medicare, New York, and Texas meet, but that is not the case at board meetings for California and the VA.

All public payers have up-to-date information about the process for decision-making on their websites except for the Medi-Cal Contract Drug Advisory Committee and the VA. Medicare is the only public payer to broadcast MEDCAC meetings live via webcast.

Medicare has the most extensive CER effort. Through its partnership with the Agency for Healthcare Research and Quality (AHRQ), Medicare has the ability to conduct professional evidence-based reviews. It can also draw upon expertise and studies underway at AHRQ's Evidence-based Practice Centers. Public testimony at these meetings can include prefiled testimony from manufacturers. The state Medicaid DUR Boards rely upon information submitted by manufacturers and Medicaid agency staff work, although Texas supplements this input with support from the University of Texas Health Sciences Center. State Medicaid DUR Boards also have support from contractors such as Xerox and First Data Bank, but these focus on drug utilization analysis services rather than comparative effectiveness input into coverage decisions. The Medicare, California, and New York bodies are

Table 10.3 Framework for decision-making with CER: openness of process

Characteristics	Medicare	Medicaid			Veterans administration
		CA	NY	TX	
Notification and number of meetings	Advance notice of all meetings published in *Federal Register*. Meetings open to the public. Meetings held 4–8 times in 2-year time period	DUR Board has advance notice of meetings sent to subscribers to Medi-Cal Subscription Service. Meetings open to the public. Meetings held once per quarter. MCDAC hold private meetings with manufacture representatives as needed	Meeting notice and agenda posted 30 days prior. Meetings held as necessary	DUR Board posts advance notice of meetings on its website. Meetings are open to the public and the public is allowed to speak with prior notification. Meetings are quarterly	None
Information available on website	Yes.[a] Also meetings broadcast live via webcast[b]	DUR Board: Yes[c] MCDAC: No	Yes[d]	Yes[e]	No
Research in-house, contracted, or submitted by manufacturers	External technology assessments conducted by the Agency for Healthcare Research and Quality (AHRQ) or an Evidence-based Practice Center (EPC) under contract. Public testimony, including manufacturer testimony, is submitted before each meeting	DUR Board: Research in-house through the MCDAC and staff of the DHCS. MCDAC relies upon manufacturer representatives and DHCS staff	Prospective DUR criteria are from the First Data Bank. Department of Health staff, specialty physicians, pharmacists, and nurses are consulted as needed	Prospective criteria are developed both in-house via contract with the University of Texas Health Science Center, and contracted pharmacy claim services vendor, Xerox, and through the First Data Bank DUR modules	Internally provided syntheses with support from academic centers

Role of CE body: advisory or prescriptive	Advisory	Both DUR Board and MCDAC are advisory to DHCS	Advisory	Prescriptive	Prescriptive
Public comment period	Notice of meetings published 2 months before meeting, including topics. Presentations, written testimony, and evidence submitted at least 20 days before the meeting. Public presentations are allowed at each meeting	DUR Board: None MCDAC: None	A public comment period is held at the beginning of each meeting. Requests to speak submitted 1 week prior to meeting	Notice of quarterly meetings published for the year. Public comment permitted on any action item. Those speaking complete a registration testimony form on the day of the meeting which includes whether the speaker is receiving compensation from any organization or manufacturer	None

[a] https://www.cms.gov/Regulations-and-Guidance/Guidance/FACA/MEDCAC.html

[b] http://www.cms.gov/live/

[c] http://files.medi-cal.ca.gov/pubsdoco/dur/dur_coe.asp

[d] https://www.health.ny.gov/health_care/medicaid/program/dur/

[e] http://www.txvendordrug.com/about/

all advisory. This means that final decisions are explicitly left to the Medicaid state agency. Comparative effectiveness review bodies at the Texas DUR and VA are prescriptive.

10.3.3 How Do Public Payers Consider Value and Budget?

Table 10.4 displays how payers consider value and budget in comparative effectiveness reviews. Topic selection is important and the first indication of how value is considered. Without a hearing, a new medical service, drug, or device can be sidelined for months or even years before receiving a favorable review that allows patient access, even if the new treatment is comparatively more effective and lower cost. All the public payers reviewed have defined processes or criteria for how they select topics for review, except New York, which does not apparently have explicit public criteria. Medicare has the most extensive list of criteria used to determine the need for national coverage determinations, ranging from significant controversy to patient advocates requesting the topic. California allows a manufacturer or a pharmacist to recommend a topic, while Texas looks to its managed care companies for topic ideas. The VA has explicit criteria similar to Medicare's.

Medicare does not have the legislative authority or the regulations in place to consider cost-effectiveness in its coverage determinations, and the Patient-Centered Outcomes Research Institute (PCORI) is specifically prohibited by law to use cost-effectiveness [7]. The other payers, Medicare and the VA, explicitly use cost-effectiveness in making coverage determinations.

Voting by the principal comparative effectiveness bodies is done by majority rule. The one exception is Medicare. In that program, MEDCAC has membership from consumer advocates and industry representatives but does not allow them to vote. Voting procedures are not known for the VA's internal deliberations. There is no explicit manufacturer appeals process for any public payer except to request a new review at a later time.

10.4 Risk-Sharing and Outcome-Based Contracting Models

While this chapter has focused so far on how comparative effectiveness decision-making bodies operate, comparative effectiveness has also played a growing and important role in the way public payers structure fee-for-service payments. These contingent payments, for which some or all payments are made based upon clinical process or outcomes measures, are becoming an increasingly important policy tool used by public payers to influence clinical and economic outcomes.

Table 10.4 Framework for decision-making with CER: how decisions are made

Characteristics	Medicare	Medicaid			Veterans administration
		CA	NY	TX	
Topic selection criteria	Topic has significant controversy. Published studies are found to be flawed, not address policy-relevant questions, or show conflicting results. A technology may have a major impact on the Medicare population or specific beneficiary groups. National coverage decisions would be better informed by viewpoints of patient advocates	A manufacturer, physician, or pharmacist may request a review	Not identified	The HHRC staff, managed care organizations, and others may request a review	P&T: Relevance to veteran population Availability of comprehensive, clinically relevant information ESP: Nominated by leadership Represents important uncertainty Not duplicative Feasible
Is cost-effectiveness used	No	Yes (both)	Yes	Yes	Yes
Voting procedures	Panel members selected for a meeting are voting members, except the consumer advocate and industry representative, who are nonvoting. Voting members score preestablished questions and scores are averaged	DUR Board: Majority MCDAC: Unknown	Majority	Majority	Not specified
Appeals process	None, except through normal process and criteria for topic selection	None	None	None	None, except through normal process and criteria for ESP topic selection

10.4.1   Hospital-Acquired Condition Reduction Program

One precedent for using comparative effectiveness to decide when to pay a fee-for-service claim is Medicare's adoption of a so-called "never event" standard to reduce payments to hospitals. In 2006, the National Quality Forum used comparative effectiveness to identify the most extreme type of ineffectiveness: when medical care leads to clearly identifiable, preventable, and serious errors with consequences for patients. Medicare incorporated eight of these National Quality Forum "never events" into its payment methodologies in 2007 and refused to pay for hospital discharges involving, for example, surgical site infections, poor control of blood sugar levels, and deep vein thrombosis following total knee replacement. To determine that a clinical event should never happen involves some level of data analysis and risk adjustment in addition to clinical judgment.

The CMS continue this practice as the Hospital-Acquired Condition Reduction Program. Payments are adjusted downward to hospitals in the worst-performing quartile of Medicare discharges with a large number of hospital-acquired conditions. In order to keep the list of these conditions current, Medicare relies upon the AHRQ to use comparative effectiveness analysis to decide whether an event meets a specific standard. This decision is important because payments to the worst performers are reduced by 1%. While seemingly small, a one percentage point reduction can mean millions of lost Medicare dollars for poor-performing hospitals, which may already have operating margins in the low single digits (or may even be negative). About half of state Medicaid programs have adopted their own versions of the Medicare Hospital-Acquired Condition Reduction Program.

10.4.2   The Medicare Accountable Care Organization (ACO)

Medicare defines accountable care organizations (ACOs) as hospitals and doctors who have joined together to receive fee-for-service payments with contingent payments based on certain joint at-risk success measures. There are multiple experimental Medicare ACO models, including the Medicare Shared Savings Program, Advanced Payment ACO Model, and Pioneer ACO Model. These models share a link between the decisions made and performance on designated success measures, which is determined by some form of comparative effectiveness analysis. There are 34 individual measures of quality performance used to decide whether Medicare ACOs will receive bonus payments under any of these programs. While they were developed primarily in dialogue with industry, the CMS determined which performance measures were to be included based upon comments and available comparative effectiveness reviews. How the list of performance measures continues to evolve – if at all – depends on CER.

10.4.3   Comprehensive Joint Replacement

Medicare's Comprehensive Joint Replacement Program, which took effect in 2016, holds hospitals in 67 geographic areas responsible for the quality of surgery for replacing hips and knees. Hospitals and physicians are paid on an ongoing basis as they have been traditionally, but at the end of a performance measurement period, quality measures are used to make bonus payments or apply penalties. Many of the measures focus on the bundle of payments associated with the procedures and their efficiency of delivery, including coordination of care. The patient outcome measures, selected based on comparative effectiveness and quality of care considerations, include metrics such as joint range of motion in degrees, use of gait aides, and patient abductor muscle strength. Thus, comparative effectiveness as a concept is being applied not only to determine whether a medical service, drug, or device should be covered and paid for but also whether specific procedures and episodes of illness should be paid for in full or only in part.

10.4.4   Medicare Advantage Star Rating System

The Medicare Advantage Star System applies comparative effectiveness considerations to beneficiary choice of Medicare Advantage health plans. Medicare introduced the Five-Star Quality Rating System in 2008 and publicly reports star ratings to help beneficiaries select a private plan during open enrollment periods each fall. In 2012, the star ratings were also used to adjust payment to plans. Plans with four or five stars receive bonus payments. Depending upon the year and based on competitive effectiveness reviews, anywhere from 26 to 80 quality measures are used to assign stars and thereby summarize each plan's quality in an easy-to-understand format for beneficiaries. The results are then published on the Medicare Plan Finder website for beneficiaries and their families to help them select a plan. An extra star can drive an additional 6–9% enrollment [8]. With over 18 million beneficiaries now enrolled in Medicare Advantage, which is more than 35% of all beneficiaries and growing, the application of quality measures based on comparative effectiveness holds important implications for plan participation and financial feasibility.

10.4.5   Medicaid Examples

States continue to struggle with how best to serve certain groups, such as participants eligible for both Medicare and Medicaid, recipients with multiple chronic conditions, and individuals in need of behavioral health services, many of whom remain in traditional fee-for-service Medicaid. These states have been experimenting since the start of the Medicaid program with the comparative effectiveness of different delivery models. Some of the prominent successes that have been adopted and are increasingly being adopted are initiatives such as patient-centered medical homes, so-called "money follows the client" models, and programs for all-inclusive care for the elderly. In a 2016 study by the Kaiser Family Foundation, Medicaid directors report that they continue to improve upon the delivery models that have worked [9]. However, they also acknowledge that they are only beginning to understand the opportunities for using comparative effectiveness in the context of population health and the social determinants of health. They also understand that they must not only improve the quality of care for their recipients through delivery models that meet policy-relevant comparative effectiveness thresholds, they must also measure and improve the outcomes themselves.

10.5   When Is CER Good Enough?

In a recent survey of 60 Medicaid medical and pharmacy directors from 46 states, researchers found overwhelming support for CER and the prospect that it can lead to better clinical decision-making and value [10]. More than 90% of those surveyed from one of the largest public payers based on persons covered felt comparative effectiveness would be used increasingly in the next 5 years and that cost-effectiveness should be an important component of the process. The same opinion survey was not administered to Medicare and VA decision-makers. The ACA, however, created the PCORI and a funding mechanism to support CER [11]. At the same time, AHRQ continues with its mandate to conduct comparative effectiveness reviews for Medicare. The VA has its Evidence-based Synthesis Program, which conducts original research and then applies it to the unique needs of the VA and veterans.

Each year, the Food and Drug Administration approves hundreds of drugs and their indications. Most of these approvals are for generic equivalents or variations on previously existing products. These approvals also cover new dosage forms for various conditions. A subset of these approvals, typically around 45, are for novel molecules and innovative drugs that advance clinical care through new mechanisms. In addition, each year, hundreds of medical devices are approved for marketing. Medical devices can be as simple as a thermometer or as complex as implantable spinal replacements or as enormous as imaging machines taking up an entire hospital room. Medical services have a different pathway to market and depend upon coverage decisions and assignment of billing codes. But the innovation and pace of change is impossible for any single payer to evaluate from a comparative effectiveness viewpoint and remain truly up-to-date [12].

For marketing approval and/or uptake by physicians, either randomized clinical trials or observational studies of real patients using these new drugs, devices, or medical services must be completed. However, even these trials and studies are often inconclusive [13]. Reviews of existing literature, conducted by experts in the field, can be helpful, but it can be difficult to gauge their ultimate impact on a payer's budget [14]. While sometimes existing research has been completed on only one demographic or age group of patients [15], the problem is that comparative effectiveness decisions can affect an entirely different group of patients.

It seems almost impossible to find a decision-making body that does not have some sort of gap in knowledge or perceived bias either for or against the comparative effectiveness evidence under review [16]. As mentioned earlier, public payers have limited resources with which to carry out their comparative effectiveness decision-making. And in some cases, legislative and political mandates can second-guess their deliberations. A notable example of second-guessing occurred when Secretary Sebelius overruled comparative effectiveness guidelines on breast cancer from the US Preventive Services Task Force within 2 weeks of their release because of an outcry from breast cancer patient groups and the media [17].

Unlike private payers, public payers are also especially subject to public scrutiny and stakeholder advocacy. Public payer decisions affect many people, sometimes millions of covered lives in programs that are specifically intended to incorporate equity considerations in their decisions in a very public way. Reviewing the framework currently in place for these public payers reveals a number of areas where decision-making is not fully transparent and open to the participation either of the public or manufacturers with an economic stake in the outcomes. Medicare can be said to have the most well-structured process and open path for participation. Medicaid programs, at least the three largest reviewed here, do fairly well at attempting to create an open process for hearing all viewpoints, though the Medi-Cal Contract Drug Program is an exception. This drug program appears to hew to a traditional government-procurement-at-the-lowest price agenda. The VA, as a closed and almost completely government-run system, relies on employee expertise without much hand-holding (or, alternatively, involvement) with the public or veterans.

It is far too soon to say where comparative effectiveness decision-making is heading for these public programs [18]. The PCORI is only a few years old, and Medicaid programs are preoccupied with implementing health-care reform and Medicaid expansion. But if comparative effectiveness holds the promise of informing coverage and payment policy decisions under an avalanche of innovation, it should play an even larger role for public payers.

References

1. Barnett, Jessica C. and Marina S. Vornovitsky (2016) Current Population Reports, P60–257(RV), Health Insurance Coverage in the United States: 2015, US Department of Commerce Economics and Statistics Administration, Bureau of the Census. https://www.census.gov/content/dam/Census/library/publications/2016/demo/p60-257.pdf
2. Gold M, et al (2014) Medicare advantage 2014 spotlight enrollment market update – firms and market structure – 8588 The Henry J. Kaiser Family Foundation.pdf. Kaiser Family Foundation. http://kff.org/report-section/medicare-advantage-2014-spotlight-enrollment-market-update-firms-and-market-structure/

3. Smith VK, Gifford K, Ellis E, Rudowitz R, Snyder L, Hinton E (2015) Medicaid reforms to expand coverage, control costs and improve care results from a 50-state medicaid budget survey for state fiscal years 2015 and 2016. Kaiser Family Foundation. http://files.kff.org/attachment/report-medicaid-reforms-to-expand-coverage-control-costs-and-improve-care-results-from-a-50-state-medicaid-budget-survey-for-state-fiscal-years-2015-and-2016
4. Atkins D, Kupersmith J, Eisen S (2010) The veterans affairs experience: comparative effectiveness research in a large health system. Health Aff (Millwood) 29(10):1906–1912. doi:10.1377/hlthaff.2010.0680
5. Levy AR, Mitton C, Johnston KM, Harrigan B, Briggs AH (2010) International comparison of comparative effectiveness research in five jurisdictions: insights for the US. Pharmacoeconomics 28(10):813–830. doi:10.2165/11536150-000000000-00000
6. Chambers JD, Chenoweth M, Cangelosi MJ, Pyo J, Cohen JT, Neumann PJ (2015) Medicare is scrutinizing evidence more tightly for national coverage determinations. Health Aff (Millwood) 34(2):253–260. doi:10.1377/hlthaff.2014.1123
7. Neumann PJ, Weinstein MC (2010) Legislating against use of cost-effectiveness information. N Engl J Med 363(16):1495–1497. doi:10.1056/NEJMp1007168
8. Reid RO, Deb P, Howell BL, Shrank WH (2013) Association between Medicare Advantage plan star ratings and enrollment. JAMA 309(3):267–274. doi:10.1001/jama.2012.173925
9. The medicare part D prescription drug benefit the Henry J. Kaiser Family Foundation.pdf. Kaiser Family Foundation (2015). Kaiser Family Foundation
10. Weissman JS, Westrich K, Hargraves JL, Pearson SD, Dubois R, Emond S, Olufajo OA (2015) Translating comparative effectiveness research into Medicaid payment policy: views from medical and pharmacy directors. J Comp Eff Res 4(2):79–88. doi:10.2217/cer.14.68
11. Selby JV, Lipstein SH (2014) PCORI at 3 years – progress, lessons, and plans. N Engl J Med 370(7):592–595. doi:10.1056/NEJMp1313061
12. Timbie JW, Fox DS, Van Busum K, Schneider EC (2012) Five reasons that many comparative effectiveness studies fail to change patient care and clinical practice. Health Aff (Millwood) 31(10):2168–2175. doi:10.1377/hlthaff.2012.0150
13. Berger ML (2013) Can the pharmaceutical industry embrace comparative effectiveness research? A view from inside. Expert Rev Pharmacoecon Outcomes Res 13(5):565–568. doi:10.1586/14737167.2013.833061
14. Glick HA, McElligott S, Pauly MV, Willke RJ, Bergquist H, Doshi J, Fleisher LA, Kinosian B, Perfetto E, Polsky DE, Schwartz JS (2015) Comparative effectiveness and cost-effectiveness analyses frequently agree on value. Health Aff (Millwood) 34(5):805–811. doi:10.1377/hlthaff.2014.0552
15. Mohr P (2012) Looking at CER from Medicare's perspective. J Manag Care Pharm JMCP 18(4 Suppl A):S5–S8. doi:10.18553/jmcp.2012.18.S8-A.S5
16. Neumann PJ (2013) Communicating and promoting comparative-effectiveness research findings. N Engl J Med 369(3):209–211. doi:10.1056/NEJMp1300312
17. Gusamo MK, Gray BH (2010) Evidence and fear: navigating the politics of evidence-based medicine. Acad Health Rep. 38. doi:citeulike-article-id:10755566
18. Pearson SD, Bach PB (2010) How medicare could use comparative effectiveness research in deciding on new coverage and reimbursement. Health Aff (Millwood) 29(10):1796–1804. doi:10.1377/hlthaff.2010.0623

Chapter 11
Patient and Stakeholder Engagement in Designing Pragmatic Clinical Trials

Anna Hung, Carole Baas, Justin Bekelman, Marcy Fitz-Randolph, and C. Daniel Mullins

Abstract The desire for patient-centeredness and more pragmatic clinical trials is increasing. Pragmatic clinical trials are conducted in normal practices to answer whether interventions work in real-world settings, whereas explanatory clinical trials are conducted under stricter settings with tighter control to answer whether an intervention can work. While both explanatory and pragmatic trials can have patient-centered elements, pragmatic trials more directly answer patient and policymaker questions.

To ensure trials are truly reflective of what is meaningful to patients, patients and stakeholders should be engaged during the entire trial process—from planning the trial through conducting the trial to disseminating the results. Identifying, recruiting, training, continually engaging, and compensating an advisory committee of patients and stakeholders to serve as a resource and guide through the trial process is one way to increase the patient-centeredness of a trial. Patients can help design recruitment and retention strategies, codevelop endpoints, review consent forms, and identify appropriate dissemination channels. Patient involvement throughout the trial benefits not only the research team but the patients themselves as well as future patients. This chapter explores how to increase patient-centeredness in clinical trials by engaging patients and stakeholders throughout the pragmatic clinical trial process.

A. Hung • C. Daniel Mullins (✉)
University of Maryland School of Pharmacy, Baltimore, MD, USA
e-mail: dmullins@rx.umaryland.edu

C. Baas
Cancer Information & Support Network, Auburn, CA, USA

J. Bekelman
University of Pennsylvania, Philadelphia, PA, USA

M. Fitz-Randolph
PatientsLikeMe, Cambridge, MA, USA

© Springer Nature Singapore Pte Ltd. 2017
H.G. Birnbaum, P.E. Greenberg (eds.), *Decision Making in a World of Comparative Effectiveness Research*, DOI 10.1007/978-981-10-3262-2_11

11.1   Patient-Centeredness in Clinical Trials

11.1.1   Patient-Centeredness

The term "patient-centered" has been widely used, especially since the establishment of the Patient-Centered Outcomes Research Institute (PCORI) in 2010. PCORI is an independent, nonprofit, and nongovernmental organization based in the United States that was established by Congress by the Patient Protection and Affordable Care Act. According to PCORI, patient-centeredness is the "extent to which the decision-making needs, preferences, and characteristics of patients are addressed in diverse settings of health care" [1]. Making clinical trials more patient-centered will increase patient participation and enrollment in trials [2], which addresses a major reason for early clinical trial discontinuation [3].

For a clinical trial to be patient-centered, patients must be asked for advice continuously throughout the trial planning, conducting, and disseminating process. From patient education materials used in recruiting patients to strategies on how and to whom to disseminate results from the trial, there are a variety of topics in which patient involvement can transform a trial to be more patient-centered.

11.1.2   Pragmatic Versus Explanatory Clinical Trials

Clinical trials may be classified by their purpose, trial design, or level of intervention. They may also be described as "pragmatic" or "explanatory," which refers to the environment in which the intervention will be studied. Pragmatic trials compare interventions under everyday clinical conditions to improve practice and inform clinical and policy decisions [4]. On the other hand, explanatory trials test a hypothesis under ideal conditions to determine causes and effects of an intervention [4]. Pragmatic versus explanatory trials were first distinguished in 1967 by Schwartz and Lellouch [5]. Both can be called randomized controlled trials (RCT) if they include randomization of trial participants to comparison arms, including a control arm.

Trials often contain trial design features that are varying degrees of "pragmatic" versus "explanatory," resulting in a continuum instead of a dichotomy [5]. To facilitate trial design, the PRagmatic-Explanatory Continuum Indicator Summary (PRECIS) 2 tool (Fig. 11.1) depicts to what degree each RCT design feature is pragmatic versus explanatory [6, 7].

It is important to determine how pragmatic any given domain of a trial should be in light of the research question being answered and the needs of the patient and provider communities in which the trial will take place.

Both explanatory trials and pragmatic trials can incorporate patient-centered elements to varying degrees (Fig. 11.2); however, the characteristics of pragmatic trials are inherently more aligned with patient-centeredness than those of explanatory trials (Table 11.1).

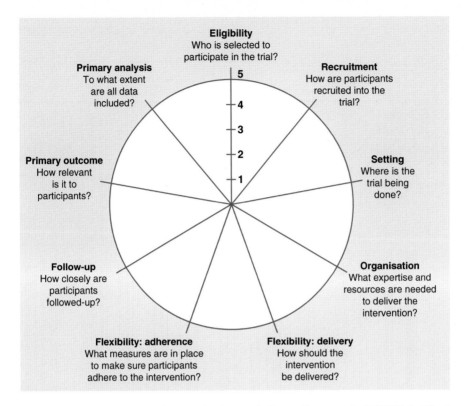

Fig. 11.1 The Pragmatic-Explanatory Continuum Indicator Summary 2 (PRECIS-2) Wheel (Abstracted from Loudon [6])

High patient involvement

Explanatory trial with high patient involvement	Pragmatic trial with high patient involvement
Example: RCT with a patient-reported outcome as the primary endpoint	Example: Patient-centered PCT comparing two previously approved drugs in which patients co-design the intervention and protocol
Explanatory trial with low patient involvement	Pragmatic trial with low patient involvement
Example: Phase II drug-dosing RCT	Example: PCT using traditional clinical outcomes but trial sites include community-based practice sites

Explanatory ←——————————————————————→ Pragmatic

Low patient involvement

Fig. 11.2 Explanatory-Pragmatic and Patient Involvement Continua in Clinical Trials. *PCT* pragmatic clinical trial, *RCT* randomized controlled trial

Table 11.1 Major differences between explanatory and pragmatic trials*

Explanatory ———————————→ Pragmatic

Feature	Explanatory	Pragmatic	How patient and stakeholder engagement can transform an explanatory RCT into a patient-centered pragmatic trial
Question	Efficacy – Can the intervention work?	Effectiveness – Does the intervention work when used in normal practice?	Ask patients and stakeholders to determine what questions are most important and meaningful to them.
Setting & organization	Well resourced, "ideal" setting (often exclusive to academic medical centers)	Normal practice (community-based practices may be included). Specifically, an identical setting to which the results are intended to be applied, and no more than the existing health care staff and resources in that setting are used.	Explain environment and resources to patients and stakeholders and solicit feedback on any discrepancies between trial settings and typical care settings (e.g., academic practices versus community-based practices). Keep in mind that different patient subpopulations may have different typical care settings.
Participants: eligibility criteria & recruitment	Highly selected. Poorly adherent participants and those with conditions which might dilute the intervention effect are often excluded, resulting in homogeneous populations.	Little or no selection beyond the clinical indication of interest. Anyone who is likely to be a candidate for the intervention if it was being provided in usual care for this condition is included. Recruitment would take place in usual care so that only people who attend a clinic with the condition of interest are recruited after they present on their own behalf without any over-recruitment effort. Recruitment would also take place in more than one clinic to increase applicability of the trial results. This results in heterogeneous populations and health care delivery settings.	Seek advice from patients and stakeholders regarding how to recruit a diverse and relevant population to the clinical question of interest. Show patients and stakeholders study recruitment materials and solicit feedback on how to make them patient-friendly and relevant to participants and community sites. Explore different recruitment channels with patients and stakeholders.
Intervention	Strictly enforced and adherence is monitored closely	Applied flexibly as it would be in normal practice, and the details of how to implement the intervention would be left up to providers	Use patient and stakeholder feedback to determine the most relevant intervention and the variation in how the intervention would be administered.
Comparison intervention	Often compare new agents to placebo treatments	Should always compare two or more currently used treatments, often for which there is clinical equipoise**	Use patient and stakeholder feedback to determine the most relevant comparators.
Intervention practitioner expertise	Often delivered by highly skilled and specialized practitioners in research settings	Often delivered by practitioners in a routine practice setting	Solicit feedback from patients regarding what types of providers are most frequently seen. Keep in mind that different patient subpopulations may have different provider types.
Participant compliance with "prescribed" intervention	Study participants' compliance with the intervention is monitored closely. Both prophylactic strategies (to maintain) and "rescue" strategies (to regain) high compliance are used.	Full flexibility in how participants adhere to the intervention. In usual care, health professionals encourage patients to take medication or follow therapy as best they can, and such encouragement would not count against a pragmatic design.	Ask patients and stakeholders about the variation in participant compliance with the intervention, as well as reasons for this variation.

Practitioner adherence to study protocol	Close monitoring of how well the participating clinicians and centers are adhering to even minute details in the trial protocol	Unobtrusive (or no) measurement of compliance or adherence, and no special strategies to maintain or improve compliance are used	Ask patients and stakeholders about the variation in practitioner compliance with the intervention, as well as reasons for this variation. Use this feedback to determine level of measurement of compliance needed.
Outcomes	Often short term surrogates or process measures (e.g., serum lipids or coronary artery calcification scores), in an attempt to decrease the time required to conduct trials	Outcomes that are of obvious importance from the patient's perspective. Outcomes selected using a pragmatic approach would also be relevant to commissioners of care (the people who decide whether to implement the intervention on the basis of its results). These outcomes are often obtainable from the electronic medical record, other extant data sources, or from simple patient surveys.	Ask patients and stakeholders what outcomes (including patient-reported outcomes) are most meaningful to them.
Analysis	"Intention-to-treat" is usually performed; however, this may be supplemented by a per-protocol analysis or an analysis restricted to "compliers" or other subgroups to estimate maximum achievable treatment effect	"Intention-to-treat" approach, recognizing that treatment crossover may be more common than in explanatory trials. Patients and practitioners are typically not blinded to treatment assignment, although the allocation of patients to different groups should be random and the process for assessing endpoints should be blinded to group allocation.	Explain to patients and stakeholders how analysis will be completed and be open to any feedback.
Follow-up	Study individuals are followed with many more frequent visits and more extensive data than would occur in routine practice	No greater frequency of follow-up of recipients than would be the case in usual care. Longer follow-up periods may be required, and administrative databases may be searched.	Using patient and stakeholder feedback, identify ways to retain participants throughout the trial.
Relevance to practice	Indirect – Little effort made to match design of trial to decision-making needs of those in usual setting in which intervention will be implemented	Direct – Trial is designed to meet needs of those making decisions about treatment options in setting in which intervention will be implemented	After explaining the research question and trial design, ask patients and stakeholders for any missing features to make the results more relevant to practice

*Adapted from Zwarenstein 2008, Rosenthal 2014, Thorpe 2011, and Loudon 2015 [6–9]

**Clinical equipoise refers to a genuine uncertainty about the preferred treatment [10]

In explanatory trials, the fundamental question being asked is whether an intervention could work in ideal settings, often by highly specialized practitioners, according to a strict protocol, under extensive follow-up, in homogeneous populations, and often being compared to placebo [7–9]. The results of explanatory trials are used to advance scientific knowledge of the efficacy of interventions. Explanatory trials are patient-centered in that providers can indirectly use explanatory trial results to determine how to manage and treat individual patients. Clinical trial researchers may also consider the patient perspective when designing the trial, for example, by considering transportation costs and other factors that impact patient burden of participation; however, patients are often not explicitly consulted in the design of an explanatory trial. Moreover, results from an explanatory trial do not directly answer patients' real-world treatment decision questions.

In contrast to explanatory trials, pragmatic trials address the fundamental question of whether an intervention works in normal, available settings, often by routine care providers, with flexible adherence to protocol to accommodate how care is usually provided, using administrative databases as follow-up, employing outcomes relevant to patients and stakeholders, and in heterogeneous populations. Pragmatic trials, which are increasingly used in health-care research [11], are aligned with patient-centeredness because the results of pragmatic trials directly answer patients', as well as providers', treatment decision questions. Furthermore, pragmatic trials can address patient preferences from diverse populations and in a myriad of health-care settings. However, early pragmatic trials (Case Study 1) focused less on patient-centeredness than contemporary pragmatic trials (Case Study 2). Today, how to ensure that the research question, trial design, and results from a pragmatic trial are authentically reflective of what is meaningful to a patient is the ongoing challenge.

11.1.3   Using Patient and Stakeholder Engagement to Make Trials Patient-Centered

Engagement in research is the "meaningful involvement of patients, caregivers, clinicians, and other health care stakeholders throughout the research process—from topic selection through design and conduct of research to dissemination of results" [12]. Stakeholders include not only patients, caregivers, and clinicians but also others who may benefit from an intervention, funders of health care and research, and researchers [6]. Patient and stakeholder engagement early and continuously [13] along every step in the clinical trial is key to drive the transition from explanatory to pragmatic trials and ensure true patient-centeredness in the clinical trial (Table 11.1).

Thinking about the patient as being at the center of decision-making is not a new concept: most clinicians believe they are patient-centered when providing care, and many researchers believe they are patient-centered in asking their questions for studies that provide health-related information. However, what is evolving is how to meaningfully incorporate patient perspectives in research through patient and stakeholder engagement. The next section will review how to use patient and stakeholder engagement to transform clinical trials to become more patient-centered.

11.2 Patient and Stakeholder Engagement in Patient-Centered Pragmatic Clinical Trials

11.2.1 Identification and Recruitment of Patient Representatives in Pragmatic Clinical Trials

One of the most important aspects of pragmatic clinical trials is the "patient-centered" focus and applicability to real-life clinical practice. To ensure these fundamental criteria are met, pragmatic clinical trials should integrate the patient voice into numerous steps in the research design and trial implementation.

The first step in this process is the creation of a Stakeholder Advisory Committee. A patient-centered trial will enroll patients as well as recruit patient representatives to serve on a Stakeholder Advisory Committee. Because the purpose of comparative effectiveness research (CER) is to assist patients, clinicians, policymakers, and health plans and other payers to make informed decisions that will improve health care, these groups should all be represented on the Stakeholder Advisory Committee and must strongly endorse the proposed study [14]. Members should also be involved with the research team throughout the study, including early involvement in reviewing proposed trial designs, active feedback on recruitment and engagement materials, regular updates during the course of the trial, and reflection on the findings and planned reporting of results. This continuous involvement ensures that research questions and outcomes are more relevant and findings more likely to be disseminated and implemented.

Patient representatives on the Stakeholder Advisory Committee are drawn from the population that is affected and may include patients as well as those who have been impacted by the disease. These representatives should reflect the patient population with regard to gender, race/ethnicity, age, and geographic and socioeconomic factors, mirroring the design of the pragmatic clinical trial to test the effectiveness of the intervention in a broad routine clinical practice. This is analogous to, but separate from, best practices on how patients should be recruited to participate in the trial. Identifying interested patient representatives often begins by reaching out to advocacy organizations associated with the disease; individuals may also be contacted through community-based health initiatives or programs specific to the population being studied. For example, patient representatives for studies involving minority populations could be recruited through local churches, which are influential institutions for health promotion in black communities [15, 16]. Prospective patient representatives from all populations may be unfamiliar with the terminology involved in clinical trial research and with the treatment parameters being studied; therefore, they may require additional education in these areas in order to fully participate as equal members of the Stakeholder Advisory Committee.

11.2.2   The Role of the Stakeholder Advisory Committee

The Stakeholder Advisory Committee serves as a guide and resource to the principal investigator and the clinical trial research team. The Stakeholder Advisory Committee should be involved in all stages of the trial: (i) planning the trial, (ii) conducting the trial, and (iii) disseminating the trial results [17].

In the design phase of the trial, Stakeholder Advisory Committee members can encourage trial participation and retention by reviewing recruitment materials and providing advice on the best methods to reach patients (Table 11.2).

The Stakeholder Advisory Committee can also codevelop endpoints important to patients and stakeholders and review how outcomes are measured. To help conduct the trial, Stakeholder Advisory Committee members can assist in engaging sites and participants in the trial.

Stakeholder Advisory Committee members can also play a key role in dissemination. Clinical trials have gained notoriety for not better informing the participants and affected community of updates and results as the clinical trial proceeds [18]. Deciding which communication channels, such as web pages, community health fairs, and clinic visits, should be used is a key to retaining participants and fostering a positive outlook on the trial by the community. In addition to channels, the types of dissemination activities and "givebacks," such as educational handouts or free screenings, are important to consider. Understanding the barriers to reaching their populations, the Stakeholder Advisory Committee should advise on the methods and activities to best reach the various target audiences. Throughout all phases of

Table 11.2 Stakeholder Advisory Committee activities

Planning the trial
Designing recruitment and retention strategies
Codeveloping patient-centered endpoints
Reviewing validity of outcome measures/instruments
Determining feasibility and patient burden of data collection
Assuring that minority and other underrepresented and hard-to-reach patients are engaged
Conducting the trial
Engaging sites and participants
Reviewing institutional review board consent forms
Participating in data safety monitoring board
Assuring that minority and other underrepresented and hard-to-reach patients are engaged
Disseminating the trial results
Assessing "givebacks" and lay reports
Determining appropriate dissemination channels
Devising community dissemination activities
Addressing barriers and facilitators to dissemination
Co-authoring manuscripts
Co-presenting results
Updating community members and providers
Assuring that minority and other underrepresented and hard-to-reach patients are engaged

the trial, the Stakeholder Advisory Committee should ensure that the hard-to-reach and underrepresented populations are involved.

Stakeholder Advisory Committee members will generally meet multiple times throughout the year. Depending on the committee members and their locations, teleconferences may suffice; however, occasional in-person meetings help foster deeper relationships. Meetings with subgroups within the Stakeholder Advisory Committee may also be beneficial for topics specific to one patient or stakeholder subpopulation, such as caregivers or providers.

Across trials, Stakeholder Advisory Committees are involved to varying degrees. Even within a trial, how much an individual member of a Stakeholder Advisory Committee contributes can differ. These levels of involvement include [19]:

 (i) Communication: A one-way flow of information from the research team to committee members
 (ii) Consultation: A one-way flow of information from committee members to the research team
(iii) Participation: A bidirectional flow of information between the research team and committee members

In some models, a potential fourth level is "user-led" research, in which research is designed and conducted by the public [20, 21]. Evaluating the degree and meaningfulness of engagement is an important area which needs further development.

It is important for investigators to demonstrate that member feedback is valued, creating a welcoming environment for ongoing dialogue. Showing appreciation for Stakeholder Advisory Committee member comments, as well as making significant changes based on feedback received, is critical not only for making members feel valued but also for maximizing potential improvements in the trial. Moreover, building authentic relationships with the Stakeholder Advisory Committee helps further engage members in their role on the committee.

Stakeholder Advisory Committee members should be offered compensation for their time. This raises the value of their participation both in their eyes and the eyes of the entire study team, confirms the importance of their participation as representatives of a larger population (of patients or other stakeholders), and allows for a wider variety of members to participate than might be possible with an entirely volunteer group. PCORI supports the concept of financial compensation and has created a framework for the financial compensation of patients, caregivers, and patient/caregiver organizations engaged in PCORI-funded research as engaged research partners [22].

While a Stakeholder Advisory Committee clearly benefits the clinical trial team by providing insight in patient values, participation in such a committee can also benefit the committee members themselves and their communities. Beyond broadening the scope of their understanding of clinical trials, participating on a Stakeholder Advisory Committee gives patients and stakeholders the opportunity to influence future policy decisions that may affect their own health as well as that of future patients (Table 11.3).

Table 11.3 Benefits of a Stakeholder Advisory Committee

	Benefits to stakeholder advisory committee members	Benefits to research team	Benefits to future patients
Planning	Voice what is important to patients and stakeholders, potentially influencing population-level decisions which could later affect themselves Learn from others on the committee, who may represent another perspective (e.g., patients and insurance companies may have different viewpoints)	Understand what is important to patients and stakeholders	Receive care that is "better" in terms of what is important to patients
Conducting	Experience how trials are conducted and implemented	Conduct a trial that is more patient-centered and relevant to payers, and perhaps more likely to be paid for Recruit patients more easily, since patients would be interested in a patient-centered trial	Are provided with an example of how they can become involved in research in a meaningful way
Disseminating	Influence how participants and the community are taught study results	Share results the way patients would like, and therefore be more likely to retain participants	Gain knowledge of trial findings more easily

11.2.3   Training

Orientation and training of patient representatives to serve on a Stakeholder Advisory Committee are crucial steps in helping them understand their role in the trial as well as the research questions [23]. Initial meetings with Stakeholder Advisory Committee members should begin with a short review of the study design as a discussion of what questions the trial is designed to answer. Group orientation may be most efficient; however, individual members of the Stakeholder Advisory Committee may need additional orientation to make them feel more comfortable with the trial and with providing feedback. Individual orientation and discussion may be as simple as a few minutes of questions and answers with an investigator or study staff before a regularly scheduled meeting.

Beyond members of the Stakeholder Advisory Committee, there are other communities who should be engaged and trained to support a patient-centered trial. These include the participants in the trials, caregivers, and immediate supporters of the participants, the community of patients not directly participating in the trials but potentially affected by trial results, and site investigators who may need support and training in engaging their own patients. Each of these groups of stakeholders will come to the Stakeholder Advisory Committee with different needs for information and orientation and different focus areas for their feedback.

11.3   Case Studies of Pragmatic Clinical Trials

The following two case studies—one of an early pragmatic trial and the other of a contemporary pragmatic trial—demonstrate the evolution of patient-centeredness in pragmatic clinical trials.

Case Study 1: ALLHAT Trial

The Antihypertensive and Lipid Lowering Treatment to Prevent Heart Attack Trial (ALLHAT) is a well-known, early pragmatic clinical trial that began in 1994 and ended in 2002 [24]. Supported by the National Heart, Lung, and Blood Institute, ALLHAT compared the effectiveness of medications to treat hypertension. In a subset of patients, ALLHAT also studied statin drug treatment versus usual care. The outcomes were mortality as well as cardiovascular events.

ALLHAT took place over 8 years in 623 North American sites, which were primarily community-based settings, and compared newer forms of antihypertensive medications to an older form of antihypertensive medication [24, 25]. ALLHAT was pragmatic because of the real-world setting, large and diverse patient population, relevant drug comparators, and long trial duration that allowed for important clinical endpoints. As a result, findings from ALLHAT were able to directly inform clinicians about how to treat their patients.

However, some have critiqued ALLHAT for a "blunted impact" due to difficulty in convincing doctors to change their treatment patterns, disagreement in the interpretation of results, and heavy pharmaceutical company marketing [26]. Some have also criticized that ALLHAT's protocol for "step-up" drugs led to a comparison of combination regimens that were less clinically relevant [25].

Moreover, while mortality and cardiovascular event endpoints should be important to patients, ALLHAT did not include patient involvement (such as through a committee of patient and stakeholder representatives) to guide the trial throughout all phases—planning, conducting, and disseminating—to ensure a patient-centered trial.

Despite these criticisms, ALLHAT demonstrated the ability to conduct RCTs in more "real-world" settings.

Case Study 2: RadComp Trial

The Pragmatic Randomized Trial of Proton vs. Photon Therapy for Patients with Nonmetastatic Breast Cancer Receiving Comprehensive Nodal Radiation, A Radiotherapy Comparative Effectiveness (RadComp) trial, is led by the University of Pennsylvania and funded by PCORI. The RadComp trial started receiving funding in 2015 and enrolled its first patient in April 2016. The RadComp trial compares two forms of radiation therapy for nonmetastatic breast cancer. The primary outcome is reduction in major cardiovascular events.

The RadComp trial is one of the few examples of clinical trials which focuses on patient-centeredness. The RadComp trial collaborated with the University of Maryland's PATIENTS Program [27] to assist with patient and stakeholder engagement. Early in the development of this trial, a Stakeholder Advisory Committee was formed. This committee consists of a wide variety of patients and stakeholders: breast cancer survivors, breast cancer advocates, insurance companies, and governmental funding agencies. In-person group and individual orientations were provided to committee members. For the entire duration of the five-year trial, the Stakeholder Advisory Committee will meet quarterly via teleconference, with in-person meetings in the first and last years of the trial.

While in the trial design phase, patients expressed in a focus group that their primary concern was major heart problems (such as heart attack) after cancer treatment for locally advanced breast cancer. Therefore, that endpoint, which was also supported by investigators, became the primary endpoint. In further discussions with patients and PCORI, investigators also learned that patients wanted the study to track cancer recurrence to make sure the new therapy "cured" breast cancer just like usual treatment. In the RadComp trial, even though investigators felt there would be no difference in efficacy between the two treatments studied, investigators altered the study design to scientifically test whether the treatments were not inferior to each other in cancer control.

The Stakeholder Advisory Committee has also provided feedback on various documents leading to numerous tangible changes. Examples of documents reviewed by the Stakeholder Advisory Committee include a summary of patient-reported outcome instruments, informed consent documents, patient handouts, the web portal, logo, videos for recruitment and education, and lists of frequently asked questions. The power of the Stakeholder Advisory Committee is illustrated by the involvement of a breast cancer survivor on the committee in the patient materials: she strongly recommended having a shorter, clearer version of the educational handout for patients and subsequently crafted her own version. This form was later approved by the Institutional Review Board and is now provided to patients enrolling in the trial.

The RadComp trial is also collaborating with PatientsLikeMe (PLM), an online data platform and social network for individuals living with chronic conditions [28]. PLM's preliminary work with other clinical trial participants yielded several areas of concern for patients in long-duration trials. This collaboration addresses the following potential issues facing patients in trials:

- Patients in trials typically are isolated from each other.

  - PLM provides a place to connect remotely and anonymously to other trial participants, creating a human frame around the trial.
  - PLM provides a portal, or hub, where participants can find links to important information about the trial.
  - PLM showcases the investigators and site staff to humanize the centers and remind participants of their place in the big picture of the trial.

- Patients have a very limited view of the progress of longer-term trials with years of follow-up, such as the study proposed by RadComp.

  - PLM can help disseminate updates by RadComp about how the study is progressing as well as any early findings and "fun facts" to help patients remain engaged.

One of the earliest ways PLM helped bring patient input into the planned online portal was by assisting in the development of the RadComp trial logo. A design consultant met with the primary team and read materials about the study that had been written by the patients on the Stakeholder Advisory Committee. From those materials, she noted the following:

"When people see the RadComp logo, we want them to feel safe, positive, and hopeful. One of the primary reasons that people enroll in randomized trials is that they feel it is gratifying to participate in generating medical knowledge to make treatment decisions and help future patients. [RadComp is] different because it will involve patients in the design and conduct of the study, as well as participating in the study."

That framing allowed her to draft a design and iterate with the RadComp team, which was pleased with the RadComp logo (Fig. 11.3), and especially the ability of the "heart bubble" to be a standalone graphic image. When asked to comment on the new logo, the Stakeholder Advisory Committee felt strongly that a tagline would help explain what RadComp meant. A quick back-and-forth exchange of ideas via email settled on a phrase suggested by one of the patient members of the Stakeholder Advisory Committee: "A study at the heart of breast cancer treatment." This logo and tagline has a real resonance with the target population because it was informed and created with patient input.

Fig. 11.3 Radcomp logo

11.4   Conclusion

The proportion of clinical trials which are considered pragmatic, as opposed to explanatory, is increasing. This growth may be due to their appeal to a wider population of patients and stakeholders in addition to clinicians. To ensure that pragmatic trials are truly relevant to all those involved with the trial, continuous patient and stakeholder engagement throughout all phases of the trial—planning, conducting, and disseminating—is necessary.

Trials also vary in the degree to which they are explanatory or pragmatic. Similarly, the degree of patient-centeredness in a trial differs across trials and depends on the level of patient and stakeholder involvement.

There are challenges in creating patient-centeredness. For instance, it can be difficult to find the "right" patients and stakeholders to inform trial design and implementation. Moreover, the heterogeneity in patients and stakeholders and their involvement means that what works in one trial to achieve patient-centeredness may not be generalizable to other trials. As we continue to learn more about how to achieve true patient-centeredness, we must remember to tailor how we engage patients and stakeholders to each study and patient population.

At the time this chapter was written, there was growing use of video, website, and other technologies to help in recruitment, retention, dissemination, and implementation. Considering that these technologies are likely to change in the future, the channels are less important than keeping in mind how they are used and incorporating the patient voice in the design of such methods of outreach.

References

1. Methodology Committee of the Patient-Centered Outcomes Research Institute (2012) Methodological standards and patient-centeredness in comparative effectiveness research: The PCORI perspective. JAMA 307(15):1636–1640. doi:10.1001/jama.2012.466
2. Mullins CD, Vandigo J, Zheng Z, Wicks P (2014) Patient-centeredness in the design of clinical trials. Value Health 17(4):471–475. doi:10.1016/j.jval.2014.02.012
3. Institute of Medicine (2012) Appendix D, Discussion Paper: The clinical trials enterprise in the United States: A call for disruptive innovation. In: Envisioning a transformed clinical trials enterprise in the United States: Establishing an agenda for 2020: Workshop summary. The National Academies Collection: Reports funded by National Institutes of Health. National Academies Press, Washington, DC. doi:10.17226/13345
4. N. I. H. Collaboratory (2014) Introduction to pragmatic clinical trials. Updated December 1, 2014. Duke University. http://sites.duke.edu/rethinkingclinicaltrials/introduction-to-pragmatic-clinical-trials/. Accessed 2 May 2016
5. Schwartz D, Lellouch J (1967) Explanatory and pragmatic attitudes in therapeutical trials. J Chronic Dis 20(8):637–648
6. Loudon K, Treweek S, Sullivan F, Donnan P, Thorpe KE, Zwarenstein M (2015) The PRECIS-2 tool: Designing trials that are fit for purpose. BMJ (Clin Res Ed) 350:h2147–h2147
7. Thorpe KE, Zwarenstein M, Oxman AD, Treweek S, Furberg CD, Altman DG, Tunis S, Bergel E, Harvey I, Magid DJ, Chalkidou K (2009) A pragmatic-explanatory continuum

indicator summary (PRECIS): A tool to help trial designers. J Clin Epidemiol 62(5):464–475. doi:10.1016/j.jclinepi.2008.12.011

8. Rosenthal GE (2014) The role of pragmatic clinical trials in the evolution of learning health systems. Trans Am Clin Climatol Assoc 125:204–216; discussion 217–208

9. Zwarenstein M, Treweek S, Gagnier JJ, Altman DG, Tunis S, Haynes B, Oxman AD, Moher D, for the CONSORT and Pragmatic Trials in Healthcare (Practihc) groups (2008) Improving the reporting of pragmatic trials: An extension of the CONSORT statement. BMJ 337(4):a2390. doi:10.1136/bmj.a2390

10. Freedman B (1987) Equipoise and the ethics of clinical research. N Engl J Med 317(3):141–145. doi:10.1056/NEJM198707163170304

11. Patsopoulos NA (2011) A pragmatic view on pragmatic trials. Dialogues Clin Neurosci 13(2):217–224

12. Patient-Centered Outcomes Research Institute (2015) What we mean by engagement. Updated 12 Oct 2015. Patient-Centered Outcomes Research Institute. http://www.pcori.org/funding-opportunities/what-we-mean-engagement. Accessed 2 May 2016

13. Mullins CD, Abdulhalim AM, Lavallee DC (2012) Continuous patient engagement in comparative effectiveness research. JAMA 307(15):1587–1588. doi:10.1001/jama.2012.442

14. Patient-Centered Outcomes Research Institute (2013) Introducing a new PCORI research funding initiative–large pragmatic clinical trials. Updated 18 Dec 2013. Patient-Centered Outcomes Research Institute. http://www.pcori.org/blog/introducing-new-pcori-research-funding-initiative-large-pragmatic-clinical-trials. Accessed 2 May 2016

15. Lancaster KJ, Schoenthaler AM, Midberry SA, Watts SO, Nulty MR, Cole HV, Ige E, Chaplin W, Ogedegbe G (2014) Rationale and design of Faith-based Approaches in the Treatment of Hypertension (FAITH), a lifestyle intervention targeting blood pressure control among black church members. Am Heart J 167(3):301–307. doi:10.1016/j.ahj.2013.10.026

16. Leone LA, Allicock M, Pignone MP, Walsh JF, Johnson LS, Armstrong-Brown J, Carr CC, Langford A, Ni A, Resnicow K, Campbell MK (2016) Cluster randomized trial of a church-based peer counselor and tailored newsletter intervention to promote colorectal cancer screening and physical activity among older African Americans. Health Educ Behav 43(5):568–76. doi:10.1177/1090198115611877

17. PCORI. Engagement rubric for applicants. http://www.pcori.org/sites/default/files/Engagement-Rubric.pdf.

18. Hudson KL, Collins FS (2015) Sharing and reporting the results of clinical trials. JAMA 313(4):355–356. doi:10.1001/jama.2014.10716

19. Deverka PA, Lavallee DC, Desai PJ, Esmail LC, Ramsey SD, Veenstra DL, Tunis SR (2012) Stakeholder participation in comparative effectiveness research: Defining a framework for effective engagement. J Comp Eff Res 1(2):181–194. doi:10.2217/cer.12.7

20. Buckland S (1994) Unmet needs for health information: A literature review. Health Libr Rev 11(2):82–95

21. Evans D, Coad J, Cottrell K, Dalrymple J, Davies R, Donald C, Laterza V, Long A, Longley A, Moule P, Pollard K, Powell J, Puddicombe A, Rice C, Sayers R (2014) Public involvement in research: Assessing impact through a realist evaluation. Health Services and Delivery Research. NIHR Journals Library Copyright (c) Queen's Printer and Controller of HMSO 2014. doi:10.3310/hsdr02360

22. Patient-Centered Outcomes Research Institute (2015) Financial compensation of patients, caregiver, and patient/caregiver organizations engaged in PCORI-funded research as engaged research partners. Updated June 10, 2015. Patient-Centered Outcomes Research Institute. http://www.pcori.org/sites/default/files/PCORI-Compensation-Framework-for-Engaged--Research-Partners.pdf. Accessed 2 May 2016

23. Katz ML, Archer LE, Peppercorn JM, Kereakoglow S, Collyar DE, Burstein HJ, Schilsky RL, Partridge AH (2012) Patient advocates' role in clinical trials: Perspectives from cancer and leukemia group B investigators and advocates. Cancer 118(19):4801–4805. doi:10.1002/cncr.27485

24. National Heart LaBI (2016) ALLHAT: Quick reference for health care providers. National Heart, Lung, and Blood Institute. https://www.nhlbi.nih.gov/health/allhat/qckref.htm. Accessed 2 May 2016

25. Nallamothu BK, Hayward RA, Bates ER (2008) Beyond the randomized clinical trial: The role of effectiveness studies in evaluating cardiovascular therapies. Circulation 118(12): 1294–1303. doi:10.1161/CIRCULATIONAHA.107.703579

26. Shih MC, Turakhia M, Lai TL (2015) Innovative designs of point-of-care comparative effectiveness trials. Contemp Clin Trials 45(Pt A):61–68. doi:10.1016/j.cct.2015.06.014

27. PATIENTS Program at University of Maryland (2014) University of Maryland School of Pharmacy. http://patients.umaryland.edu/. Accessed 31 May 2016

28. PatientsLikeMe (2016) PatientsLikeMe. https://www.patientslikeme.com/. Accessed 31 May 2016

Chapter 12
Policy Considerations: Ex-US Payers and Regulators

Jipan Xie, Kalipso Chalkidou, Isao Kamae, Rebecca E. Dittrich, Rifaiyat Mahbub, Arjun Vasan, and Cinzia Metallo

Abstract In response to a growing demand to demonstrate the value of health technologies, comparative effectiveness research (CER) is gaining importance worldwide. CER-based evidence is key to informing the process of health technology assessment (HTA), a policy-oriented form of systematic research that aims to assess the clinical, economic, social, and ethical effect of health technologies. In many countries around the world, the results of HTA are increasingly used to inform healthcare policy decisions, shaping care delivery and treatment strategies. In this chapter, HTA practices outside the USA are described, with a focus on the policy implications of HTA. To exemplify how CER is integrated into the HTA appraisal process and how HTA can be implemented within the context of different healthcare systems and reimbursement structures, the experience of England, Japan, and the Global Fund is presented. While England has a well-established and renowned HTA program that includes detailed CER guidelines, Japan has only recently started a two-year HTA pilot program with no CER guidelines, and the Global Fund has yet to embrace HTA practices. These three case studies serve as excellent examples of

J. Xie (✉) • C. Metallo
Analysis Group, Inc, Boston, MA, USA
e-mail: jipan.xie@analysisgroup.com

K. Chalkidou
Institute of Global Health Innovation, Imperial College London, London, UK

I. Kamae
University of Tokyo, Tokyo, Japan

R.E. Dittrich
Georgetown University Law Center, Washington, DC, Johns Hopkins Bloomberg School of Public Health, Baltimore, MD, USA

R. Mahbub
Results for Development (R4D), Washington, DC 20005, USA

A. Vasan
U. S. Department of Treasury, Washington, DC 20220, USA

© Springer Nature Singapore Pte Ltd. 2017
H.G. Birnbaum, P.E. Greenberg (eds.), *Decision Making in a World of Comparative Effectiveness Research*, DOI 10.1007/978-981-10-3262-2_12

how HTA can be adopted to meet a country's unique healthcare needs and outline some of the challenges to be overcome during the HTA implementation phase, including the generation of CER-based evidence.

12.1 Introduction

As healthcare costs continue to rise and outpace inflation [1], the pressure to efficiently allocate available funds and resources intensifies. As a result, there exists a growing demand to demonstrate the clinical and economic value of health technologies. Accordingly, comparative effectiveness research (CER) is gradually gaining importance worldwide. CER is a type of systematic research that compares healthcare interventions (e.g., pharmaceuticals, procedures, medical, and assistive devices) in order to determine which one is the most effective and/or which one results in the best possible health outcomes. In the USA, CER is defined by the Institute of Medicine (IOM) as follows:

> CER is the generation and synthesis of evidence that compares the benefits and harms of alternative methods to prevent, diagnose, treat, and monitor a clinical condition or to improve the delivery of care. The purpose of CER is to assist consumers, clinicians, purchasers, and policy makers to make informed decisions that will improve healthcare at both the individual and population levels [2]

To generate evidence, CER necessitates the use and development of a wide range of data sources and methods [3, 4]. Among the most commonly used methods are randomized trials (including pragmatic trials, cluster randomized trials, and adaptive trials), observational cohort studies and case control studies, retrospective studies involving a variety of data sources (such as medical records, insurance claims, and surveys), and systematic reviews and meta-analyses using data from both randomized trials and observational studies [3, 4]. Recently, new comparison-based methods have also been developed, including matching-adjusted and simulated indirect treatment comparisons [5, 6].

While CER does not directly incorporate cost implications, the evidence it generates is key to informing the process of health technology assessment (HTA), a multidisciplinary and policy-oriented form of systematic research aimed to assess the clinical, economic, social, and ethical properties and effects of health technologies [7]. By applying CER and other various health economics methodologies, HTA evaluates the safety, efficacy, real-world effectiveness, patient-reported outcomes, cost, and cost-effectiveness of health technologies in order to inform policy decisions [8]. The results of HTA feed into clinical guidelines and policy decisions, which are crucial to shaping care delivery and treatment strategies.

Since this chapter focuses on healthcare policy making outside the USA, it should be noted that the definition of CER not incorporating economic evaluation – and thus the concept of cost-effectiveness – is largely unique to the USA. Indeed, not only do most other countries consider CER and HTA to be one and the same but also prefer to use the term HTA regardless of whether economic evaluation is included or not. This difference is believed to have its roots in the distinctive features of the US multi-payer healthcare system, which is likely to require a different approach to economic valuation from that employed in the single-payer systems characterizing the vast majority of countries that have implemented HTA. To avoid confusion and be consistent with the rest of the book, in this chapter, the terms CER and HTA will be used according to the US-based definitions provided above. For a more comprehensive review of the definitions of CER and HTA from a US and non-US point of view, the study by Luce et al. [9] can be consulted.

12.2   Brief History of HTA

Over the past three decades, several countries have established agencies to carry out HTA within the context of their unique healthcare and reimbursement systems. In Europe, the first HTA-dedicated organizations were set up as early as the 1980s in several European countries such as France, Spain, and Sweden. In England, the National Institute for Health and Care Excellence (NICE) was established in 1999, followed by the German Institute for Quality and Efficiency in Healthcare (IQWiG) in 2004 and the French High Health Authority (HAS) in 2005. Since then, most Western and Eastern European countries have formed national and/or regional HTA agencies. Outside Europe, the Canadian Agency for Drugs and Technologies in Health (CADTH) was set up in Canada in 1990 and the Pharmaceutical Benefits Advisory Commission (PBAC) in Australia in 1993. In Asia, the number of countries with HTA agencies is on the rise and includes South Korea, Taiwan, Thailand, and Japan. In Latin America, there are several countries with institutionalized HTA, including Chile, Mexico, Brazil, and Argentina.

In the USA – where, as previously mentioned, HTA mostly comprises safety, effectiveness, and CER [10] – the Agency for Healthcare Research and Quality (AHRQ) has been historically the largest federal funder of technology assessments, which inform national coverage decisions for the Medicare and Medicaid programs [11]. In 2009, the US government allocated $1.1 billion to foster the uptake of CER [12] and, in 2010, established the Patient-Centered Outcomes Research Institute (PCORI) to fund and support CER [13].

12.3   Key Features of HTA

12.3.1   HTA Methodologies

The clinical benefits of a health technology remain the most important criterion to inform coverage decision making. As such, CER is the centerpiece of HTA, particularly CER evidence based on randomized controlled trials (RCTs) and systematic reviews. However, randomized trials are not always feasible or appropriate to measure all the desired clinical outcomes. In the absence of head-to-head trials, HTA agencies frequently rely on indirect comparisons to assess the comparative effectiveness of health technologies, as is the case in England, Scotland, Australia, and Canada. To this end, a wide array of methodologies is being used, including pairwise meta-analysis, mixed treatment comparisons (or network meta-analyses), Bucher analysis, and matching-adjusted and simulated indirect comparisons. NICE, for example, prefers mixed treatment comparisons based on systematic literature reviews of clinical trials and relies on the Decision Support Unit (DSU) to support and guide the development of CER [14].

When assessing the economic value of health technologies, different HTA agencies have adopted and place different emphasis on different methodologies. For instance, cost-effectiveness analysis (CEA) is an integral part of the HTA process in several countries. In CEA, cost-effectiveness is typically measured as incremental cost per quality-adjusted life year (QALY) and is used to determine value for money, coverage status, or reimbursement levels [15]. England, the Netherlands, Sweden, South Korea, Taiwan, and Thailand are some of the countries that explicitly use cost-effectiveness, albeit with slightly different cost-per-QALY thresholds. Other countries such as Germany, however, very rarely consider cost-effectiveness in their health technology appraisal process, choosing instead to rely more on CER-based evidence, particularly evidence from RCTs.

12.3.2 HTA Governance

Because of the differences in healthcare systems, the organization, scope, and authority of HTA agencies vary across countries. Some HTA agencies have a direct impact on decision making, as is the case with England's NICE, while others have more of an advisory role, as is the case with France's HAS and Germany's IQWiG. Besides recommending coverage decisions, some HTA agencies also develop clinical guidelines and provide guidance on interventional procedures and public health programs, as is the case in Australia, Canada, and England.

The way HTA topic selection is conducted varies across countries. Some countries such as Australia consider every new drug and indication; other countries are more selective, as is the case with Germany, where HTA topic selection is triggered by payers [16].

While a comprehensive review of the methodological and organizational differences among HTA agencies across the world is beyond the scope of this chapter, three case studies are presented below to illustrate the range of possible HTA organizations and functions.

12.4 Case Studies

Three case studies – England, Japan, and the Global Fund – were selected to describe HTA at different implementation stages. While England has a well-established HTA program, Japan has only recently started a two-year HTA pilot program, and the Global Fund has yet to embrace HTA. These case studies well exemplify how CER is incorporated into HTA practices and how HTA is implemented within the context of different healthcare systems and reimbursement structures to meet a country's unique needs. They also outline some of the challenges to be overcome during the HTA implementation phase and provide an overview of how HTA might prove beneficial in enabling a more efficient use of resources within budget-constrained healthcare systems, such as those of the Global Fund's recipient countries.

12.4.1 England

12.4.1.1 Overview of the Role and Responsibilities of NICE

England's HTA agency NICE was established in 1999 to support the National Health Service (NHS), a publicly funded single-payer system that is funded out of general taxation and that requires few patient co-payments. NICE issues guidance to the NHS on the adoption and appropriate use of drugs and medical devices, diagnostic techniques, surgical procedures, social care, and health promotion activities – though the latter is being increasingly dealt with by Public Health England. The appropriateness and relevance of comparator technologies, clinical and cost-effectiveness, health-related factors, and social value are all part of the decision-making process at NICE. In most cases, technology appraisal is based on evidence synthesis and economic analyses outsourced to research organizations and universities, which are also often tasked with critiquing the evidence presented by industry sponsors.

A positive recommendation from NICE means that a drug or a medical device must be covered under the NHS and must be offered to patients when clinically indicated and requested by a provider. In 2004, a program was specifically implemented within NICE to support and monitor guidance uptake. Health technologies that receive a negative recommendation can be offered when deemed clinically appropriate, but their reimbursement often requires a special agreement with the local purchasing agency, mostly because of the financial constraints that NHS is currently facing. Although the Department of Health has authority to override NICE recommendations, to date, this has never happened.

NICE is internationally recognized for its methodological rigor and often serves as an example for countries considering the adoption of HTA or in the process of adopting it. Furthermore, as demonstration of its international standing, any positive or negative recommendation issued by NICE has a strong influence on other countries' decisions [17, 18]. As one of the first HTA agencies to be established, NICE is also recognized to have played a key role in prompting pharmaceutical and medical device companies to conduct economic studies and thus prove the clinical and cost-effectiveness of their products.

12.4.1.2 Methodologies Used by NICE

CER Methodologies

As previously discussed, CER is an integral part of the HTA assessment at NICE. Like most HTA agencies, NICE considers evidence on comparative effectiveness gathered directly from RCTs as the gold standard for evidence. However, since this type of evidence is not always attainable or available, NICE also relies heavily on indirect treatment comparisons. Indeed, NICE has published a series of guidelines detailing the key methodologies for systematically synthesizing available clinical information and conducting indirect treatment comparisons [19]. In particular, when no head-to-head RCTs exist to compare healthcare technologies, NICE recommends "data from a series of pairwise head-to-head randomized controlled trials [...] be presented together with a network meta-analysis, if appropriate" [19]. However, in some cases, not enough data exist to conduct pairwise or network meta-analysis, as is the case with many novel oncology drugs for which only single-arm trials are available. Although NICE has been traditionally cautious regarding the use of evidence based on observational studies, its DSU published a technical support document in 2015 intended to summarize accepted methods for the analysis of patient-level data from nonrandomized trials and issued a series of recommendations aimed at improving the quality of this type of CER analysis. The DSU document also lists several methods to be used to minimize the risk of bias, including matching methods, inverse probability weighting, regression adjustment, multivariate regression, propensity score, and difference-in-differences. In addition, the DSU document includes an algorithm and a checklist to assess the overall quality of the analysis based on the applied methods.

Economic Evaluation

As part of its appraisal process, NICE explicitly considers cost-effectiveness and relies heavily on CEA data to issue its recommendations. In particular, a key factor in the decision-making process is the incremental cost-effectiveness ratio (ICER), which measures the incremental cost per unit of benefit gained from using one technology versus another. The strength of the recommendation is closely dependent on the ICER value: when ICER is lower than £20,000, the recommendation is likely to be positive; when ICER is between £20,000 and £30,000, the recommendation is likely to be positive if supported by additional evidence; when ICER is higher than £30,000, the recommendation is likely to be negative. In the case of end-of-life treatments and interventions, including cancer drugs, recommendations are typically based on a QALY threshold, which has been increased to £40,000–50,000 per QALY. However, NICE's threshold has recently come under scrutiny for being too generous [20].

When a health technology is found to be too expensive or its cost-effectiveness is unclear, patient access schemes with manufacturers may be implemented, particularly for cancer drugs. These schemes include pricing or outcome assessment agreements and are aimed to reduce the transaction price of a product and, hence, increase its cost-effectiveness. In addition, since 2011, English patients can obtain cancer drugs not available on the NHS through the Cancer Drugs Fund (CDF). In March 2016, NICE implemented a new appraisal process in line with the newly minted CDF operating model, which involves a managed entry scheme whereby the CDF can pay for cancer drugs before they are approved or rejected by NICE [21]. As Sir Andrew Dillon, NICE chief executive, explained: "In a first of its kind approach, NICE will issue draft recommendations on the use of cancer medicines before they receive their license, with funding from [NHS] available if approved" [21]. The newly reformed CDF is expected to feature a strong role for NHS since, for the first time in its history, NICE will share decision making with the NHS – a clear departure from the traditional one-way decision-making model that saw NICE issuing direct recommendations to the NHS. This might be taken as reflection of how important budgetary constraints are becoming.

12.4.1.3   The Future of HTA in England

The way HTA influences decision making in England is evolving. NICE seems to be concentrating more on medical technologies and pharmaceuticals and less on disease prevention and health promotion, which are being more prominently managed by Public Health England [22]. For instance, the involvement of NICE in developing clinical guidelines and setting quality standards is being diluted over time as such activities gradually devolve to regional authorities.

The future of NICE will likely be shaped by the financial pressures that the NHS is currently experiencing. In the present setting, quality is becoming less relevant – as evidenced by the government decision in 2015 to revoke a NICE program aimed to develop evidence-based guidelines on safe staffing, including nursing staff. By contrast, increasing emphasis is placed on managing *all* pharmaceuticals, particularly expensive new drugs and cancer drugs. At the same time, there are growing calls by industry, academics, and patients to revise the affordability threshold used by NICE and allow for price negotiations and price-volume agreements [20]. In the near future, the push for managed entry agreements – which allow payers and manufacturers to share financial risks and thus are believed to help curb drug price hikes – might lead NICE to prioritize the budgetary implications of its coverage decisions over quality and clinical benefits.

12.4.2   Japan

12.4.2.1   The Japanese Healthcare System

In Japan, the health coverage system is universal and heavily regulated by the government [23]. Currently, no pharmacoeconomic evidence is required for the approval of new health technologies, including drugs [24]. Nevertheless, since the early

1990s, pharmacoeconomic evidence is recommended – but not mandatory – for pricing decisions, which are made by the Central Social Insurance Medical Council ("Chu-i-kyo") based on empirical evidence as well as political considerations.

Given that the fee schedule regulating pricing and reimbursement is almost entirely, and unilaterally, controlled by the government, there has been, so far, little to no incentive for pharmaceutical companies to conduct health economics studies that might complement and support clinical trial data [17, 25, 26]. Indeed, after being listed in the National Formulary, all new health technologies are reimbursed at a fixed 70% rate of the listed price, which is reduced every 2 years regardless of value for money or effectiveness [24, 27]. Exceptions to this rule are considered only in the case of products that have shown exceptional clinical benefit in the post-market setting. The result of this fixed-fee and strictly regulated pricing structure is that pharmaceutical companies tend to develop and bring to market the largest possible number of new health technologies – to maximize profit during the first 2 years after approval – instead of generating evidence for cost-effectiveness or conducting indirect comparison studies [26, 29].

12.4.2.2 HTA Pilot Program

April 2016 marked the official start of Japan's first HTA pilot program, aimed to include cost-effectiveness in setting the prices of pharmaceuticals and medical devices. The program will initially include only a limited number of target products designated by the government for price reassessment – seven drugs and seven medical devices due for repricing in 2018. Despite its initial small scale, the program is expected to move Japan toward fully adopting HTA practices in the near future.

The main reason behind Japan's choice to embrace HTA finds its roots in the country's rapidly aging population, which has put a significant strain on the healthcare system at a time when the country's economy was slowing [23, 29, 30]. HTA is now seen as a necessity to curb rising medical costs and avoid investing in technologies whose clinical and economic benefits are not properly supported by evidence-based data. As part of the HTA pilot program, the cost-effectiveness of a drug or a device must be reported using an ICER, with no specified threshold. To measure the value and clinical outcomes associated with a health technology, the use of QALY is recommended but other clinical indicators, such as progression-free survival, will also be considered by the "Chu-i-kyo." Three main issues remain to be solved: (1) how will the "Chu-i-kyo" make a determination on cost-effectiveness without an ICER threshold? (2) how, and to what extent, will clinical indicators affect decisions (which will be in the form of either a "yes" or a "no," as announced by "Chu-i-kyo")? and (3) how will a final decision issued by the "Chu-i-kyo" affect fair pricing based on value?

While, so far, no clear guidelines have been issued regarding CER, RCTs are expected to continue to be regarded as the most important source of evidence for efficacy and safety, as it happens in most countries around the world. At the same time, real-world studies are also being encouraged. Indirect comparisons are not required but they might be considered to support clinical trial data or in the absence of head-to-head clinical trials.

In the case of all the new drugs that pharmaceutical companies intend to submit for approval in 2017, the government has been actively soliciting cost-effectiveness evidence in addition to clinical trial data to gradually expand the scope of the pilot program. As HTA processes and practices become more developed in Japan, the approval of most, if not all, new drugs might gradually come to include cost-effectiveness data.

12.4.2.3 Challenges Ahead

Japan started the HTA pilot program with the goal to inform and optimize coverage and pricing decisions and, thus, decrease healthcare spending. However, incorporating HTA into Japan's existing healthcare system while balancing the interests of patients, healthcare providers, and the pharmaceutical industry will not be an easy task for the government. To successfully build long-term HTA capacity and extend the current pilot program, Japan must resolve several issues. In particular, the fixed-fee reimbursement schedule presents unique challenges. For instance, if a currently covered medical device or drug is not found to be cost-effective during the ongoing HTA-based reassessment process, how should its price be discounted? Should a new reimbursement scheme be developed? One of the most pressing issues is indeed to clearly define the extent to which HTA should be used to issue guidance on coverage, pricing, and reimbursement decisions.

Because of the strong control that the government has on the coverage and reimbursement decision-making process – representatives of pharmaceutical companies are even prohibited from communicating with the "Chu-i-kyo" or government offices [26] – the adoption of HTA is likely to initially face limited opposition. However, the downside of such tight government control is that the group supervising the implementation of HTA in Japan might struggle to retain its independence. For this reason, it will be important for Japan to comprehensively define the governance of HTA, particularly in light of the growing skepticism among pharmaceutical companies. In addition, as many countries in the process of adopting HTA, another challenge that Japan faces is a lack of in-country expertise to conduct health economics research, review and interpret evidence, and ultimately make final recommendations regarding the clinical and economic value of new and existing health technologies [23, 27]. Building this type of expertise will be crucial to establishing and maintaining strong HTA capacity in the event the government decides to establish a self-sustaining HTA entity at the end of the pilot program. Besides building HTA knowledge and expertise, the key to the success of HTA in Japan will be the ability of the government to solicit evidence for HTA from pharmaceutical companies while ensuring that they have the capacity to generate it. To this end, the nationwide electronic medical record system that is under construction will likely play a critical role in generating the real-world evidence that is needed for HTA.

12.4.3 Global Fund and Developing Countries

12.4.3.1 Brief Overview of the Global Fund

The Global Fund is an international organization whose goal is to attract and disburse resources to prevent and treat acquired immune deficiency syndrome (AIDS), tuberculosis (TB), and malaria in low- and middle-income developing countries. Every year, nearly $4 billion are mobilized and invested by the Global Fund to support programs run by local experts in countries such as Mali, Pakistan, Bangladesh, Ghana, and Indonesia. Per the Global Fund website, "[p]rograms…have put 9.2 million people on antiretroviral treatment for HIV, provided 15.1 million people with TB treatment, and distributed 659 million mosquito nets to protect families from malaria" [31]. In September 2016, donors – including the USA, the UK, France, Germany, Japan, the Bill & Melinda Gates Foundation, and private donors – pledged approximately $13 billion to support the Global Fund over a three-year period [31].

The Global Fund, whose secretariat is based in Switzerland, is a financing rather than an implementation agency. The implementation of the funded programs is supervised by committees comprising representatives from the local government, nongovernmental organizations, the United Nations, faith-based organizations, private sector, and patient groups.

12.4.3.2 HTA and the Global Fund

The need to effectively allocate healthcare resources is particularly evident in developing countries, where efforts to bring infectious and noninfectious diseases under control are dampened by inadequate healthcare system capacity and budget shortcomings. Implementing the principles and methods of HTA could thus prove particularly beneficial to the Global Fund. In particular, given how important value for money and justification for costs are for the Global Fund, assessing the cost-effectiveness of health technologies and implementation strategies in the context of each country's healthcare system could be especially helpful in maximizing the clinical and economic impact of the Global Fund's investments.

The limited success of the Global Fund's investment in the GeneXpert test system represents a good example of how HTA-based evidence could better inform disbursement decisions. Between 2013 and 2015, the Global Fund invested about $8 million in GeneXpert – a molecular test for the diagnosis of TB and the resistance to the anti-TB antibiotic rifampicin. However, despite the superior accuracy of GeneXpert over traditional smear tests, this investment did not lead to any meaningful reduction in TB-related mortality and morbidity, likely because of a less-than-expected uptake of GeneXpert in local clinical practices and a high number of patients lost to follow-up. This experience suggests that CER-based evidence alone might not always be enough to inform policy decisions. Instead, HTA-based studies should take into account local healthcare systems and their budget and human resource constraints. Pragmatic trials, for instance, could generate evidence on the comparative effectiveness of treatments and diagnostics and thus inform decisions on the use of aid money. Moreover, incorporating the evaluation of cost-effectiveness evidence into the Global Fund's decision-making processes has the potential to help make smarter spending decisions by maximizing value for money and return on investment.

12.4.3.3 Implementation of HTA in Developing Countries

To help countries manage their healthcare needs without falling back on aid money, more investments are needed to support the implementation of in-country HTA programs. In this way, clinical and policy decisions could be made within the context of each country's healthcare priorities, preferences, and budgetary limitations. Besides financing therapies and interventions to eradicate diseases, the Global Fund is in a unique position to help developing countries build HTA capacity. By generating their own HTA evidence, local governments could more efficiently allocate their limited resources toward the most promising health technologies. This would also allow them to provide more robust evidence to the Global Fund and other organizations when requesting aid money for healthcare interventions. As the working group on Priority-Setting Institutions for Global Health stated, "[r]eallocating a portion of public and donor monies toward the most cost-effective health interventions would save more lives and promote health equity. The obstacle is not a lack of knowledge about what interventions are best, but rather that too many low- and middle-income countries lack the fair processes and institutions needed to bring that knowledge to bear on funding decisions. […] The group calls for an interim secretariat to incubate a global health technology assessment facility designed to help governments develop national systems and donors get greater value for money in their grants" [32]. In 2013, this led to the creation of the International Decision Support Initiative – a partnership between governments, academic institutions, and nongovernmental organizations which received funding, among others, from the Bill & Melinda Gates Foundation – aimed to assist developing countries in setting up HTA programs and activities [33]. More recently, in September 2016, the UK's government and the Global Fund signed a performance agreement in which evidence of value for money is explicitly considered as a prerequisite for allocating resources to interventions [34].

As evidence of the increasing importance of HTA in shaping policy decisions in developing countries, several upper-middle-income countries such as Brazil, Chile, and Thailand have recently established HTA agencies. In Thailand, following the advent of universal health coverage, the Health Intervention and Technology Assessment Program (HITAP) was formed in 2007 as a semiautonomous research unit under Thailand's Ministry of Public Health. Drawing from its own experience, in 2013, HITAP established an international unit with the goal of working with foreign governments and organizations to help low- and middle-income countries build in-house HTA knowledge and capacity.

As it becomes more widely accepted worldwide, HTA is posed to greatly benefit developing countries. While several initiatives are in place to support low- and middle-income countries in establishing local HTA programs, more needs to be done to select and finance technologies and interventions that yield the best possible health outcomes. HTA-based evidence is likely to become increasingly important in guiding decisions of donors and organizations such as the Global Fund on how to distribute aid money and resources in the most efficient way possible.

12.5 Concluding Remarks

Demand for HTA arises from the need that many countries have to provide affordable healthcare amidst budget constraints and rising costs. CER-based evidence is an integral part of the HTA process, though different countries accept it to varying degrees. While evidence from RCTs is nearly universally accepted as key to HTA, other types of CER-based evidence such as indirect treatment comparisons and observational studies are less widely accepted. When real-world evidence is taken into account, HTA agencies typically require that possible biases be identified and adjusted for, as is the case with NICE. As HTA continues to flourish, real-world CER-based evidence is gradually gaining more recognition, as demonstrated by the growing number of countries that include it in their technology appraisal process, including Australia, England, and Scotland.

Outside the USA, CER is almost always complemented by economic evaluation as part of the HTA process. Indeed, more and more countries are recognizing *value for money* as an important factor in the assessment of the relative benefits of health technologies, effectively integrating comparative- and cost-effectiveness analyses – as is the case with England, Australia, France, Japan, and many of the countries supported by the Global Fund. However, as exemplified by the Japanese experience, setting up an HTA program is no small feat given that it requires incorporating HTA principles and methodologies into existing – and often complex – coverage, reimbursement, and pricing schemes. To facilitate the implementation and development of HTA practices abroad, NICE International – now the Global Health Group at Imperial College London – was established in 2008 with the goal of offering strategic advice and technical support to governments and organizations interested in building capacity for HTA [35].

To sustain the growth of HTA efforts over time, it will be of paramount importance to provide evidence for its effectiveness in clinical practice – something that has yet to be convincingly demonstrated. Indeed, first and foremost, countries want to quantify the impact of HTA on improving clinical outcomes, care delivery, and quality of life, both at the national and local level. Real-world studies are ideally suited to address this issue. However, how – and the extent to which – evidence from these studies should be incorporated into HTA and policy decision making remains unclear. In light of the significant progress that is being made in the development of CER and other HTA methodologies as well as the larger availability of higher quality real-world data, the robustness and reliability of real-world studies is posed to grow over time, increasing the likelihood that real-world results will be gradually – and more efficiently – integrated into HTA. This is particularly important since, in the next few years, the ultimate goal of HTA organizations will be to measure the effects of their technology appraisal processes on the quality of the healthcare that is administered in everyday clinical practice.

References

1. Fortune (2016) Rising healthcare costs. Available from http://fortune.com/2016/06/21/healthcare-rising-costs/. Accessed Sept 2016.
2. Institute of Medicine (2009) Initial National Priorities for Comparative Effectiveness Research. Washington, DC
3. Sox HC, Goodman SN (2012) The methods of comparative effectiveness research. Annu Rev Public Health 33:425–445. doi:10.1146/annurev-publhealth-031811-124610
4. Concato J, Peduzzi P, Huang GD, O'Leary TJ, Kupersmith J (2010) Comparative effectiveness research: what kind of studies do we need?. J Investig Med Off Publ Am Fed Clin Res 58 (6):764–769. doi:10.231/JIM.0b013e3181e3d2af
5. Signorovitch JE, Sikirica V, Erder MH, Xie J, Lu M, Hodgkins PS, Betts KA, Wu EQ (2012) Matching-adjusted indirect comparisons: a new tool for timely comparative effectiveness research. Value Health J Int Soc Pharmacoecon Outcomes Res 15(6):940–947. doi:10.1016/j.jval.2012.05.004
6. Ishak KJ, Proskorovsky I, Benedict A (2015) Simulation and matching-based approaches for indirect comparison of treatments. Pharmacoeconomics 33(6):537–549. doi:10.1007/s40273-015-0271-1
7. World Health Organization (2016) Health technology assessment. Available from http://www.who.int/medical_devices/assessment/en/. Accessed Sept 2016.
8. Sullivan SD, Watkins J, Sweet B, Ramsey SD (2009) Health technology assessment in healthcare decisions in the United States. Value Health J Int Soc Pharmacoecon Outcomes Res 12(Suppl 2):S39–S44. doi:10.1111/j.1524-4733.2009.00557.x
9. Luce BR, Drummond M, Jonsson B, Neumann PJ, Schwartz JS, Siebert U, Sullivan SD (2010) EBM, HTA, and CER: clearing the confusion. Milbank Q 88(2):256–276. doi:10.1111/j.1468-0009.2010.00598.x
10. Sullivan SD, Watkins J, Sweet B, Ramsey SD (2009) Health technology assessment in healthcare decisions in the United States. Value Health 12:S39–S44. doi:http://dx.doi.org/10.1111/j.1524-4733.2009.00557.x
11. Agency for Healthcare Research and Quality (2016) Technology Assessment Program. Available from: http://www.ahrq.gov/research/findings/ta/index.html. Accessed Sept 2016.
12. Ali R, Hanger M, Carino T (2011) Comparative effectiveness research in the United States: a catalyst for innovation. Am Health Drug Benefits 4(2):68–72
13. Barksdale DJ, Newhouse R, Miller JA (2014) The Patient-Centered Outcomes Research Institute (PCORI): information for academic nursing. Nurs Outlook 62(3):192–200. doi:10.1016/j.outlook.2014.03.001
14. NICE (2016) Decision Support Unit. Available from: http://www.nicedsu.org.uk/Evidence-Synthesis-TSD-series(2391675).htm. Acessed Oct 2016.
15. Sorenson C (2010) Use of comparative effectiveness research in drug coverage and pricing decisions: a six-country comparison. Issue Brief (Commonw Fund) 91:1–14. http://www.commonwealthfund.org/~/media/Files/Publications/Issue%20Brief/2010/Jul/1420_Sorenson_Comp_Effect_intl_ib_71.pdf
16. AcademyHealth (2015) Improving quality and efficiency in health care through comparative effectiveness analyses: an international perspective. Available from: http://ah.cms-plus.com/files/publications/2014CERImprovingQuality.pdf. Accessed Oct 2016.
17. Ikegami N, Drummond M, Fukuhara S, Nishimura S, Torrance GW, Schubert F (2002) Why has the use of health economic evaluation in Japan lagged behind that in other developed countries? Pharmacoeconomics 20(Suppl 2):1–7. doi:10.2165/00019053-200220002-00001

18. Hernández-Villafuerte K, Garau M, Devlin N (2014) Do NICE decisions affect decisions in other countries? Value Health 17(7):A418. doi:http://dx.doi.org/10.1016/j.jval.2014.08.1020

19. NICE (2013) Guide to the methods of technology appraisal 2013. Available from: https://www.nice.org.uk/process/pmg9/resources/guide-to-the-methods-of-technology-appraisal-2013-pdf-2007975843781. Accessed Oct 2016

20. Financial Times (2015) Expensive drugs cost lives, claims report. Available from: https://www.ft.com/content/d00c4a02-b784-11e4-981d-00144feab7de. Accessed Oct 2016.

21. NICE (2016) Cancer Drugs Fund. Available from: https://www.nice.org.uk/about/what-we-do/our-programmes/nice-guidance/nice-technology-appraisal-guidance/cancer-drugs-fund. Accessed Sept 2016.

22. GOV.UK (2016) Public Health England. Available from: https://www.gov.uk/government/organisations/public-health-england. Accessed on Sept 2016.

23. Kamae I (2010) Value-based approaches to healthcare systems and pharmacoeconomics requirements in Asia: South Korea, Taiwan, Thailand and Japan. Pharmacoeconomics 28(10):831–838. doi:10.2165/11538360-000000000-00000

24. Hisashige A (2009) History of healthcare technology assessment in Japan. Int J Technol Assess Health Care 25(Suppl 1):210–218. doi:10.1017/s0266462309090655

25. Oliver A (2003) Health economic evaluation in Japan: a case study of one aspect of health technology assessment. Health policy (Amsterdam, Netherlands). 63(2):197–204. doi:http://dx.doi.org/10.1016/S0168-8510(02)00066-0

26. Dittrich R, Asifiri E (2016) Adopting health technology assessment. International Decision Support Initiative. Available from: http://www.idsihealth.org/wp-content/uploads/2016/11/Final-6-2016_HTA-Adoption_Dittrich-Asifiri_Updated-Frameworks.pdf

27. Kennedy-Martin T, Mitchell BD, Boye KS, Chen W, Curtis BH, Flynn JA, Ikeda S, Liu L, Tarn YH, Yang B-M, Papadimitropoulos E (2014) The Health Technology Assessment Environment in Mainland China, Japan, South Korea, and Taiwan—implications for the evaluation of diabetes mellitus therapies. Value Health Reg Issues 3:108–116. doi:http://dx.doi.org/10.1016/j.vhri.2014.03.001

28. Ikegami N, Ikeda S, Kawai H (1998) Why medical care costs in Japan have increased despite declining prices for pharmaceuticals. Pharmacoeconomics 14(Suppl 1):97–105. doi:10.2165/00019053-199814001-00012

29. Tatara K, Okamoto E (2009) Health systems in transition. Available from: http://www.euro.who.int/__data/assets/pdf_file/0011/85466/E92927.pdf. Accessed Oct 2016.

30. Tokuyama M, Gericke CA (2014) Health technology assessment in Japan: history, current situation, and the way forward. Value Health 17 (7):A798. doi:10.1016/j.jval.2014.08.480

31. Global Fund (2016) Global Fund Donors Pledge Nearly $13 Billion to Help End Epidemics. Available from: http://www.theglobalfund.org/en/news/2016-09-17_Global_Fund_Donors_Pledge_Nearly_$13_Billion_to_Help_End_Epidemics/. Accessed Sept 2016.

32. Center for Global Development (2016) Priority-setting in health: building institutions for smarter public spending. Available from: http://www.cgdev.org/publication/priority-setting-health-building-institutions-smarter-public-spending. Accessed Sept 2016

33. International Decision Support Initiative (2016) Available from: http://www.idsihealth.org/. Accessed Sept 2016

34. Department of International Development (2016) Performance Agreement, United Kingdom and The Global Fund to fight aids, tuberculosis and malaria. Available from: https://www.gov.uk/government/uploads/system/uploads/attachment_data/file/552983/perf-agreement-global-fund.pdf. Accessed Oct 2016

35. NICE International (2016) About NICE International. Available from: https://www.nice.org.uk/about/what-we-do/nice-international/about-nice-international. Accessed Sept 2016.

Chapter 13
Perspectives on the Common Drug Review Process at the Canadian Agency for Drugs and Technologies in Health

Patrick Lefebvre, Marie-Hélène Lafeuille, and Sean Tiggelaar

Abstract The Common Drug Review (CDR) is a federal review process that provides funding and adoption recommendations to Canadian provinces and territories on non-oncological drugs. This chapter will begin with providing an introduction to the Canadian Agency for Drugs and Technologies in Health (CADTH) and its role within the Canadian health-care system and will then describe and provide a commentary on the intricacies of the CDR process. The pathway of the CDR process is then outlined, from manufacturer submission, to the formation and evaluation of that submission by a review team, to the dissemination and publication of final recommendations from a pan-Canadian Drug Expert Committee. In addition to the CDR process pathway, details on key factors considered and desired in HTA submissions are outlined (large disease burden or an unmet need), as well as the recommended methodology manufacturers should consider when conducting clinical trials and cost-effectiveness models. This chapter then discusses CADTH's performance, as reviewed by other organizations against fellow international HTA agencies. Based on the discussed strengths and limitations, the chapter concludes with providing future direction, encouraging CADTH's continued focus on improved transparency and responsiveness while also urging them to conduct continued reviews (past the adoption milestone) that manage obsolescence and facilitate evidence translation.

13.1 Introduction

Canada is a federation of ten provinces and three territories, governed in a relatively decentralized fashion compared to most European systems (e.g., the United Kingdom (UK) or Sweden). This is also true with regard to the many aspects of health-care decision making and is particularly illustrated within the scope of pharmaceutical treatments. The path from pharmaceutical innovation to consumer use can be traced through

P. Lefebvre (✉) • M.-H. Lafeuille • S. Tiggelaar
Analysis Group, Inc., Montreal, Quebec, Canada
e-mail: patrick.lefebvre@analysisgroup.com

© Springer Nature Singapore Pte Ltd. 2017   167
H.G. Birnbaum, P.E. Greenberg (eds.), *Decision Making in a World of Comparative Effectiveness Research*, DOI 10.1007/978-981-10-3262-2_13

multiple decision makers: to start, the federal government directs pharmaceutical premarket approval and price regulation [1, 2]. The evaluation of emerging and existing drug technologies is conducted by a pan-Canadian evaluating body, the Canadian Agency for Drugs and Technologies in Health (CADTH) [3]. CADTH has a unique place as it acts at a centralized level to support decision making by federal/provincial/ territorial jurisdictions (except Quebec), but multiple other agencies exist at the local (provincial, regional, district, municipal, or even hospital) level that also assess emergent technologies. The decision to reimburse for specific drugs occurs at the individual federal (e.g., Non-Insured Health Benefits)/provincial/territorial administration level and at the private insurance provider level. Canada has a publicly funded health-care system that provides universal coverage for medically necessary hospital and physician services [4]; health-care services are mostly provided by private entities and primarily reimbursed by each governing province or territory through a public insurance plan run as a not-for-profit entity [5]. The federal government also plays a small role in reimbursement [5], usually covering particularly vulnerable populations and other groups of people under direct federal jurisdiction [2]. For additional medical care needs, private health insurance can also be purchased, or patients may pay directly. Up to 30% of all health-care costs are estimated to be covered by the private sector (includes private health insurance and out-of-pocket payments) [5, 6], and this percentage of private sector coverage increases to 57.4% for pharmaceutical costs [6].

As technology develops ever faster, pharmaceutical innovation results in an ever growing number of new drugs that need to be assessed (for efficacy and safety) against the current standard of care. In recognition of this need, the Canadian Coordinating Office for Health Technology Assessment (CCOHTA), an independent not-for-profit agency, was first formed in August of 1990. CCOHTA was part of an initial three-year pilot program funded by Canadian federal, provincial, and territorial governments to support medical device purchasing and coverage decisions. In 1993 it was given permanent funding, and the agency's scope expanded to include pharmaceutical reviews. In 2002–2003 the agency implemented the Common Drug Review (CDR) for all eligible pharmacological evaluations. The CDR is a single, uniform process for reviewing the cost-effectiveness and therapeutic benefits of drugs and offers listing recommendations to all participating publicly funded drug benefit plans.[1] Its aims were threefold: to streamline the drug review process at the national level; to reduce duplication of efforts at the provincial, regional, and local level; and to increase the standardization and improve the quality of the drug review process. In 2005, the agency added the Health Technology Inquiry Service (HTIS) to its portfolio, offering a rapid response service (24 h–30 days) to inquiries about drugs, devices, and procedures. In 2006, the CCOHTA was renamed CADTH. Finally, as of April 2014, the pan-Canadian Oncology Drug Review Process (pCODR) was transferred to CADTH to ensure its governance and

---

[1] This process excludes the Province of Quebec, which has its own independent health technology assessment (HTA) body (l'Institut national d'excellence en santé et en services sociaux, INESSS) and which independently administers the health-care insurance plan of its inhabitants. INESSS requires the same methodological approach as the one used by CADTH in its pharmacoeconomic assessments [7].

long-term sustainability [8]; previously, most oncological drug reviews were done separately from CADTH, since pCODR's inception as the interim Joint Oncology Drug Review (iJODR) in 2007. Currently, CADTH not only performs formal HTAs and rapid reviews of drugs, devices, and diagnostics but also performs other forms of health technology evaluations (environmental and horizon scans) and offers scientific advice to manufacturers regarding their drug development plans in order to ensure that the health technologies that manufacturers are interested in pursuing coincide with the needs of the Canadian population. To illustrate, in the 2013–2014 fiscal year, CADTH had completed more than 250 reports, including 36 Common Drug Review reports, four Optimal Use reports, 240 Rapid Response reports, four Environmental Scans, and five Horizon Scans [9]. Although CADTH's drug review process includes pCODR and CDR, in the current chapter, we will primarily focus on the CDR process, describing the standard submission and review process, as the many other facets of CADTH's work have been previously described elsewhere [1, 10–13].

13.2 Methodological Considerations

Before a drug's eligibility for public reimbursement can be evaluated, all new medicines must pass regulatory scrutiny by the federal health agency—Health Canada—in terms of demonstrating a favorable safety and efficacy profile relative to the existing standard of care or placebo for the condition in question [14]. Once market approval is granted, another governmental committee within the federal health department—the Patented Medicine Prices Review Board (PMPRB)—ensures that the price is not considered "excessive," as per their guidelines, which also considers the perceived innovative contribution of the new medicine to the targeted disease area [15]. In addition to these two steps, the manufacturer (or in limited cases a drug plan) submits a request for evaluation of the product of interest to CADTH. If the drug in question has an oncological indication, it is reviewed through the pCODR process; otherwise, it is reviewed through the CDR process. Drugs reviewed through the CDR include new drugs, drugs with new indications, combination drugs, and subsequent entry biologics (biologics for a disease where there already is a relatively similar biologic treatment). Line extensions of already marketed drugs and generics are not reviewed by CADTH but rather by the individual provincial, territorial, and federal (e.g., Department of Defense) drug plans.

13.2.1 Overview: How Does the CDR Process Function?

CADTH reviews new medicine evaluation requests on a first-come, first-served basis, with a priority review process in place in case of a backlog of CDR applications [16]. Approximately 1 month before the targeted date for the receipt of a new drug submission, CADTH solicits patient input—regarding the condition that is being assessed,

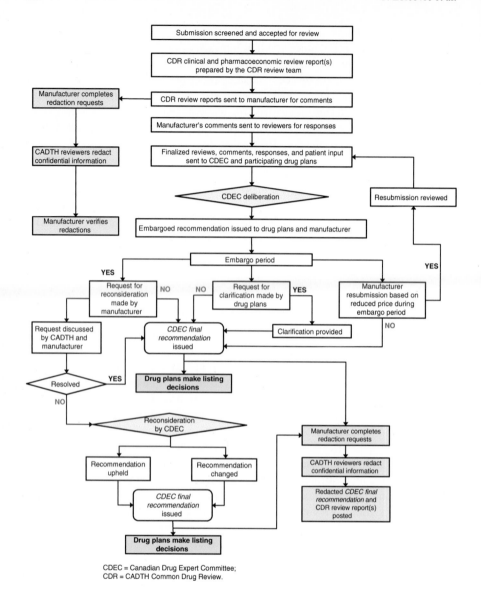

Fig. 13.1 CADTH Common Drug Review process. *CDEC* Canadian Drug Expert Committee, *CDR* CADTH Common Drug Review

patient experience with existing treatments, and patient expectations of incoming treatment options. During this time, a team of internal and external reviewers, as well as clinical experts in the field, is assembled. The review team then examines clinical and economic evidence of the benefits (in terms of clinical effectiveness, safety, and cost-effectiveness) of the new drug relative to the existing standard of care [16] (see Fig. 13.1). The evidence used in the review is from manufacturer-provided material, patient input, and information obtained by the review team from a literature search. The draft report produced by the review team is then shared with the manufacturer, who is given the opportunity to comment and make revisions.

Once the reports are complete, the information is reviewed by the Canadian Drug Expert Committee (CDEC), who provides recommendations regarding reimbursement. CDEC is an appointed, pan-Canadian, 11–14-member advisory committee, composed of members of the public, physicians, pharmacists, and other health-care professionals with expertise in drug therapy, drug evaluation, drug utilization, or health economics. After a draft assessment is released by the CDEC, manufacturers are given the opportunity to comment, contribute further information, and/or request reconsideration of the CDEC's decision during a short embargo period (10 days). Within the embargo period, CDEC reviews any comments. If, after discussion between CADTH and the manufacturer, the issue is not resolved, the review may be reconsidered by CDEC, and then the final recommendation—reimburse, reimburse with clinical criteria and/or conditions, or do not reimburse—is publicly released [17]. At the same time, a final CDR Clinical and Pharmacoeconomic Review Report(s)—containing clinical and pharmacoeconomic evidence (from the manufacturer) and the review team's economic model and cost-effectiveness assessment—is released by the review team, and a final report is posted on CADTH's website. Drug plans then independently decide whether or not to adopt CADTH's recommendations.

13.2.2   Submission Process Details: What Factors Are Considered in an HTA Submission for a New Drug Evaluation?

CADTH has posted detailed instructions on the submission process on its website [18]. In brief, six main factors are considered as part of a drug review: the context of the assessment, the audience, the needs for the new treatment, the benefits and harms of the new drug relative to existing treatments, the drug's economic impact when it enters the market, and patient preferences. These considerations are split into two types of evidence: category 1 and category 2 requirements. Category 1 requirements are used by CADTH in the review process and include:

1. Background information on the need for review and the drug of choice (the requestor of the HTA, prior regulatory history from Health Canada, submission history of similar drugs reviewed by the CDEC, international context in terms of the regulatory and reimbursement status outside Canada).
2. Safety and efficacy/effectiveness of the drug in question relative to existing treatments (from published and unpublished clinical trials, including a measure of the validity of the outcomes reported [19]).
3. Economic information (pharmacoeconomic evaluation of the cost-effectiveness of the new drug compared to the current care for the patient population indicated

by Health Canada and any pertinent subpopulations, the economic model underlying the analysis, and the number of patients in Canada currently accessing the drug).

4. Planned pricing and distribution information (the submitted price per unit and the planned method of distribution to pharmacies).

The CDR contains the manufacturer's comments on the clinical and pharmacoeconomic report, the reviewer's responses, as well as patient perspectives on the impact of the new drug. As previously mentioned, patient input is solicited from the public by CADTH and then integrated into the CDR process.

Category 2 requirements—which only includes the submission of a budget impact analysis (and any supporting documentation)—are used by the drug plans and are not considered part of the CDR review/recommendation process. However, these requirements are mandatory for issuance of the CDEC final recommendation, as the availability of this information is critical for territory/provincial authorities as they develop an informed reimbursement decision.

13.2.3   Review Process Details: The CADTH Experience

As with any technology assessment process, not all of the required evidence has equal weight in the deliberative process. Historically, key decision drivers for a positive reimbursement recommendation have included whether it demonstrated clinical benefit/unmet clinical need and if it was cost effective [20–22]. To demonstrate a clear clinical benefit, in terms of both efficacy and safety over currently used therapies (generally, the most commonly used of available options), CADTH requires RCT of sufficient size and length to appropriately model the disease and the new drug's effect against the standard of care. Other types of evidence (from observational or nonrandomized studies) may not be sufficient to establish clinical benefit.

Regarding cost-effectiveness analyses, CADTH encourages that pharmacoeconomic models use the perspective of the publicly funded health-care system (and, ideally, a lifetime horizon) [23]. This perspective includes direct costs for a publicly funded health-care system and publicly funded services outside the health system (such as home care), but does not include direct costs to patients and their relatives or lost productivity costs (these can be modeled separately). CADTH encourages the use of real-world evidence in developing the economic model underlying the cost-effectiveness analysis. Examples of useful real-world information to be included into the model are known rates of patients' adherence to treatment, screening and diagnostic accuracy, and health-care providers' compliance and skill [23]. Whenever possible, evidence incorporated into the model should be country specific (should apply to the Canadian population). For modeling long-term outcomes (the lifetime perspective for a chronic disease), short-term (intermediate) outcomes observed in clinical trials should be robustly linked to long-term primary outcomes observed over the patient's lifetime. Regarding patient quality of life/preferences/

utilities, CADTH encourages the use of a generic measurement instrument (such as the HUI or the EQ-5D) because such standardized information is easier to obtain, compare, and interpret than that acquired with a disease-specific instrument. The use of expert judgment to establish utilities associated with various conditions is discouraged, since it is thought that the preferences of the general public would best serve to produce an accurate cost-effectiveness estimate—both because the ultimate payer is the general public and because it includes the patient population. Discount rates (5%) are to be used when modeling cost data and sensitivity analyses are required to model any uncertainties around model inputs.

However, even in the context of an essentially standardized process, multiple factors (a large disease burden and a corresponding unmet need, the lack of availability of information, patient input, etc.) can influence the review process. For example, in the case of orphan drugs, clinical benefit can be hard to establish due to limited available evidence—small number of clinical studies, limited number of participants, absence of comparator groups, insufficient evidence on meaningful clinical end points, etc. [24]. A significant unmet need may prompt the recommendation that the drug be conditionally reimbursed even with limited evidence of clinical or pharmacoeconomic benefit. CADTH has also established criteria for defining a significant unmet need, which must be established beyond the rarity of the condition that is to be treated [24]. However, even with this recognition, the large uncertainty around the clinical benefit offered by the orphan drug can negatively affect estimates of cost-effectiveness, and the combined evidence may not be sufficient to clearly determine a benefit of the new drug's use. CADTH is encouraging the use of real-world evidence of clinical effectiveness in such cases [24], in order to reduce existing uncertainty with respect to potential clinical benefit of the new treatment.

13.3 Perspective on and Implications of the HTA Process

The CDR process follows most of the key principles for conducting a rigorous HTA process, as outlined by Drummond et al. [25] and as reinforced by the International Working Group for HTA Advancement (INAHTA)—a 52-member body representing HTA agencies from all over the world [26]. Various assessments of CADTH's performance over the years [27–30] have praised the agency for its methodological rigor (using cost per quality-adjusted life-year and cost-effectiveness thresholds (undisclosed) for a fair comparison, considering a wide range of evidence and outcomes, defining uncertain clinical benefit and unmet need standards, including patient perspectives), transparency (publicly available assessments, reports generally submitted for peer-reviewed journal publication, commitment to monitor the impact of HTA decisions), and fairness (including all relevant stakeholders throughout the HTA process, availability of an appeals process). Over the years, CADTH has implemented several initiatives that have aided in its mission to represent a pan-Canadian process, remove duplication of efforts at various regional levels, and provide access to health technology reviews to reimbursing bodies that lack infrastructure for performing standardized/comprehensive HTAs. In 2003, at the behest of the Canadian Ministry of Health, a special task force reviewed the performance of CADTH and recommended the creation of a Policy Forum, to develop shared policy advice across multiple national, provincial, and regional HTA bodies for non-drug health technologies. In addition, a Health Technology Analysis Exchange—a network of HTA producers across the nation—was established, to "coordinate providing research evidence and policy advice (when requested) to specific decision points determined by the Forum" [31]. In response to criticism/feedback regarding patient access to technology, CADTH has modified its assessment program to include patient input [1]. Furthermore, to streamline access to HTAs produced at the national, regional, and local level, CADTH, in collaboration with HTA agencies from Ontario, Quebec, and Alberta, and with the National Institute for Health Research's Centre for Reviews and Dissemination, has created a common repository and search tool for Canadian HTA reports—the HTA Database Canadian search interface.

However, even with all these initiatives, there are still several areas where there is room for improvement. From a methodological perspective, process transparency could be improved by publicly disclosing ways in which they determine whether a drug's cost-effectiveness is reasonable or excessive. Furthermore, it is important for CADTH to continue pushing for incorporating effectiveness information into clinical benefit [1], using a societal point of view [27], and using non-pharmacologic comparators if appropriate [1, 27], as it produces more robust and credible conclusions on cost-effectiveness. From an organizational perspective, the drug review processes would benefit from timelier assessments [27], and if a backlog of CDR applications occurs again due to capacity limitations, CADTH could improve transparency by disclosing a clear and systematic prioritization of drug reviews [1]. Both of these factors have been repeatedly criticized as sources of effort duplication by other HTA agencies [27, 32, 33], thus creating overall inefficiency in the system. From a translational perspective, it is important to know how well CADTH has fulfilled its main mandate—achieving savings to the public health-care system (through removal of duplicative services, providing access to high-quality evidence at the national level and recommending the replacement of less effective technologies with newer alternatives)—as well as how has the disseminated evidence affected funding decisions by drug plans or the revision of clinical guidelines by medical societies.

Multiple stakeholders have analyzed CADTH's track record relative to international HTA agencies (in terms of volume of applications analyzed, the rate of recommended to non-recommended drugs and their respective characteristics [22, 30, 34–36]) and relative to performance within the Canadian health-care system (changes in review times after the implementation of the CDR [27], translation of recommendations into drug reimbursement decisions [20, 37–39] and the economic impact of such decisions [40]). These reviewers have concluded that:

- The implementation of the CDR has resulted in a more streamlined process, with the duration of an evaluation (from Health Canada approval to drug plan listing decision) estimated at 4–6 months [27]. However, some provinces report that the review time has decreased, while others report an increase [27]. Compared to international HTA agencies, this time is among the slowest review times (ranked 16th of the 18 OECD countries reviewed in a recent report by Canada's Research-Based Pharmaceutical Companies, Rx&D) [41]. However, these metrics on review times can be misleading as they are not directly linked to the HTA review process itself. Manufacturers can delay submissions throughout the process, and once the CDR funding recommendation is released, the provinces still need to negotiate a funding agreement.

- Since inception (and until 2011), 153 drugs (for 175 indications) were reviewed through the CDR process (this excludes oncological drugs, which are reviewed through pCODR); 51% of the drugs reviewed by the CDEC received positive recommendations [42]. Compared to similar reviews in other countries (the United Kingdom, Australia, Korea), Canada is more conservative and restrictive in technology recommendations [34, 35, 41, 42], though this may be due to differences in submitted drugs and review processes [36] and the bigger role reimbursement decision makers have in adoption [34].

- Since the implementation of the CDR process (up to 2009), drug coverage by public plans has decreased with time; however, this may not necessarily be due to HTA recommendations [37]. The total positive reimbursement average across Canada was 51–55% in 2009–2012 assessments [42–44], but it must be noted that only 23% of new medicines received public reimbursement across all Canadian provinces (during 2009–2013) [41]. By contrast, on average, among 29–30 OECD countries, the public reimbursement rate was over 80% [43, 44]—though it must be noted that, unlike CADTH, some of these HTA agencies have the ability to enforce their recommendations to health service distributors.

- Drug plans generally are in agreement with CDR's recommendations [38]: in 2011, 54% of drugs had received a positive recommendation from the CDEC, and 46% of drugs reviewed through CDR were being reimbursed (including those that had restrictions) [44]. In addition, most provincial formularies agree on reimbursement decisions for most of the top-selling medicines [39].

- Technology adoption—listing on the provincial formulary—has occurred for 35–59% of CDR recommended cases [38]. This is relatively low compared to international standards [42], but a contributing factor to this is the advisory (as opposed to regulatory) role the agency plays into the Canadian health-care system, and CADTH, unlike other HTA bodies, reviews all new drugs.

- Although no report has yet detailed the cost savings associated with the CDR process, several reports have estimated that technology adoption (at the local level) has resulted in significant cost savings (on the order of millions of CAD in a five-year or eight-year period) to the health-care system [45, 46].

Overall, the CDR process has achieved some of its goals, with setting a clear, mostly transparent, rigorous HTA process that produces assessments in a reliable fashion. However, prioritization and timeliness could be further improved, and implementation by drug plans is still largely unmeasured/unreported. The ongoing Canadian HTA/Payer Agency Metrics Assessment Project at the Centre for Innovation in Regulatory Science (CIRS)—a neutral, independent, international organization conducting research on global regulatory and HTA policies—aims to measure and report on the impact of recommendations issued by CADTH on provincial payer decision making [47], but this information is yet to be forthcoming.

13.4   Future Directions/Ongoing Initiatives

The goal of CADTH remains to centralize the HTA process and to facilitate access to and interpretation of accumulated evidence by all relevant stakeholders. As such, in 2015, the Pan-Canadian HTA Collaborative has proposed adding structured decision-relevant summaries to all listed HTAs within its database in order to facilitate timely decision making [48].

However, the mission of CADTH may need to evolve even beyond its currently expanded scope in order to better fulfill societal needs. One area that has been relatively neglected is the continued review of technology past the adoption milestone—in terms of technology management and of clinical pathway management. This strategic view is important in maintaining health-care system efficiency, managing obsolescence [49, 50], and facilitating evidence translation across the reimbursement and clinical practice landscapes. As a first step on this new path, future CADTH efforts will focus on developing an effective technology management program.

13.5   Conclusion

CADTH's main mandate was to implement rigorous HTA procedures in order to increase health-care system efficiency (reduce effort duplication at the provincial, regional, and local level), allow access to systematic evidence analysis to interested groups across Canada (regardless of local resources), and produce cost savings to the health-care system by identifying the current technologies that are most efficacious and cost effective for any given condition. On the whole, CADTH has fulfilled that mandate, even without direct reimbursement authority. In addition, CADTH has undertaken several initiatives in response to stakeholder interest, including patient outcomes in assessments, increasing process transparency, creating a national HTA repository, and establishing a gathering place for all Canadian HTA agencies (the pan-Canadian HTA collective). Future efforts at CADTH will continue to focus on improving transparency and responsiveness of the HTA process, as well as developing the current system into a technology management process and eventually pathway management.

Acknowledgments Editing assistance was provided by Ana Bozas, PhD, a salaried employee of Analysis Group, Inc.

Disclaimer This research was not sponsored. The views and opinions expressed in this article are those of the authors and do not necessarily reflect the official policy or position of (their) current or past employers, or any related entities, or those of scientific collaborations of which they are members.

References

1. Menon D, Stafinski T (2009) Health technology assessment in Canada: 20 years strong? Value Health 12(Suppl 2):14–19
2. Government of Canada Strategic Policy Branch, Health Care Policy Directorate HC (2011) Canada's Health Care System [Health Canada, 2011]
3. Canadian Agency for Drugs and Technologies in Health (2016) About CADTH. In: CADTH. ca. Canadian Agency for Drugs and Technologies in Health. Available from: https://www.cadth.ca/about-cadth
4. Canada Health Act, C. 6, s. 1 (1985) Available from: http://laws-lois.justice.gc.ca/eng/acts/c-6/fulltext.html
5. Canadian Institute for Health Information (2005) Exploring the 70/30 split: how Canada's health care system is financed, pp 1–133
6. Labrie Y (2015) Setting the record straight on health care funding in Canada. Notes Heal Care Ser
7. (Québec) C du médicament (2007) Selecting medication for coverage in Quebec: a responsible, transparent process. Information on the scientific evaluation process of medication. Government of Quebec, Quebec
8. Canadian Agency for Drugs and Technologies in Health (2016) Frequently asked questions about pCODR. In: CADTH.ca. Canadian Agency for Drugs and Technologies in Health. Available from: https://www.cadth.ca/pcodr/faqs
9. Canadian Agency for Drugs and Technologies in Health (2014) 25 years of supporting informed health care decisions. 2013–2014 Annual Report, pp 1–13. Available from: https://www.cadth.ca/sites/default/files/pdf/CADTH-Annual-Report-13-14.pdf
10. Rocchi A, Chabot I, Glennie J (2015) Evolution of health technology assessment: best practices of the pan-Canadian Oncology Drug Review. Clinicoecon Outcomes Res 7:287–298
11. Battista RN, Côté B, Hodge MJ, Husereau D (2009) Health technology assessment in Canada. Int J Technol Assess Health Care 25(Suppl 1):53–60
12. Dubois A, Dubé M-P (2014) Are payers ready to assess the combined value of drugs with a companion diagnostic? Cognoscienti 31–6
13. Ferrusi IL, Ames D, Lim ME, Goeree R (2009) Health technology assessment from a Canadian device industry perspective. J Am Coll Radiol 6(5):353–359
14. Health Canada (2015) How Drugs are Reviewed in Canada. Available from: http://www.hc-sc.gc.ca/dhp-mps/prodpharma/activit/fsfi/reviewfs_examenfd-eng.php
15. Paris V, Belloni A (2014) Value in pharmaceutical pricing country profile: Canada, p 1–16. Available from: https://www.oecd.org/health/Value-in-Pharmaceutical-Pricing-Australia.pdf
16. Canadian Agency for Drugs and Technologies in Health (2013) Procedure for Common Drug Review. CADTH Common Drug Review. Canadian Agency for Drugs and Technologies in Health, pp 1–68. Available from: http://www.cadth.ca/media/cdr/process/CDR_Procedure_e.pdf
17. Canadian Agency for Drugs and Technologies in Health (2015) Consultation on recommendation framework for CADTH Common Drug Review and pan-Canadian Oncology Drug Review programs, pp 1–4. Available from: https://www.cadth.ca/sites/default/files/cdr/consult/recommendations_consultation_document.pdf

18. Canadian Agency for Drugs and Technologies in Health (2014) Submission guidelines for the CADTH Common Drug Review. Canadian Agency for Drugs and Technologies in Health, pp 1–113. Available from: https://www.cadth.ca/media/cdr/process/CDR_Submission_Guidelines.pdf

19. Wells G, Shukla VK, Bak G, Bai A (2012) Quality assessment tools project report. Canadian Agency for Drugs and Technologies in Health (CADTH), Ottawa, pp 1–139

20. Mills F, Poinas AC, Siu E, Wyatt G (2014) Consistency in reimbursement decisions at Canadian HTA agencies: INESSS versus CDR. Value Health 17(3):A28–A28

21. Rocchi A, Miller E, Hopkins RB, Goeree R (2012) Common drug review recommendations: an evidence base for expectations? PharmacoEconomics 30(3):229–246

22. Solon C, Kanavos P (2015) An analysis of HTA decisions for orphan drugs in Canada and Australia. Working paper No 42/2015, pp 1–36

23. Canadian Agency for Drugs and Technologies in Health (2006) Guidelines for the economic evaluation of health technologies. 3rd edn. Ottawa, pp 1–75

24. Canadian Agency for Drugs and Technologies in Health. Recommendation Framework for CADTH Common Drug Review and pan-Canadian Oncology Drug Review Programs: guidance for CADTH ' s Drug Expert Committees (2016) Available from: https://www.cadth.ca/media/cdr/templates/pre-sub-phase/pCODR_CDR_recommendations_framework.pdf

25. Drummond MF, Schwartz JS, Jönsson B, Luce BR, Neumann PJ, Siebert U et al (2008) Key principles for the improved conduct of health technology assessments for resource allocation decisions. Int J Technol Assess Health Care 24(3):244–248

26. Neumann PJ, Drummond MF, Jönsson B, Luce BR, Schwartz JS, Siebert U et al (2010) Are key principles for improved health technology assessment supported and used by health technology assessment organizations? Int J Technol Assess Health Care 26(1):71–78

27. Canadian Diabetes Association (2012) In the balance: a renewed vision for the common drug review. Canadian Diabetes Association, pp 1–33. Available from: https://www.diabetes.ca/CDA/media/documents/publications-and-newsletters/advocacy-reports/renewed-vision-for-the-common-drug-review-english.pdf

28. Stephens J, Hanke J, Doshi J (2012) International survey of methods used in health technology assessment (HTA): does practice meet the principles proposed for good research? Comp Eff Res 2:29–44

29. Wilsdon T, Fiz E, Haderi A (2014) A comparative analysis of the role and impact of health technology assessment. Charles River Associates, pp 1-132. Available from: http://www.efpia.eu/uploads/documents/cra-comparative-analysis.pdf

30. Levy AR, Mitton C, Johnston KM, Harrigan B, Briggs AH (2010) International comparison of comparative effectiveness research in five jurisdictions. PharmacoEconomics 28(10):813–830

31. Health Canada, Health Technology Assessment Task Group (2004) Health Technology Strategy 1.0 Final Report June 2004 Information and Emerging Technologies. Federal/Provincial/Territorial Advisory Committee on Information and Emerging Technologies, pp 1–17

32. Blomqvist Å, Busby C, Husereau D (2013) Capturing value from Health Technologies in Lean Times. CD Howe Inst 396:2–28

33. Spitz S (2013) A decade of the Common Drug Review. CMAJ 185(7):E277–E278

34. Bae G, Bae EY, Bae S (2015) Same drugs, valued differently? Comparing comparators and methods used in reimbursement recommendations in Australia, Canada, and Korea. Health Policy 119(5):577–587

35. Clement FM, Harris A, Li JJ, Yong K, Lee KM, Manns BJ (2009) Using effectiveness and cost-effectiveness to make drug coverage decisions: a comparison of Britain, Australia, and Canada. JAMA 302(13):1437–1443

36. Lexchin J, Mintzes B (2008) Medicine reimbursement recommendations in Canada, Australia, and Scotland. Am J Manag Care 14(9):581–588

37. Gamble J-M, Weir DL, Johnson JA, Eurich DT (2011) Analysis of drug coverage before and after the implementation of Canada's Common Drug Review. CMAJ 183(17):E1259–E1266

38. Milliken D, Venkatesh J, Yu R, Su Z, Thompson M, Eurich D (2015) Comparison of drug coverage in Canada before and after the establishment of the pan-Canadian Pharmaceutical Alliance. BMJ Open 5(9):e008100–e008100
39. Morgan S, Hanley G, Raymond C, Blais R (2009) Breadth, depth and agreement among provincial formularies in Canada. Healthc policy Polit santé 4(4):e162–e184
40. Paradis PE, Mishagina N, Carter V, Raymond V (2012) Economic impact of delays in listing decisions by provincial drug plans after a positive Common Drug Review recommendation: the case of a smoking-cessation treatment. Healthc Q 15(2):52–60
41. (Rx&D) CR-BPC (2015) Access to new medicines table of contents. Canada's Research-Based Pharmaceutical Companies (Rx&D). Available from: http://innovativemedicines.ca/wp-content/uploads/2015/05/RxD_6580_AccessToMedicinesReport_WEB.pdf
42. (Rx&D) CR-BPC (2012) The Rx&D international report on access to medicines. Canada's Research-Based Pharmaceutical Companies (Rx&D). Available from: file: http://www.wyatthealth.com/wp-content/uploads/IRAM/iram-2011-2012-English.pdf; Healthcare/Canada's Research-Based Pharmaceutical Companies (Rx&D) - 2012.pdf
43. (Rx&D) CR-BPC (2010) The Rx&D international report on access to medicines. Canada's Research-Based Pharmaceutical Companies (Rx&D). Available from: http://www.wyatthealth.com/wp-content/uploads/IRAM/iram-2009-2010.pdf
44. (Rx&D) CR-BPC (2011) The Rx&D international report on access to medicines. Canada's Research-Based Pharmaceutical Companies (Rx&D), pp 1–26. Available from: http://www.wyatthealth.com/wp-content/uploads/IRAM/iram-2010-2011-English.pdf
45. Gagnon M-P, Desmartis M, Poder T, Witteman W (2014) Effects and repercussions of local/hospital-based health technology assessment (HTA): a systematic review. Syst Rev 3:129
46. Husereau D, Cameron CG (2011) The use of health technology assessment to inform the value of provider fees: current challenges and opportunities. CHSRF series of reports on cost drivers and health SYSTEM efficiency: paper 6. Canadian Health Services Research Foundation, pp1–33. Available from: http://www.cfhi-fcass.ca/Libraries/Commissioned_Research_Reports/Husereau-Fee-Schedule-E.sflb.ashx
47. Science TC for I in R (2016) Canadian HTA/Payer Agency Metrics Assessment Project. The Centre for Innovation in Regulatory Science. Available from: http://www.cirsci.org/canadian-htapayer-agency-metrics-assessment-project/.
48. Polisena J, Lavis JN, Juzwishin D, McLean-Veysey P, Graham ID, Harstall C et al (2015) Supporting the use of health technology assessments by decision-makers. Healthc Policy Polit santé 10(4):10–15
49. (HTAi) HTA international (2016) HTAi: disinvestment and early awareness. Health Technology Assessment international (HTAi). Available from: http://www.htai.org/interest-groups/disinvestment-and-early-awareness.html.
50. Canadian Agency for Drugs and Technologies in Health. Policy Perspectives on the Obsolescence of Health... In: CADTH.ca. Canadian Agency for Drugs and Technologies in Health (CADTH). Available from: https://www.cadth.ca/collaboration-and-outreach/advisory-bodies/policy-forum/discussion-papers/policy-perspectives-obsolescence-health#5

Chapter 14
Evaluating Non-Pharmaceutical Technologies at the Canadian Agency for Drugs and Technologies in Health

Patrick Lefebvre, Marie-Hélène Lafeuille, and Sean Tiggelaar

Abstract The medical device industry is an innovative and fast-paced field that necessitates frequent technology assessments and post-market reviews from health technology assessment (HTA) agencies. In this chapter we will describe the Canadian regulatory pathway for medical devices and provide a commentary. First we provide a summary of the regulatory assessment conducted by Health Canada's Medical Devices Bureau, which evaluates the quality, efficacy, and safety of the device before it is reviewed by Canada's federal HTA agency, the Canadian Agency for Drugs and Technologies in Health (CADTH). We then discuss CADTH's Health Technology Expert Review Panel (HTERP), which provides final recommendations for a medical device's management (adoption, replacement, or disinvestment). Specifically within the HTERP process, we provide a summary of the requirements for a manufacturer's submission, the HTERP process for priority ranking medical devices and how it differs from a pharmaceutical review, and how HTERP comes to developing their recommendations. We conclude with providing opportunities for improvement within CADTH's device recommendations and discuss the future challenges of pushing disinvestment, particularly in a field with a limited device life span that advances through iterative improvements.

14.1 Introduction

The value of the medical device industry in Canada was estimated at $8 billion in 2014, ranking it as the ninth largest market in the world [1]. It represents roughly one-third of the total revenue of the pharmaceutical market [2], but the medical device technology field is, arguably, more innovative and faster paced: in 2008, between 6000 and 8000 medical devices were submitted for licensing, and nearly

---

P. Lefebvre (✉) • M.-H. Lafeuille • S. Tiggelaar
Analysis Group, Inc., Montreal, Quebec, Canada
e-mail: patrick.lefebvre@analysisgroup.com

© Springer Nature Singapore Pte Ltd. 2017   181
H.G. Birnbaum, P.E. Greenberg (eds.), *Decision Making in a World of Comparative Effectiveness Research*, DOI 10.1007/978-981-10-3262-2_14

half of those technologies were nonexistent a decade earlier [3]. Relative to the pharmaceutical industry, the time from innovation to market access for medical devices is shorter, ranging between 2 and 5 years [4]. Given this field's dynamics, it necessitates frequent technology assessments and post-market reviews for technology disinvestment.

Although the Canadian Agency for Drugs and Technologies in Health (CADTH) is mostly known for its drug reviews [5–7], this chapter will focus on the lesser known, but equally important role of CADTH in reviewing non-pharmacological technologies—medical devices, diagnostics, medical/dental/surgical procedures, and programs.

14.2 Methodological Considerations

14.2.1 Regulatory Assessment

Before a new medical technology is reviewed by CADTH, the medical device or diagnostic test is evaluated for quality, efficacy, and safety by Health Canada's Medical Devices Bureau (MDB). The MDB operates under the authority of Medical Devices Regulations and the Food and Drugs Act. Under the act, medical devices (including diagnostic tests and technologies) are classified into four categories, based on the perceived risk (in terms of invasiveness, dependence on a power source for use, and interaction with cardiovascular or nervous system) associated with the use of the device [4]. Class I devices are the lowest-risk items (e.g., bandages, toothbrushes, microbiological diagnostics) and do not require a product license application from the manufacturer, only an application for medical device establishment license. Class II–IV devices require a medical device license, which must be annually renewed [8]. Class II devices are items that interact with the human body but are minimally invasive (e.g., contact lenses, gloves, magnetic resonance imaging equipment, patient management diagnostics). Class III devices are invasive (absorbed into the body or inserted into the body for at least 30 days) and/or may pose some safety risk to the patient (e.g., genetic tests, home glucose test kits, hip replacement implants, mammographies). Finally, class IV devices are items that may pose serious risk of harm due to their invasive nature or long-term use (e.g., implantable defibrillators, breast implants, carotid stents, HIV test kits). For class I/II devices, only preclinical tests are required, while for class III/IV devices, clinical trials are required to demonstrate safety and efficacy. However, unlike in the pharmaceutical review process, where randomized clinical trials are the standard requirement for safety and efficacy [9], randomized controlled trials, nonrandomized trials, and observational studies are acceptable clinical trials for a medical device [9, 10]. This is in large part due to the fact that the emphasis in the evaluation process is placed not on the performance of the device but rather on the entire episode of care [9]. Typically, only class III/IV devices undergo a health technology assessment (HTA) by CADTH after marketing approval is granted by Health Canada, because devices must be associated with patient-related health outcomes or their use must require evidence for coverage, policy, or clinical practice decisions [11].

14.2.2 Medical Device HTA Overview

CADTH has separate processes for reviewing either pharmaceutical or nondrug products (medical devices). Pharmaceuticals are processed through either the pan-Canadian Oncology Drug Review (pCODR) (cancer treatments) or the Common Drug Review (CDR) (all other pharmacological treatments), while medical devices (which include devices, diagnostics,[1] and surgical/medical/dental procedures) are reviewed in a separate process, which is overseen by the Health Technology Expert Review Panel (HTERP). HTERP consists of five appointed core members: a public member representative, an ethicist, a health economist, and two health-care practitioners and/or HTA specialists. For each topic under assessment, expert members—specialists with subject matter expertise—may additionally be appointed to serve on the panel on a per-project basis [14]. Technology management recommendations require a quorum vote, where five core members and at least three expert members must be present [14]. HTERP meets quarterly to develop recommendations for technology management (adoption, replacement, or disinvestment) and to provide input on draft reports for HTAs and other documents requiring non-pharmacological expert review (scoping briefs and protocols).

When compared to the federal process, HTAs on medical devices proceed organically at the provincial level, as they occur only when an increase in adoption at the local (hospital) level is noted across the province [8]. However, at the federal level, CADTH may start this process earlier. CADTH routinely flags HTA queries that originate from its horizon and environmental scans, from the Council of the Federation's priority list, or from queries originating from interested individuals or entities (e.g., hospitals or provinces without medical device HTA processes) for prioritization [15]. The rest of the topic screening process, consisting of identification, priority ranking, and final determination of topics to be assessed, occurs on a quarterly basis. After the potential review topics are identified, CADTH assesses them for fit within its mandate, by taking into consideration the technology's approval status, potential impact on patient care, and possible duplication of work with other HTA agencies [15]. Once appropriateness has been determined, the topic is then ranked using weighted scores and put on a prioritization list. High-priority topics meet most of the following criteria: have great potential to improve clinical outcomes, are expected to produce large cost savings, and affect a large segment of the Canadian population (e.g., 1–5%); there is multi-jurisdictional interest in the technology (it is expected to affect multiple provincial, territorial, or federal programs), and the technology has a high potential to affect health status equity [15]. The ranked prioritization list is publicly disseminated on CADTH's website and is also shared internally with the CADTH executives, liaison officers, committees, and other relevant groups. In order to determine the topics to be included in the current review cycle, CADTH considers the resources needed and available for the task, as well as their processing capacity for each topic on the weighted list. Topics that are ranked lower on the priority list may be addressed instead under the rapid response process or may be kept on

---

[1] Of note, companion diagnostics (categorized as class III devices) are reviewed under either the CDR or pCODR, depending on the indication of their companion pharmacological [12, 13]. There is currently no clear joint assessment process for companion diagnostics.

the list for reassessment in the next review cycle. Topics that become obsolete (due to either technological progress or a duplicative review by another HTA agency) are removed from the prioritization list at the next review cycle.

Once a topic is assigned for review, a CADTH research team is assembled [11]. The research team develops the research protocol (research questions and analysis plan) based on the brief summary and policy-related information collected for the earlier topic screening process. This protocol is shared with and reviewed by HTERP members, external experts, and methodologists. Once approved, the protocol is publicly disseminated on the CADTH's website and registered with PROSPERO, the international database of prospectively registered systematic reviews in health care and social care.

During the evidence collection and review step, the research team assembles information from the manufacturer and published literature regarding clinical outcomes, economic impact, patient preferences, and other important factors, such as ethical, legal, and environmental or social implications of technology adoption. If applicable, patient input is requested on the scope of the project and patient experiences and for clinical outcome validation [11]. This initial report is reviewed by the HTERP committee, and additional feedback is sought from peer reviewers, methodologists, and other stakeholders. The final report incorporates all feedback and is posted on CADTH's website.

When issuing recommendations for a technology's adoption and implementation, HTERP considers not just the evidence at hand (final report) but also the target audience, the type of decision sought, and implementation-related factors. These points are all listed in the recommendation report, including the rationale for the recommendation and implementation and other considerations (such as selection of the appropriate population for technology use, the optimal use of the technology, or evidence gaps requiring further investigation) [11]. Finally, in order for a recommendation to be issued, a majority vote must occur.

14.2.3   Medical Device HTA Submission Process

The information required from a manufacturer for HTERP's HTA process is detailed in CADTH's 2006 Guideline on economic evaluations of health technologies [9]. The following essential elements are required from the manufacturer (or the requestor of the HTA) [9]:

1. Context/background (on the need addressed and the reasons for requesting a technology assessment).
2. Evidence on clinical safety—including adverse events where they affect the clinical or economic aspects of the evaluation, "based on meaningful differences between the intervention and alternatives" [9].
3. Clinical effectiveness—which can include randomized controlled trials, nonrandomized trials, and observational studies. Since medical devices tend to use more observational or registry data, other factors that affect efficacy (such as diagnostic/screening accuracy, patient adherence, operator's skill, certain patient subpopulation differences) are taken into account when submitting evidence of

clinical effectiveness. For all such additional real-world considerations, it is important to consider the strength of the available evidence linking these factors with important patient outcomes, before deciding on effectiveness inputs for the analysis. Finally, organizational infrastructure must be taken into account as it can impact the overall effectiveness of the device (e.g., accessibility issues) [16].

4. Economic impact—this includes considerations of cost-effectiveness of the new technology, infrastructure support costs, and budget impact. The perspective for these evaluations should be that of the public payer, which only includes direct costs to patients and their families, the publicly funded health-care system, and other publicly funded services [9]. The health-care system perspective (which includes time costs to patients and their families) and the societal perspective (which includes productivity costs) are suggested as sensitivity analyses, but are not mandatory considerations. Health-related quality of life (HRQoL) is essential to analyses of cost-effectiveness. CADTH recommends using generic HRQoL instruments (the EQ-5D, the SF-6D, or the 15D), for easier comparability across disease areas and patient populations, which helps represent the experiences/preferences of the "informed" general public [9].

The HTERP committee also considers the following additional factors in all HTA reviews, as applicable [11]:

1. Ethical implications—these include considerations of possible inequity of access to treatment or resource allocation after the technology's introduction. Equity considerations in the economic assessment include choice and valuations of outcomes such as life years gained vs. quality-adjusted life years, larger or smaller discount rates, and deciding on choosing a clinical trial eligible population or a real-life affected population.

2. Legal implications—specific to medical device's post-marketing surveillance and/or recall issues.

3. Organizational impact—these include factors affecting the integration of the new technology into existing infrastructure and workflow, future training/competency requirements for use, and future repair/maintenance issues and costs.

4. Social impact—such as the environmental impact for certain technologies.

These additional factors are generally provided by CADTH's research team through a systematic review of the evidence, but information provided by the manufacturer or other stakeholders are also taken into consideration. CADTH stresses that when providing information on all of these factors, the focus for medical devices should be on the entire episode of care and not strictly on the technical performance of the device [9].

14.2.4   Differences in the Medical Device HTA Review Process

Although the same main factors of effectiveness and safety are considered critical for a rigorous HTA, there are several key differences between evaluating a medical device and a pharmaceutical.

At a regulatory level, the stringency of the requirements for clinical evidence differs among the two types of technologies. As several authors have noted [3, 17] and as the CADTH 2006 Economic Analysis Guidelines also acknowledged [9], the standards of evidence for clinical effectiveness in medical devices are typically lower than that for pharmaceutical products (e.g., class I/II devices do not need clinical trials, and class III/IV can use nonrandomized, observational studies—for which HTA agencies have been criticized given the accepted methodology's diminished rigor [17]). However, this evidentiary standard is lower due to several practical reasons. Conducting randomized controlled trials (RCT) is problematic due to the importance that experience with the device has on device effectiveness, particularly for clinicians in surgical devices [18]. Randomization in RCTs can also pose challenges, such as attempting to blind physicians when they are often actively involved with the device or finding patients who agree to randomization for an invasive procedure when other minimally invasive treatments are available [3, 18]. Additionally, device manufacturers in Canada tend to be small-sized firms with relatively small budgets. These manufacturers have difficulty producing the clinical/ economic evidence at the large scale or of the high-quality and rigorous design as that produced in large pharmacological trials. This is due to high costs of investment/clinical trials (relative to company size), limited patent protection (therefore increased competitive pressure), and a short technological life span (an estimated 18 months) for medical devices [3, 19]. Given the criticism, post-marketing surveillance is used to mitigate the inherent evidence gap [10, 16].

Medical devices are more closely tied within the hospital infrastructure. As such, cost-effectiveness is, in some ways, not as important as budgetary impact when taking the perspective of a technology adopter [10]. Therefore, cost-effectiveness calculations are made more complex than those for drug evaluations [7] because of the organizational impact [17] that must be accounted for in addition to the device-specific impact.

Expert opinion plays a larger role in determining the use and applicability of a technology [10]. Physicians are usually decision-makers in terms of establishing the need for a technology and passing judgment on its future uses [8, 20].

14.3   Perspective on the HTA Process

CADTH's HTA process has been compared to several international HTA agencies, and CADTH was found to meet or exceed expectations for methodological rigor, fairness (including patient views and other stakeholders), and transparency [10, 21]. However, there is room for improvement on several issues, such as increased methodological transparency (actual incremental cost-effectiveness ratio threshold used for decision-making), fairness (consistently including a societal perspective in analyses), and transparency (little is currently published with regard to device-specific requirements). Furthermore, CADTH's track record with respect to technology adoption subsequent to an affirmative decision is currently unknown. While it is clear that CADTH's mandate is that of an advisory nature, offering recommendations and suggesting tools for implementation without the power to enforce adoption, it is nevertheless important to evaluate how this role has:

1. Helped hospitals decide on technological implementation and integration
2. Helped provinces in terms of overall budget management
3. Helped Canada improve its overall health, technological standard, and reduced health-care system inefficiencies and waste

14.4 Future Directions and Ongoing Initiatives

Technological change is estimated to represent over 37% of the growth in non-pharmacological health-care spending (from 1996 to 2008), totaling an estimated $23 billion in additional expenses [22]. While past and current HTA efforts are aimed at reviewing emergent technologies, it becomes increasingly clear that technology management that pushes disinvestment in the event of obsolescence is an equally important part of a well-functioning health-care system—particularly in a field where a device's life span is an estimated 18 months [3], and advances in technology occur through iterative improvements in current devices [19]. Future efforts should focus not just on technology management but also on pathway management.

14.5 Conclusion

With the growth of medical devices and ballooning costs, CADTH holds an important informative role conducting HTAs that provide decision-makers with the evidence they need to make informed decisions. Given the rapid movement of the medical device market, it is important for CADTH to promote continual reviews for technological disinvestment to provincial and local HTA groups. Although CADTH's HTA process has been praised for its methodological rigor, improvements in transparency, such as publishing device-specific requirements, need to be addressed.

Acknowledgments Editing assistance was provided by Ana Bozas, PhD, an employee of Analysis Group, Inc., at the time of support.

References

1. US Department of International Trade and Commerce (2016) 2016 top markets report medical devices: Canada. U.S. Department of Commerce, International Trade Administration, Washington, DC
2. Government of Canada (2016) Life science industries: pharmaceutical industry profile. canada.ca. Government of Canada, Ottawa
3. Ferrusi IL, Ames D, Lim ME, Goeree R (2009) Health technology assessment from a Canadian device industry perspective. J Am Coll Radiol 6:353–359
4. Snowdon A, Zur R, Shell J (2015) Transforming Canada into global centre for medical device innovation and adoption. University of Western - Centre for Health Innovation and Leadership, pp 1–52. Available from: http://sites.ivey.ca/healthinnovation/files/2011/06/ICHIL_Medical_Devices_White_Paper_FINAL2.pdf
5. Rocchi A, Chabot I, Glennie J (2015) Evolution of health technology assessment: best practices of the pan-Canadian Oncology Drug Review. Clinicoecon Outcomes Res 7:287–298
6. Canadian Diabetes Association (2012) In the balance: a renewed vision for the Common Drug Review. Canadian Diabetes Association, pp 1–33. Available from: http://www.diabetes.ca/CDA/media/documents/publications-and-newsletters/advocacy-reports/renewed-vision-for-the-commondrug-review-english.pdf
7. Battista RN, Côté B, Hodge MJ, Husereau D (2009) Health technology assessment in Canada. Int J Technol Assess Health Care 25(Suppl 1):53–60
8. Husereau D, Arshoff L, Bhimani S, Allen N (2015) Medical device and diagnostic pricing and reimbursement in Canada. The Institute of Health Economics (IHE), Edmonton
9. Canadian Agency for Drugs and Technologies in Health (2006) Guidelines for the economic evaluation of health technologies, 3rd edn. Canadian Agency for Drugs and Technologies in Health, Ottawa
10. Stephens J, Hanke J, Doshi J (2012) International survey of methods used in health technology assessment (HTA): does practice meet the principles proposed for good research? Comp Eff Res 2:29–44
11. Health Technology Expert Review Panel (2015) Process for developing recommendations. Canadian Agency for Drugs and Technologies in Health, pp 1–3. Available from: https://www.cadth.ca/sites/default/files/pdf/HTERP_Process.pdf
12. Canadian Agency for Drugs and Technologies in Health (2013) Pharmaceuticals requiring companion diagnostics background and context. Environ Scan. Canadian Agency for Drugs and Technologies in Health, Ottawa
13. Dubois A, Dubé M-P (2014) Are payers ready to assess the combined value of drugs with a companion diagnostic? Cognoscienti 31–6
14. Canadian Agency for Drugs and Technologies in Health (2013) CADTH Health Technology Expert Review Panel: Terms of Reference. Canadian Agency for Drugs and Technologies in Health (CADTH), Ottawa
15. Canadian Agency for Drugs and Technologies in Health (2015) Health technology assessment and optimal use: medical devices; diagnostic tests; medical, surgical, and dental procedures. Top identify prioritization. Canadian Agency for Drugs and Technologies in Health, Ottawa
16. Tsoi B, O'Reilly D, Masucci L, Drummond M, Goeree R (2015) Harmonization of HTA-based reimbursement and regulatory approval activities: a qualitative study. J Popul Ther Clin Pharmacol 22:E78–E89
17. Ciani O, van Giessan A, Taylor R, Wilcher B (2014) Comparative assessment of HTA reports on drugs and medical devices for the treatment of cardiovascular disease. Work Package 1 – Deliv. D 1.3
18. Drummond M, Griffin A, Tarricone R (2009) Economic evaluation of medical devices and drugs—same or different? Value Health 12:402–406
19. World Health Organization (2011) Health technology assessment of medical devices, WHO Medical device technolgy series. World Health Organisation, Geneva

20. Hogue S, Brogan A, Fernandez M, Hong L (2014) Point of care tests: the long and winding road to reimbursement in the United States and Canada. Value Health 17:A14
21. Levy AR, Mitton C, Johnston KM, Harrigan B, Briggs AH (2010) International comparison of comparative effectiveness research in five jurisdictions. Pharmacoeconomics 28:813–830
22. Grootendorst P, Nguyen VH, Constant A, Shim M (2011) Health technologies as a cost-driver in Canada. Final report to strategical policy Branch, Office. Pharm. Manag. Strateg. Heal. Canada. Health Canada, Toronto

Chapter 15
Challenges and Opportunities in the Dissemination of Comparative Effectiveness Research Information to Physicians and Payers: A Legal Perspective

Mary E. Gately and Paul D. Schmitt

Abstract This article examines the underutilization of FDAMA-114 by pharmaceutical manufacturers from its passage in 1997 until the present and the reasons for its under-use. While the original intention of Congress in passing FDAMA-114 was to expand the range of information that manufacturers could legally disseminate regarding their products, as a practical matter FDAMA-114 has not been used in that way. Despite lowering the bar for the type of health care economic information ("HCEI") that manufacturers could share – from the more stringent "substantial evidence" standard to "competent and reliable scientific evidence" – the statute has been restrictively interpreted by the FDA in informal statements, and was not the subject of formal guidance from 1997-2016, which led to it not being used expansively. That dynamic might change with the recent passage of the 21st Century Cures Act, which amended FDAMA-114 to broaden the meaning of HCEI and loosened restrictions on the dissemination of HCEI as well as draft guidance issued by the FDA. Previously, restrictions on manufacturers' ability to provide HCEI and other scientific information contributed to an asymmetry of information between manufacturers and third parties and entities responsible for making formulary and coverage decisions, who are free to disseminate the very same scientific information. The amendments to FDAMA-114, as well as recent decisions by U.S. federal courts, could end or alleviate this dynamic for manufacturers and enable them to more freely participate in the market for information regarding pharmaceutical products.

M.E. Gately (✉) • P.D. Schmitt
DLA Piper LLP, Washington, DC, USA
e-mail: mary.gately@dlapiper.com

© Springer Nature Singapore Pte Ltd. 2017   191
H.G. Birnbaum, P.E. Greenberg (eds.), *Decision Making in a World of Comparative Effectiveness Research*, DOI 10.1007/978-981-10-3262-2_15

15.1   Introduction

In 1997, Congress created a safe harbor to permit dissemination of certain health care economic information ("HCEI") to formulary committees or similar entities even if such information is not included in the medication label and is not supported by substantial evidence. Until December 2016, Section 114 of the Food and Drug Administration Modernization Act of 1997, Pub. L. No. 105–115 (FDAMA 114) defined HCEI as any "analysis that identifies, measures, or compares the economic consequences, including the costs of the represented health outcomes, of the use of a drug to the use of another drug, to another health care intervention, or to no intervention." As discussed further below, that definition has changed with the passage of the 21st Century Cures Act, which broadened the meaning of HCEI under the statute and added several provisions designed to loosen restrictions on the dissemination of HCEI. Prior to the amendment to the statute, FDAMA 114 authorize[d] the dissemination of "written information concerning the safety, effectiveness, or benefit of a use not described in the approved labeling of a drug," provided that:

(1) The information provided is HCEI;
(2) The HCEI is provided to a "formulary committee or other similar entity in the course of the committee or the entity carrying out its responsibilities for the selection of drugs for managed care or other similar organizations";
(3) The HCEI "directly relates" to an approved indication for the medication;
(4) The HCEI is based on "competent and reliable scientific evidence."

Comparative effectiveness research ("CER") falls squarely under FDAMA 114.

FDAMA 114 is limited in scope. It does not apply to regulatory standards that cover industry-supported scientific and educational activities, pharmaceutical manufacturers' responses to unsolicited requests for information about their products, and dissemination of medical journal articles.[1] A goal of FDAMA 114 was to increase the flow of HCEI between pharmaceutical manufacturers and those who make formulary decisions in health care plans. But nearly two decades after its passage, FDAMA 114 has fallen short. Until January 17, 2017, when the

---

[1] Final Guidance on Industry Supported Scientific and Educational Activities, FDA, 62 Fed. Reg. 64,074, (Dec. 3, 1997); Good Reprint Practices for the Distribution Of Medical Journal Articles and Medical or Scientific Reference Publications on Unapproved New Uses of Approved Drugs and Approved or Cleared Medical Devices, FDA (Jan. 2009), http://www.fda.gov/RegulatoryInformation/Guidances/ucm125126.htm; Guidance for Industry—Responding to Unsolicited Requests for Off-Label Information about Prescription Drugs and Medical Devices, FDA (Dec. 2011), http://www.fda.gov/downloads/Drugs/GuidanceComplianceRegulatoryInformation/Guidances/UCM285145.pdf; Notice of Availability of Draft Guidance, FDA, 76 Fed. Reg. 82,303 (Dec. 30, 2011), http://www.gpo.gov/fdsys/pkg/FR-2011-12-30/pdf/2011-33550.pdf; Guidance for Industry – Distributing Scientific and Medical Publications on Unapproved New Uses – Recommended Practices, FDA (Feb. 2014), http://www.fda.gov/downloads/Drugs/GuidanceComplianceRegulatoryInformation/Guidances/UCM387652.pdf.

FDA issued draft guidance in the waning days of the Obama Administration, regulators had never issued official guidance regarding the scope of the statute and had limited their interpretation to informal (and restrictive) pronouncements. In large part due to this widespread uncertainty, FDAMA 114 has been an underutilized vehicle for manufacturers to disseminate key information. Instead, manufacturers have relegated most of their sharing of HCEI to the Academy of Managed Care Pharmacy ("AMCP") Format for Formulary Submissions, which permits manufacturers to compile scientific data regarding their medications and share it with third parties only after an unsolicited request for information. Once seen as a modest breakthrough for commercial speech, today FDAMA 114 has become a virtually dormant outlet for manufacturers. It remains to be seen whether the amendments contained in the 21st Century Cures Act will change this dynamic.

As this article discusses below, the limited use of FDAMA 114 – and the FDA's restrictive view of what the statute permits – is not fully consistent with Congress's intentions when it passed the statute in 1997. Notably, the statute allows manufacturers to disseminate HCEI based on "competent and reliable scientific evidence," a standard that established a bar arguably lower than the "substantial evidence" standard relied upon by the FDA for dissemination of scientific information. Prior to the passage of the 21st Century Cures Act, the FDA, acting in its capacity to protect the public health, interpreted the statute – at least on an unofficial basis – as requiring the "substantial evidence" standard for the data underlying HCEI disseminated by manufacturers. This stance was likely due to the FDA's emphasis on safety/efficacy and focus on how the medication will perform in a controlled setting as opposed to payers' concerns, which include cost containment and real-world effectiveness. The FDA's restrictive view eventually led to the desire to amend the statute. Given these recent amendments to the statute and several judicial decisions, FDAMA 114's period of underuse may be drawing to a close.

15.2   An Overview of FDAMA 114

When FDAMA 114 was enacted, Congress made it clear during drafting and passage that it intended for information to be distributed pursuant to FDAMA 114 to face a less onerous standard than what manufacturers faced for the standard drug approval process. The FDA, however, subsequently interpreted FDAMA 114 to require the more restrictive standard. This discrepancy has led to the underutilization of FDAMA 114 as well as to efforts to amend the legislation, which culminated in the passage of amendments to FDAMA 114 in the 21st Century Cures Act.

15.2.1   The Legislative History of FDAMA 114

The legislative history of FDAMA 114 makes it clear that Congress limited the provision of HCEI to formulary committees or similar entities because those entities would be able to understand and appreciate HCEI.[2] The House Report noted that FDAMA 114 was not intended to provide HCEI to "medical practitioners who are making individual patient prescribing decisions nor is it intended to permit the provision of such information in the context of medical education."[3] Further, the legislative history shows that Congress intended that the standard ("competent and reliable scientific evidence") set forth in the HCEI provision was to be lower than what was required in the standard drug approval process – i.e., adequate and well-controlled clinical studies, the so-called "substantial evidence" standard.[4] At the time, the FDA had been applying the substantial evidence standard to HCEI and Congress explicitly noted that FDAMA 114 was to correct the approach taken by the FDA.[5]

---

[2] See, e.g., H.R. Rep. No. 105–310, 1997 WL 633083 at 65 (1997), the House Report on the proposed legislation that became FDAMA 114. The Report noted that the purpose of the legislation was to allow "drug companies to provide information about the economic consequences of the use of their products to parties that are charged with making medical product selection decisions for managed care or similar organizations," which were described as "formulary committees, drug information centers, and other multidisciplinary committees within health care organizations that review scientific studies and technology assessments and recommend drug acquisition and treatment guidelines." The Report noted that the provision was limited to these entities "because such entities are constituted to consider this type of information through a deliberative process and *are expected to have the appropriate range of expertise* to interpret [HCEI] presented to them to inform their decision-making process, and to distinguish facts from assumptions." (emphasis added). See also S. Rep. No. 105–43 (1997), 1997 WL 394244 at 42–43, which noted that "[pharmaceutical] companies typically have the best and most comprehensive information about the cost, effectiveness, and safety of their products. The FDA should not unduly impede the flow of that information to *experts* who need it for patient and health plan decisions." (emphasis added).

[3] H.R. Rep. No. 105–310 (1997); 1997 WL 633083 at 65.

[4] See Food, Drug, and Cosmetic Act 21 U.S.C . § 355(d) ("...the term 'substantial evidence' means evidence consisting of adequate and well-controlled investigations, including clinical investigations, by experts qualified by scientific training and experience to evaluate the effectiveness of the drug involved, on the basis of which it could fairly and responsibly be concluded by such experts that the drug will have the effect it purports or is represented to have under the conditions of use prescribed, recommended, or suggested in the labeling or proposed labeling thereof."); see also 21 CFR Chap. 1, Section 314.126 ("Adequate and well controlled studies").

[5] The FDA Division of Drug Marketing, Advertising and Communications (DDMAC) issued draft non-binding guidance in 1995 that stated that pharmacoeconomic claims used in promotion had to be supported by substantial evidence. "Principles for the Review of Pharmacoeconomic Promotion (Draft)," (Mar. 20, 1995), http://ehoganlovells.com/cv/b7a0ea32dc480661cdb4b7e36bd84e47bf6f2 4fd, n. 2 (last visited Jul. 27, 2016).

Congress then proposed legislation that later became FDAMA 114 to address this issue. See "Pharmacoeconomic Claims Based on Non-Clinical Studies Would Be Allowed Under Senate Bill; Language Being Prepared for Bill Introduction, Cmte. Says," PINK SHEET (Jun. 2, 1997), https:// pink.pharmamedtechbi.com/PS030296/PHARMACOECONOMIC-CLAIMS-BASED-ON-NONCLINICAL-STUDIES-WOULD-BE-ALLOWED-UNDER-SENATE-LABOR-BILL-LANGUA ("FDA is regulating health economic claims as if they are new efficacy labeling, subject

The "competent and reliable scientific evidence" standard was a standard that had been used by the Federal Trade Commission and its use was intended to increase access to HCEI, including information on cost, effectiveness and safety, by formulary committees and similar entities by lowering the standard those kinds of claims would need to meet.[6] Congress never clearly defined the meaning of "competent and reliable scientific evidence," however, its discussion in the legislative history of the distinction between the substantial evidence standard for clinical claims versus the competent and reliable standard for economic claims leaves open the possibility that Congress intended that clinical claims underlying economic claims must meet the substantial evidence standard.[7]

Following the adoption and implementation of FDAMA 114, the FDA opened a docket to receive comments from relevant stakeholders on the proper interpretation and implementation of the statute and issued a notice that it was considering issuing guidance on the issue. The FDA, however, never issued any guidance on FDAMA 114 or its interpretation of FDAMA 114.[8] Under FDAMA 114(b), the U.S. Comptroller General was to conduct a study related to the implementation of the statute by May 21, 2002, but that study also was never done.

15.2.2   Informal Guidance on FDAMA 114 from FDA Officials

While there was no official FDA guidance on FDAMA 114, Dr. Robert Temple, Deputy Center Director for Clinical Science, FDA Center for Drug Evaluation and Research, made statements during public speeches and conferences in which he expressed his views on FDAMA 114. In February 2012, Dr. Temple in a presentation explained his views on FDAMA 114 and reiterated that it is "*not* intended to provide a path for promoting new off-label indications or claiming clinical

---

to the requirement of clinical trials… This is an inappropriate standard for claims which discuss economic issues related to approved uses of drugs or devices. FDA believes it should require clinical data for health economic claims to protect doctors who cannot understand this information. This overlooks the nature of health economic claims and studies and ignores the target audience for health economic claims which are formulary committees and managed care plan operators.").

[6] S. Rep. No. 105–43 (1997), 1997 WL 394244 at 3, 42–43.

[7] The House Report provided an example of economic claims based on observational studies for the prevention of fractures due to osteoporosis. The Report concluded that this would be permissible because the medication evaluated was already approved for prevention of fractures. This supports the argument that Congress intended that the economic analysis be done only after the underlying clinical claim, prevention of fractures, is supported by substantial evidence. Similarly, the House Report also focused on the requirement that the HCEI "directly relate" to the approved indication for the medication and gave specific examples of the types of information that would and would not be appropriate. In one of those examples, it suggested that economic claims based on disease progression would not be considered to be directly related to the approved indication unless the medication was approved to prevent disease progression. H.R. Rep. No. 105–310 (1997); 1997 WL 633083 at 65–66.

[8] 63 Fed. Reg. 40,719 (Jul. 30, 1998).

advantages of one drug over another when these claims do not satisfy FDA's evidentiary standards for the claims being made."[9]

Dr. Temple went on to interpret FDAMA 114 narrowly and stated that with respect to "[e]conomic consequences directly related to [the] labeled effect of the drug," those consequences must be supported by the label and "[a]ny comparative claims must be supported by substantial evidence that directly compares the treatments in question (i.e., head-to-head clinical trials that provide a valid comparison of two drugs)." He further explained that supporting evidence for HCEI depends on which component of the analysis is involved with "*economic costs* and *consequences* used to construct HCEI" … based on standards widely used by economic experts" while "[e]stablishing the *clinical outcome assumptions* used to construct HCEI would be based on the evidence of effectiveness used to support effectiveness."[10]

In summary, although the statutory language and legislative history of FDAMA 114 made it clear that Congress intended HCEI claims to be substantiated by the more relaxed standard of "competent and reliable scientific evidence," the lack of clear guidance from the FDA coupled with the narrow interpretation of the statute by FDA officials like Dr. Temple led to the underutilization of FDAMA 114.

15.2.3   21st Century Cures Act – Amendment to FDAMA 114

On December 13, 2016, President Obama signed the 21st Century Cures Act into law, which, inter alia, included an amendment to FDAMA 114.[11] The amendments changed FDAMA 114 in the following ways: (1) removed "directly" from the statute which previously provided that HCEI claims must "directly relate" to an approved indication for the medication; (2) added a provision that pharmaceutical companies provide a "conspicuous and prominent statement describing any material differences between the HCEI and the labeling approved..."; (3) modified the definition of HCEI to broaden it;[12] and (4) broadened those who can receive HCEI

---

[9]Communication of CER Findings, Asymmetry in the Ability to Communicate CER Findings: Ethics and Issues for Informed Decision Making, (Feb. 9, 2012), (http://www.npcnow.org/system/files/conferences/download/rtemple_asym12.pdf) at slide 11 (emphasis in original). This is similar to a concern raised by Congress in the legislative history of FDAMA 114. See House Report, H.R. Rep. No. 105–310 (Oct. 7, 1997) at 66 ("This provision also is not intended to provide manufacturers a path for promoting off-label indications or claiming clinical advantages of one drug over another when such claims do not satisfy FDA's evidentiary standards for such claims.").

[10]Communication of CER Findings, Asymmetry in the Ability to Communicate CER Findings: Ethics and Issues for Informed Decision Making, (Feb. 9, 2012), (http://www.npcnow.org/system/files/conferences/download/rtemple_asym12.pdf) at slide 12–14 (emphasis in original).

[11]*See* 21st Century Cures Act, Pub. L. No. 114–255, 130 Stat. 1033 (2016).

[12]21st Century Cures Act, Pub. L. No. 114–255, § 3037, 130 Stat. 1033, 1105 (2016) (HCEI "means any analysis (including the clinical data, inputs, clinical or other assumptions, methods, results, and other components underlying or comprising the analysis) that identifies, measures, or

information to include payers.[13] While prior versions of the proposed amendment to FDAMA 114 included a provision that the Department of Health and Human Services issue draft guidance on "facilitating the responsible dissemination of truthful and non-misleading scientific and medical information not included in the approved labeling of drugs…"[14], the amendment that passed did not provide for the issuance of draft guidance by HHS. In the waning days of the Obama Administration, however, the FDA issued draft guidance addressing manufacturers' communications with payers including under FDAMA 114.[15] According to the draft guidance, the Agency will not consider the HCEI to be false or misleading or use it as evidence of a new intended use if the HCEI is distributed in a manner consistent with the guidance.[16] The draft guidance also explains additional types of persons or entities who can receive HCEI (e.g., drug information centers, technology assessment panels and pharmacy benefit managers),[17] and clarifies what FDA will consider as "related to an approved indication."[18] Finally, the draft guidance provides a broader

---

describes the economic consequences, which may be based on the separate or aggregated clinical consequences of the represented health outcomes, of the use of a drug. Such analysis may be comparative to the use of another drug, to another healthcare intervention, or to no intervention.").

[13] 21st Century Cures Act, Pub. L. No. 114–255, § 3037, 130 Stat. 1033, 1105 (2016) (The definition of those who can receive HCEI include "a payor, formulary committee, or other similar entity with knowledge and expertise in the area of healthcare economic analysis, carrying out its responsibilities for the selection of drugs for coverage or reimbursement.").

[14] H.R. 2414, 114th Cong. (2015).

[15] U.S. DEP'T. OF HEALTH & HUMAN SERV., Drug and Device Manufacturer Communications with Payors, Formulary Committees, and Similar Entities – Questions and Answers, Draft Guidance (last visited Feb. 2, 2017), http://www.fda.gov/downloads/Drugs/GuidanceCompliance RegulatoryInformation/Guidances/UCM537347.pdf.

[16] U.S. DEP'T. OF HEALTH & HUMAN SERV., *supra* note 15, at Q.A.3 ("If a firm disseminates to an appropriate audience HCEI that is the type of information within the scope of section 502(a) (i.e., HCEI that relates to an approved indication and is based on competent and reliable scientific evidence (CARSE), as each of these elements is described in this guidance), FDA does not intend to consider such information false or misleading.").

[17] U.S. DEP'T. OF HEALTH & HUMAN SERV., *supra* note 15, at Q.A.2 ("This audience includes payors, formulary committees (e.g., pharmacy and therapeutics committees), drug information centers, technology assessment panels, pharmacy benefit managers, and other multidisciplinary entities that review scientific and technology assessments to make drug selection, formulary management, and/or coverage and reimbursement decisions on a population basis for health care organizations.") (footnotes omitted).

[18] U.S. DEP'T. OF HEALTH & HUMAN SERV., *supra* note 15, at Q.A.4. The draft guidance provides that "HCEI analyses should relate to the disease or condition, manifestation of the disease or condition, or symptoms associated with the disease or condition in the patient population for which the drug is indicated in the FDA-approved labeling." Examples of HCEI analyses that relate to the approved indication include duration of treatment, practice setting, burden of illness, dosing, patient subgroups, length of hospital stay, validated surrogate endpoint and clinical outcome assessments. Examples of HCEI that FDA considers not to be "related" to an approved indication include HCEI on patient populations outside the labeled population and HCEI on use of a drug for prevention of a disease when the drug is approved for treatment of symptoms. *Id.* at 8.

definition of the "competent and reliable scientific evidence standard for HCEI.[19] The recent amendment did not address the differing standards for clinical and economic claims, but the removal of the word "directly" coupled with the use of disclosures related to the evidence supporting the HCEI may expand the use of FDAMA 114.[20] The draft guidance, together with the recent amendment to FDAMA 114, may actually promote widespread use of the statute by removing some of the ambiguities that have led to FDAMA 114 being underutilized.

15.3   Issues and Challenges Regarding FDAMA 114

The evolution of scientific research and health care information exchange has only further underscored the importance of manufacturers' ability to disseminate information under FDAMA 114. To date, manufacturers have faced an asymmetry of information whereby they are restricted by FDA regulations from freely disseminating HCEI regarding their products, while payers and third parties may freely speak about HCEI regarding those same products. These entities often have financial interests in the dissemination of HCEI and other health care information, and in the case of payers and other entities, can impact formulary acceptance and use of a particular product.

15.3.1   *Manufacturers' Ability to Disseminate Information as Compared to Payers and Other Interested Entities in the Health Care Field*

The uncertainty regarding, and consequent underutilization of, FDAMA 114 have been particularly problematic given the broader challenges facing pharmaceutical manufacturers in the realm of communicating scientific information regarding their

---

[19] U.S. DEP'T. OF HEALTH & HUMAN SERV., *supra* note 15, at Q.A.5 ("FDA considers HCEI to be based on CARSE if the HCEI has been developed using generally-accepted scientific standards, appropriate for the information being conveyed, that yield accurate and reliable results. In evaluating whether the amount and type of evidence that forms the basis for a particular communication of HCEI meets the generally-accepted scientific standards for such information, FDA will consider the merits of existing current good research practices for substantiation developed by authoritative bodies (e.g., International Society for Pharmacoeconomic and Outcomes Research (ISPOR), Patient-Centered Outcomes Research Institute). For example, when evaluating HCEI based on indirect treatment comparisons in the absence of data from head-to-head controlled clinical trials, FDA may refer to guidelines issued by external expert bodies regarding current rigorous methodologies and best practices for such comparisons (e.g., network meta-analyses).").

[20] 162 Cong. Rec. E1602–02, 2016 WL 7107406 (daily ed. Dec. 6, 2016) (statement of Hon. Billy Long) ("in particular, the omission of the word "directly" from the requirement in existing law that information be "directly related" to an approved indication means that information that is consistent with an approved use, but not in the labeling itself, falls within the scope of information that can be communicated to payors . . .").

medications. Perhaps the biggest issue faced by manufacturers (both prior to passage of the 21st Century Cures Act and continuing in the present), including within the context of HCEI under FDAMA 114, is providing such information to physicians and payer entities.[21] Evolving health care models, as well as current FDA regulations regarding HCEI and other scientific information that does not satisfy the FDA's "substantial evidence" standard, leaves manufacturers disadvantaged in freely communicating the full range of information regarding their products. This is particularly important because the communication of HCEI (which is theoretically facilitated by FDAMA 114, but in practice prior to the passage of the 21st Century Cures Act amendment and issuance of draft guidance in January 2017, chilled by the FDA's restrictive interpretation of FDAMA 114) can help demonstrate that a medication is effective and/or economically more feasible when compared to other medications – information that is crucial to formulary decisions.

Pharmaceutical manufacturers' need to speak more freely regarding their medications in the context of these decisions – including as to HCEI – is underscored by payers' financial stake in formulary decisions. Payer decisions impact how much a payer will reimburse for a particular prescription medication. Additionally, large payers have extraordinary power in the prescription medication market to control or influence the coverage and reimbursement of approximately 90 percent of all outpatient prescriptions dispensed in the United States.[22] Because one of their financial objectives is lowering reimbursement costs, payers can leverage their market power to encourage physicians to prescribe and patients to use particular prescription medications. Additionally, the rise in prescription medication costs has led to the growth of "value-based" arrangements, through which payers and manufacturers contractually agree to link reimbursement to a pre-defined outcome measurement, relying on data from how the medication performs in the population covered by the payers. Finally, as discussed further below, payers are already free to conduct and disseminate their own HCEI studies using data they have independently collected from their own coverage pools, and the ability of payers to do so continues to grow rapidly.

Other scientific groups, unlike manufacturers, are also free to discuss research and data, including HCEI, and are not restricted by the boundaries of FDAMA 114 or the FDA's relatively narrow interpretation of the regulation. This dynamic has only accelerated because of recent exponential changes in the managed care environment and the expansion of care to previously uninsured patients through the Affordable Care Act; together these have placed a greater emphasis on the effectiveness of medications. First, the government is providing funding to scientific organizations to sponsor this type of research. For example, the Agency for Healthcare Research and

---

[21] The payer community includes private and public health insurance plans, as well as pharmacy benefit managers (third-party corporate entities which manage prescription benefits for both private and public health plans). In the US market, there are several hundred payers, ranging in size from relatively small local plans that cover a few thousand patients, to large, Fortune 100-size payers covering tens of millions of patients each.

[22] See Adam J. Fein, Pembroke Consulting, Inc. and Drug Channels Institute, *2012–13 Economic Report on Retail, Mail, and Specialty Pharmacies*, (Jan. 2013) at 45, http://www.drugchannels.net/2013/01/new-2012-13-economic-report-on-retail.html.

Quality (AHRQ) funds comparative effectiveness research (CER), which evaluates the relative effectiveness of different medications.[23] In addition, the 2010 Patient Protection and Affordable Care Act established the Patient-Centered Outcomes Research Institute (PCORI). PCORI funds clinical effectiveness research focused on evaluating certain outcomes of medical therapies ("outcomes research") to improve quality of care nationwide.[24] Much of the research conducted or sponsored by AHRQ and PCORI does not satisfy the FDA's substantial evidence standard.

Finally, Accountable Care Organizations (ACOs) are free to discuss HCEI and other scientific information, all of which is relevant to evaluating treatment pathways, the cost effectiveness of medications, and preventive care options, even if that information does not meet the FDA's substantial evidence standard. ACOs are networks of physicians and providers who manage health care costs by sharing financial risk and savings and incentivizing the delivery of quality care. Through ACOs, health care providers attempt to deliver higher quality care to patients while reducing unnecessary visits, procedures and other resources. In some cases, ACOs even establish their own formularies [1].[25] Accordingly, ACOs pay careful attention to data regarding which medications are most effective. This includes both clinical and "real world" data on clinical efficacy, effectiveness, and safety.

This imbalance in the ability to share information has broad implications for manufacturers. As has been recently documented in a study by Tufts University, the average cost to develop and gain marketing approval for a new medication is over $2.5 billion, and the timeline from commencement of development to approval typically takes 10 years [2].[26] Given this investment and the high stakes of payer decisions with regard to prescription medications, it is particularly important for a manufacturer to be able to share information with payers – including HCEI – that demonstrates that its product is comparable or superior to a prescription medication made by another manufacturer. Otherwise, the manufacturer risks the possibility of its product being unreimbursed, and thus not available, for a significant portion of the U.S. population, thus resulting in reduced use and potential failure of a medication in which the manufacturer has made a considerable investment.

Since its passage, FDAMA 114 has not been used by pharmaceutical manufacturers to provide HCEI to payers [3–5].[27] Indeed, one recent study found that "representatives of pharmaceutical manufacturers rarely mentioned FDAMA Section

---

[23] See Agency for Healthcare Research and Quality, "Comparative Effectiveness Research Grant and ARRA Awards," http://effectivehealthcare.ahrq.gov/index.cfm/comparative-effectiveness-research-grant-and-arra-awards/ (providing link to database of AHRQ grants for comparative effectiveness research).

[24] Patient-Centered Outcomes Research Institute, "National Priorities and Research Agenda", http://www.pcori.org/research-results/research-we-support/national-priorities-and-research-agenda.

[25] See Wendy Diller, "ACO Initiatives Test Pharma's Traditional Sales Model" Forbes, (Jun. 17, 2014).

[26] See Joseph A. DiMasi, Henry G. Grabowski, Ronald W. Hansen, "Innovation in the pharmaceutical industry: New estimates of R&D costs" Journal of Health Economics, (Mar. 2016) at 20-33.

[27] See, e.g., Eleanor M. Perfetto, Laurie Burke, Elisabeth M. Oehrlein, and Mena Gaballah, "FDAMA Section 114: Why the Renewed Interest?" Journal of Managed Care & Specialty Pharmacy May 2015; Peter J. Neumann, "What Ever Happened to FDAMA Section 114? A Look back after 10 Years" Value in Health, (2009), at189-190.

114 when they presented scientific data" and "presenting data with FDAMA disclosure is rarely exercised and thus far has been considered of limited value."[28]

Rather, manufacturers have communicated CER and/or HCEI through the AMCP Format for Formulary Submissions, also known as the AMCP dossier.[29] Through the dossier, the AMCP provides a standardized format that allows pharmaceutical companies to include summaries of studies (including CER and other HCEI) based on clinical-trial data, meta-analyses, off-label data, published and unpublished studies supporting labeled and off-label indications, and observational studies.[30] The AMCP recognizes that such evidence is "a necessary component of a comprehensive product dossier."[31] Such evidence need not satisfy the "substantial evidence" standard because manufacturers do not proactively offer it to payers; rather, they may only provide the information contained therein (typically available for download by the payer online) after "*an authentic, validated unsolicited request*" from a payer.[32] As one set of researchers recently explained, the AMCP dossier format "minimizes legal quandaries" and "circumvents" FDAMA 114 "because it relates only to unsolicited requests from a health care system to a pharmaceutical manufacturer."[33]

However, for that reason, the dossier is an imperfect communication tool, as manufacturers may only passively provide the data contained in the dossier – i.e., in response to a specific request. Further, the dossier is filled with an extremely cumbersome amount of data which is difficult to navigate absent specific interest by the payer in a particular issue. And payers typically review the dossier only when they have a specific reason to do so – for example, during a periodic review of a class of prescription medications. In short, while the dossier provides the manufacturer with "the opportunity to present economic evidence to justify the price of a new [prescription medication] in terms of its overall value to the health system[,]"[34] it is a far cry from the type of affirmative and targeted communications that manufacturers would prefer to have with payers and other entities concerning the complex data contained within the dossier.

---

[28]Yoonyoung Choi and Robert P. Navarro, "Assessment of the Level of Satisfaction and Unmet Data Needs for Specialty Drug Formulary Decisions in the United States," Journal of Managed Care & Specialty Pharmacy, Vol. 22, No. 4 (Apr. 2016), at 7.

[29]AMCP, *The AMCP Format for Formulary Submissions, Version 3.1*, Dec. 2012 ("AMCP dossier"), http://www.amcp.org/practice-resources/amcp-format-formulary-submisions.pdf.

[30]*Id.* at 16, section 5.1.1. *See also* Richard N. Fry, "AMCP Format for Formulary Submissions," in Robert P. Navarro, ed., Managed Care Pharmacy Practice (2d ed. 2009), at 273–286.

[31]*AMCP Format for Formulary Submissions* at 33.

[32]*Id.* at vii (emphasis in original).

[33]Eleanor M. Perfetto, Laurie Burke, Elisabeth M. Oehrlein, and Mena Gaballah, *"FDAMA Section 114: Why the Renewed Interest?"* Journal of Managed Care & Specialty Pharmacy, May 2015 at 371; *see also* Guidance for Industry—Responding to Unsolicited Requests for Off-Label Information about Prescription Drugs and Medical Devices, FDA (Dec. 2011), http://www.fda.gov/downloads/drugs/guidancecomplianceregulatoryinformation/guidances/ucm285145.pdf; Guidance for Industry – Distributing Scientific and Medical Publications on Unapproved New Uses – Recommended Practices, FDA (Feb. 2014), http://www.fda.gov/downloads/Drugs/GuidanceComplianceRegulatoryInformation/Guidances/UCM387652.pdf.

[34]Fry at 277.

15.3.2   Payers' Ability to Freely Generate and Discuss HCEI

The effect of the restrictions on prescription medication manufacturers in communicating HCEI is exacerbated by the growing ability of payers to generate research and
freely communicate the very same information in the marketplace. When presented
with information contained in manufacturers' dossiers, payers are free to disregard
the information contained therein and instead conduct their own HCEI studies using
data they have independently collected.[35] The ability of payers to fulfill these functions has grown exponentially in recent years; large payers increasingly employ
highly trained and specialized on-staff scientists and professionals, including health
researchers, economists, physicians, pharmacists, bio-statisticians and methodologists, all of whom have the training and expertise to evaluate scientific data and information related to prescription medications, and to disregard data that do not meet the
minimum standards of reliable and credible evidence. Additionally, larger payers
have an inside edge on the data market, as they possess vast databases (i.e., so-called
"big data") of medical and pharmacy claims gathered from their covered members,
which they can then use to conduct their own research and make formulary decisions.
Moreover, payers are not bound by FDAMA 114, and may therefore speak about and
disseminate HCEI in a variety of settings, including journals and conferences.

15.3.3   Asymmetry in Communication Regarding HCEI

As a result of FDA restrictions on pharmaceutical manufacturers and payers' leverage of their considerable resources and comparative freedom to speak, there is an
increasing "asymmetry" in communication regarding relevant scientific evidence
that does not satisfy FDA's "substantial evidence" standard. Pharmaceutical companies are strictly bound by FDA regulations and the constraints of FDAMA 114 with
regard to the dissemination of HCEI, while other entities, including payers, the government, and academics, are allowed to conduct research and freely communicate
HCEI in the marketplace. This asymmetry in communication undermines the manufacturer's ability to provide a complete and accurate picture of the body of evidence
that currently exists regarding a particular prescription medication, because it places
the manufacturer at a disadvantage in a variety of areas, most notably formulary and
drug reimbursement decisions, which depend in part on the strength of the medication's cost effectiveness and its benefit-risk profile. Payers, on the other hand, are
free to develop and communicate with other entities about this evidence.[36] While the

---

[35] For a study as to the current state of payors' reliance on data and how it is evolving, *see*
Yoonyoung Choi and Robert P. Navarro, *"Assessment of the Level of Satisfaction and Unmet Data
Needs for Specialty Drug Formulary Decisions in the United States,"* Journal of Managed Care &
Specialty Pharmacy, Vol. 22, No. 4 (Apr. 2016).

[36] Additionally, the development of ACOs and other risk and shared savings agreements has led to
a focus on cost-effective quality care. Accordingly, ACOs analyze data, including HCEI, about
which prescription medications are most efficient and cost-effective. Like payers, they are free to
discuss HCEI for prescription medications, even if that information does not meet FDA's "substantial evidence" standard.

amendments to FDAMA 114 contained in the 21st Century Cures Act may loosen the restrictions on manufacturers, it does not change the essential dynamic that disadvantages manufacturers' speech vis-à-vis payers and other entities.

15.4 Developments in the Law – Uncertainty and Opportunities

While prescription medication manufacturers are restricted in disseminating HCEI, developments in the law may suggest that the FDA's ability to regulate the dissemination of HCEI, as well as a broader range of research, may be more limited than originally believed. As discussed above, prior to January 2017, uncertainty existed related to FDAMA 114 and precisely what information manufacturers were able to disseminate, because the FDA had never released guidance on FDAMA 114. Further, the FDA "has never taken any action in the form of a warning letter or notice of violation against a company that specifically mentions a violation" of FDAMA 114 [6].[37] Consequently, thus far manufacturers have been reluctant to use the statute to disseminate information regarding their products.[38] However, three recent cases, while not specifically addressing FDAMA 114, suggest that the FDA's ability to regulate commercial speech may be eroding. In the short term, this can lead to increasing uncertainty for manufacturers as to what HCEI (and other research) they may communicate; in the long term, together with the 21st Century Cures Act amendments, it may provide an opportunity to challenge previous boundaries of FDA-approved commercial speech.

Under the modern commercial speech doctrine, commercial speech (such as the type of speech that pharmaceutical manufacturers employ in discussing their products) receives a lower protection ("intermediate scrutiny") than other forms of expression, such as political expression (which receives "strict scrutiny").[39] Although the Supreme Court had recognized for years that the government did not have "complete power to suppress or regulate commercial speech,"[40] it permitted such regulation of speech provided that the government asserted a substantial interest to be achieved, the restriction directly advanced the interest asserted, and there was no less restrictive alternative than the one chosen by the government.[41]

That calculus changed in *Sorrell v. IMS Health Inc.*, in which the Supreme Court struck down a Vermont law that restricted the sale, disclosure, and use of pharmacy records of doctors' prescribing practices.[42] The Vermont legislature had attempted to regulate manufacturers' efforts to use this information to promote prescription

---

[37] Peter J. Neumann and Cayla Saret, *"When Does FDAMA Section 114 Apply? Ten Case Studies"* Value in Health, (2015), at 1-2.

[38] *Id.*

[39] See *Cent. Hudson Gas & Elec. Corp. v. Pub. Serv. Comm'n of New York*, 447 U.S. 557 (1980).

[40] *Id.* at 562.

[41] *Id.* at 563–64.

[42] *Sorrell v. IMS Health Inc.*, 131 S. Ct. 2653 (2011).

medications to doctors whom manufacturers believed were likely to prescribe their products. Additionally, the law prohibited pharmaceutical companies from using that same information for marketing purposes. However, the law allowed other entities to acquire and use the same information. The Court found that these regulations were both content- and speaker-based, because it targeted manufacturers and their communications to physicians. Because of these content and speaker-based restrictions, the Court imposed a form of "heightened judicial scrutiny" (which it did not fully explain) whereby, given the content and speaker based restrictions, a statute was "presumptively invalid" unless the government could demonstrate otherwise. Because the State essentially had restricted pharmaceutical manufacturers' speech because it was persuasive and conflicted with the government's objectives, while allowing others to engage in the same speech, the Court found the statute unconstitutional.[43]

The United States Court of Appeals for the Second Circuit further bolstered protections for commercial speech in *United States v. Caronia*.[44] In that case, a sales representative who detailed doctors challenged his prosecution and conviction under the FDCA for dissemination of off-label information for a prescription medication. The representative was found guilty of misbranding under the FDCA after the prosecutor made certain statements suggesting to the jury that Caronia's speech, by itself, was proscribed by the FDCA.[45] The jury found Caronia guilty of misbranding, and Caronia appealed the conviction, which the Second Circuit overturned, holding that the Government's prosecution for Caronia's promotional speech alone violated the First Amendment.[46] The court reasoned that Caronia's prosecution amounted to both a content- and speaker-based speech restriction, as it targeted off-label promotion and pharmaceutical manufacturers who engaged in such promotion, while allowing "physicians and academics . . . to speak about off-label use without consequence . . ."[47] Given these circumstances, the Court found that *Sorrell*'s heightened scrutiny standard applied, and noted that "[t]he government's construction of the FDCA essentially legalizes the outcome – off-label use – but prohibits the free flow of information that would inform that outcome."[48]

Most recently, in *Amarin Pharma, Inc. v. U.S. Food & Drug Admin.*, 119 F. Supp. 3d 196 (S.D.N.Y. 2015), a federal court in New York held that biopharmaceutical manufacturer Amarin could not be prevented by the FDA from disseminating truthful, non-misleading information regarding off-label use of one of its medications to reduce triglyceride levels in patients with persistently high levels.[49] Amarin had sought to make specific and truthful statements, and provide peer-reviewed literature (which did not meet the "substantial evidence" standard), about the clinical

---

[43] *Id.* at 2672.

[44] 703 F.3d 149 (2d Cir. 2012).

[45] *Caronia*, 703 F.3d at 158–59.

[46] *Id.* at 160.

[47] *Id.* at 165.

[48] *Id. at 167.*

[49] *Amarin Pharma, Inc. v. U.S. Food & Drug Admin.*, 119 F. Supp. 3d 196 (S.D.N.Y. 2015).

trial data to medical professionals, and planned to include disclaimers in these state-ments that the medication had not been FDA-approved for the specific off-label use. In finding in favor of *Amarin*, the court held that, in line with Caronia, the FDA could not bring action against the manufacturer "based on truthful promotional speech alone" – even if the promotional speech concerns an off-label indication.

These cases suggest that the ground for commercial speech regarding prescrip-tion medication research – including HCEI – is shifting.[50] *First*, Sorrell's focus on "heightened scrutiny" makes clear that commercial speech restrictions that are both content- and speaker-based are subject to higher scrutiny than previously applied. *Second*, in Sorrell and subsequent cases, courts are increasingly expressing skepti-cism as to restrictions that subject two different groups of speakers (e.g., manufac-turers and payers) to different standards when discussing the same information.[51] *Third*, as the Supreme Court explained in *Sorrell*, paternalistic regulations "that seek to keep people in the dark for what the government perceives to be their own good" are especially concerning in situations where the audience is sophisticated and experienced.[52] This provides a basis for manufacturers to argue that physicians and payers (which routinely produce, disseminate, and review HCEI as part of their decision-making process) – are "sophisticated and experienced" and therefore the type of audience from whom the government should not paternalistically shield information simply because of its persuasiveness. In short, *Sorrell, Caronia,* and *Amarin* reflect an increasing skepticism by the courts toward commercial speech restrictions, and have expanded the opportunity for potential First Amendment chal-lenges by manufacturers as to the dissemination of HCEI.

Nevertheless, these rulings do not necessarily provide manufacturers with an untrammeled ability to disseminate scientific information regarding their products. First, *Caronia* and *Amarin* were decided within the Second Circuit, and only have

---

[50] Additionally, the US House of Representatives' Committee on Energy and Commerce, Subcommittee on Health (which is the FDA's authorizing committee), recently wrote to HHS and, citing the decisions in *Sorrell, Caronia,* and *Amarin*, stated that it was "increasingly perplexed by the agency's unwillingness or inability to publicly clarify its current thinking" on the exchange of scientific information, May 26, 2016 from the Hon. Fred Upton and Joseph Pitts to the Honorable Sylvia Burwell. The subcommittee noted its disfavor with the FDA issuing guidance on these issues through "non-binding policy statements" and instead encouraged FDA to address lingering uncertainties through "comprehensive guidance," including draft amendments to the FDCA to that effect.

[51] *See also* Erin E. Bennett, *Central Hudson-Plus: Why Off-Label Pharmaceutical Speech Will Find Its Voice*, 49 Hous. L. Rev. 459, 461 (2012) (predicting that the Court eventually will overturn FDA regulations on off-label pharmaceutical marketing because, like the regulation in Sorrell, "the FDA's regulatory policies prohibit certain speakers – pharmaceutical manufacturers – from freely disseminating information that others may lawfully express.").

[52] *Sorrell*, 131 S. Ct. at 2671; *see also* Nat Stern & Mark Joseph Stern, *Advancing an Adaptive Standard of Strict Scrutiny for Content-Based Commercial Speak Regulation*, 47 U. Rich. L. Rev. 1171, 1186 (2013) (noting "the Court's intolerance of restrictions on truthful, nonmisleading com-mercial speech" in *Sorrell*); Kate Maternowski, *The Commercial Speech Doctrine Barely Survives Sorrell*, 38 J.C. & U.L. 629, 648 (2012) (arguing that a law that prohibits "the free flow of truthful information based on [a] belief that the public is better off without that information . . . fails the means-ends inquiry and is therefore invalid.").

persuasive value outside of that Circuit. Second, the *Amarin* decision explicitly limited the scope of the ruling and noted that "[a] statement that is fair and balanced today may become incomplete or otherwise misleading in the future as new studies and new data is acquired."[53] Consequently, the court warned manufacturers that they "bear[] the responsibility, going forward, of assuring that . . . communications to doctors" remain truthful and non-misleading.[54] Third, the Court's ruling as to what constitutes "misleading" information relied upon a fact-intensive determination as to the particular statements at issue in the case (and regarding a product which had a well-documented regulatory history), thus rendering future judicial evaluation of other statements unpredictable and forcing manufacturers to seek FDA approval for promotional statements. Fourth, the FDA still has authority to pursue misbranding charges against pharmaceutical companies under the False Claims Act, and States' Attorneys-General have the ability to bring state law claims. In short, it remains to be seen what continuing effect *Amarin* will have on First Amendment protections and whether it will be informative to statements regarding other products presenting different facts and circumstances.

15.5   Conclusion

As this article has emphasized, pharmaceutical manufacturers' ability to share HCEI, including CER, remains both restricted and unclear, despite the passage of FDAMA 114 almost two decades ago. Although one purpose of the statute was to permit manufacturers to more freely share such information, the lack of clear guidance regarding the statute, as well as informal statements from the FDA suggesting a restrictive approach, has led to an underutilization of FDAMA 114 by manufacturers. Manufacturers' inability to provide HCEI and other scientific information as FDAMA 114 intended exacerbates an already stifling dynamic in which manufacturers' speech regarding their medications is restrained, while third parties and entities responsible for making formulary and coverage decisions are free to disseminate the very same scientific information. It remains to be seen whether constraints on manufacturers regarding HCEI and other scientific information will be alleviated in practice through the amendments to FDAMA 114 contained in the 21st Century Cures Act or judicial challenges. The FDA's issuance of draft guidance, however, provides an opening for a more liberal use of FDAMA 114 that should encourage the use of FDAMA 114.

---

[53] *Amarin Pharma, Inc.*, 119 F. Supp. 3d at 236.
[54] *Id.*

References

1. Diller W (2014) ACO Initiatives test pharma's traditional sales model. Forbes Media, LLC, Jersey City, NJ
2. DiMasi JA, Grabowski HG, Hansen RW (2016) Innovation in the pharmaceutical industry: New estimates of R&D costs. J Health Econ 47:20–33. doi:10.1016/j.jhealeco.2016.01.012
3. Perfetto EM, Burke L, Oehrlein EM, Gaballah M (2015) FDAMA section 114: Why the renewed interest? J Manag Care Spec Pharm 21(5):368–374. doi:10.18553/jmcp.2015.21.5.368
4. Neumann PJ (2009) What ever happened to FDAMA Section 114? A look back after 10 years. Value Health 12(2):189–190. doi:10.1111/j.1524-4733.2008.00429.x
5. Choi Y, Navarro RP (2016) Assessment of the level of satisfaction and unmet data needs for specialty drug formulary decisions in the United States. J Manag Care Spec Pharm 22(4):368–375. doi:10.18553/jmcp.2016.22.4.368
6. Neumann PJ, Saret CJ (2015) When does FDAMA section 114 apply? Ten case studies. Value Health 18(5):682–689. doi:10.1016/j.jval.2015.02.013

Chapter 16
Legal Considerations in a World of Comparative Effectiveness Research

Paul Kalb, Paul E. Greenberg, and Crystal T. Pike

Abstract In this chapter, Dr. Paul Kalb, Esq. draws on his expertise in the area of health care fraud and abuse in discussing legal considerations relating to comparative effectiveness research (CER). He describes barriers to CER dissemination by manufacturers as well as use of CER in various litigation settings. Dr. Kalb talks about these issues with Paul Greenberg and Crystal Pike, economists at Analysis Group with whom he has worked on many legal cases involving off-label promotion and kickback allegations. They discuss recent developments concerning: (a) what can be communicated by manufacturers based on different standards of scientific evidence (e.g., substantial, competent and reliable, truthful and non-misleading); (b) to whom different types of product claims can be communicated (i.e., where the sophistication of the audience may matter); and, (c) who can do the communicating (i.e., where the expertise of the spokesperson may matter). In addition, they talk about managing company risk in an environment where some communications may be seen as off label and others may be seen as false or misleading. Finally, they talk about specific examples in which CER has been used effectively in litigation.

Paul Greenberg As you know, the issue of whether Comparative Effectiveness Research (CER) evidence can be properly disseminated by pharmaceutical companies has been disputed in litigation. The US Food and Drug Administration (FDA) and the US Department of Justice (DOJ) have taken the position that pharmaceutical companies cannot include information such as CER in their promotional materials if it is not based on direct results of clinical trials, or what they call "substantial evidence," as that would constitute off-label promotion. The companies have argued that dissemination of this type of scientific information, under appropriate circumstances, is protected by First Amendment rights. This is one area where CER has

P. Kalb (✉)
Sidley Austin, LLP, Washington, DC, USA
e-mail: pkalb@sidley.com

P.E. Greenberg • C.T. Pike
Analysis Group, Inc., Boston, MA, USA

© Springer Nature Singapore Pte Ltd. 2017
H.G. Birnbaum, P.E. Greenberg (eds.), *Decision Making in a World of Comparative Effectiveness Research*, DOI 10.1007/978-981-10-3262-2_16

been emerging as a hot topic outside of academic circles. Related to this is the debate over whether the relative sophistication of the audience should figure in to what is deemed acceptable in terms of information sharing. In this context, reference has been made to Section 114 of the Food and Drug Administration Modernization Act of 1997 (FDAMA 114) and recent amendments in the 21st Century Cures Act – which permit companies to distribute "health care economic information" based on "competent and reliable scientific evidence" to payers, for example, for purposes of formulary decision making.

Since the passage of FDAMA 114 many years ago, it is not obvious just how much dissemination of CER-type of information has made its way to payers, because of all the concerns that have been expressed by FDA and other watchdog groups. But given the growing emphasis on CER by a variety of stakeholders, I expect that views on this topic will continue to evolve and the need for clarity in terms of what information can and cannot be shared will become even more critical.

So, with that as a backdrop, it is really those two strands that we had in mind to discuss with you today – CER in the context of litigation specifically, as well as the more general surveying of the environment in which CER has the potential to be used.

Paul Kalb Let me actually start at the more general level and then we'll get to litigation more specifically.

It seems to me that at the highest level of abstraction that pharmacoeconomic information, or more broadly information about the economics of treatment, is going to become even more important over time. It has become increasingly important over the last 5–10 years, or maybe even longer than that, and there is no doubt that it will become more and more important. I think we are probably at an inflection point, as we saw the major presidential candidates talking about the cost of drugs as a major political issue and President Trump has continued the discussion. So this debate has risen from the modest obscurity of academic pharmacoeconomic journals to the macro stage of presidential politics, and I don't see the debate about the cost of drugs, biologics, and devices going away any time soon. We continue to move into an era of highly specialized drugs that are often highly effective, and therefore, in a free market could potentially bear a very high price. So at the highest level I see an enormous focus on this set of issues and therefore, an enormous demand for good data about the cost, cost-effectiveness, and comparative effectiveness of drugs and devices.

Within that context, I absolutely agree with you that there is a need for a political and regulatory solution to the question of what information can be shared. You're absolutely right that we are in a strange world in which it's not entirely clear what can lawfully be shared and with whom. These two points are key. The first concerns the content and what standard you would apply to it. Are we talking about *any* pharmacoeconomic information? Does the nature of the information need to be disclosed? Does the information need to rise to the level of some standard and if so, what is that standard? Is it "truthful and non-misleading," whatever that means, or "competent and reliable," whatever that means? Those are all competing standards.

At the moment, all of that is unclear. And for the second point, the audience for CER information dissemination is unclear. I've heard about some draft legislation float-ing around which would create a safe harbor for communication of some universe of that information – really to payers, meaning formulary committees or those responsible for formulary decisions. But that leaves open the entire physician world, notwithstanding the fact that physicians increasingly are being pressured to take into account the costs and cost-benefit assessments of drugs and devices.

So, absolutely, what can lawfully be communicated by manufacturers, and the related question of to whom, are both very real questions, very important questions, all against the backdrop of what is permissible under the First Amendment.

Greenberg And I'll add that there's also the question of "by whom." In particular, is there a meaningful distinction to be made between a person representing the drug or device company, such as a more traditional sales representative, and a medical science liaison.

Kalb Right. I think the "by whom" is really only an issue with respect to manufac-turers. If you're an academic or you're otherwise in the business of generating this kind of information, you have seemingly complete latitude to share it. It's only the rules governing manufacturer communications that are really an issue. But you're absolutely right that within the manufacturer world, there is a question of "by whom?" Do people have to have certain backgrounds or credentials to be on the scientific side of the organization? Can you communicate CER through sales reps?

We've been living for a decade or so in a kind of weird self-imposed world in which the manufacturer community has generally accepted the notion that scientific information, including pharmacoeconomic information, somehow needs to come from specialized people within an organization. Manufacturers have concluded that such information generally should not be disseminated by sales people. There are good reasons for that – sales people are often not trained to communicate highly technical or nuanced information. It's not clear, however, that that's a constitution-ally mandated rule. That seems to be much more of a self-imposed rule. But whether that makes sense and whether that should be continued – those are open questions as well.

Crystal Pike Turning to litigation for a moment, what is your sense of how dis-semination of CER and CER-like information can trigger a lawsuit against a manu-facturer and what are the corresponding legal underpinnings?

Kalb I think that the best way of thinking about litigation stemming from the use of this kind of information is first through the prism of the Food, Drug and Cosmetic Act, and the risk there really is twofold: one is that the information is often off-label, so you run some risk that it's actionable simply on that basis; and the other, perhaps a more global, risk is that regardless of whether it's on- or off-label, it nonetheless may be viewed as false or misleading in some respect. So you can trigger either of the two principal enforcement theories under the Food, Drug and Cosmetic Act by communicating that kind of information. Secondarily, if the communication of that information by a manufacturer causes, or is alleged to cause, providers to submit claims that are not reimbursed – that is, they are off label and off compendia – then you run into False Claims Act risk as well. Those are the traditional types of cases that we have seen; that is the enforcement risk that accompanies the communication of this sort of information.

Greenberg One of the classic scenarios in which CER has been applied – quite beneficially to the audiences that have reviewed the results – is as follows:

The drug or device company has clinical-trial results for Product A versus placebo. There are also a competitor product's clinical-trial results for Product B versus placebo. But there's no head-to-head trial that either of the two manufacturers ever did that looked at the comparative effectiveness of Product A versus Product B. Fortunately, there are methodologies that have now been developed to try to overcome that information problem. In the absence of that methodological fix, or plugging of the gap, you would not be able to compare Product A to Product B directly, which is ultimately both what payers really want to know about and also, I think, what prescribing physicians and their patients would want to know about.

It turns out that there are not that many direct comparative results of traditional randomized controlled-trial studies involving head-to-head trials of one product versus another. So, in that environment, there have been methodological advances with respect to CER that have resulted in good insights – in some therapeutic areas at least – regarding this Product A versus Product B question.

So, now the issue is that we have all this great CER-type of information, but what can we do with it – particularly in light of the risks that have been identified today of manufacturers running afoul of the Food, Drug and Cosmetic Act, the False Claims Act, or maybe even the Anti-Kickback Statute. So, how do you, as counsel for a manufacturer, advise companies to navigate in an environment where they are, at times, sitting on really valuable information like that, but are puzzled as to how to proceed in a way that won't invite regulatory scrutiny?

Kalb I don't think there is a particularly good answer to that question. It's ultimately a risk judgment. The contours of that discussion are "here's what's permitted by statute. Here's what we believe is permitted constitutionally, which would be the communication of truthful and non-misleading information, even if not statutorily permitted." And then the client makes a judgment within that context as to what level of risk it is willing to take. This is the classic approach to counseling in this environment.

Pike What are some of the things that might change the level of risk? For example, would you have to look at the particulars of the study or the analysis that they are sitting on and see what methods were used or what data were used – whether or not they paid physicians to participate in the study or paid other people to participate in the study? Are those some of the things that you would look at when you're evaluating the risk? Or is it just a matter of looking at the statutes versus what you think is constitutionally allowed?

Kalb I would consider the factors that you articulated plus several other factors. One factor is the sophistication of the audience. The more sophisticated the audience, the more latitude you should have. Another factor is the manner in which data are presented. If data are presented in a straightforward manner without conclusions, that is safer. Finally, manufacturers can include "disclaimer" language, which really should best be thought of as "contextual" language. The more you describe about the methodology, the more you caveat that methodology, the more you provide the audience with an understanding of the limitations of the methodology, the safer the communication is. It seems to me that the safest way of communicating such analysis would be to the kinds of people who could really understand the data, to communicate the data without conclusory language, and to communicate the data along with both specific information about how that study was done and with some discussion of whatever is known about the accuracy of such an analysis versus an actual head-to-head trial. That would be the key: how good is this methodology at mimicking an actual head-to-head trial?

So it seems to me if you could provide all of that information, you'd be in a much safer position than if you just provided the data straight up.

Greenberg   Are there any other considerations? For instance, would it matter if the companies have recently or previously entered into a Corporate Integrity Agreement ("CIA") with the government as a result of past alleged improprieties?

Kalb   I guess I always have a little bit more worry about companies under a CIA. Most companies under CIAs tend to be in a more conservative posture and so the same counseling may lead to a different decision.

Greenberg   And, that conservative posture may become even more conservative – hyperconservative – in an environment where CEOs now are in the crosshairs.

Kalb   That's correct. It's ultimately a risk-based assessment that would include attention to all of these things.

Pike   And if the methodology is in line with published guidelines on best practices for CER, does that help with the risk?

Kalb   I think it does, because, if nothing else, it establishes the truthful and non-misleading nature of it. It certainly seems to me to inform that question, particularly if there is consensus among experts that has informed the methodological approach.

Greenberg   Let's just run through a stylized example to draw out some of these points.

Suppose a manufacturer commissioned a CER study along the lines I just described: Product A, Product B, each versus placebo to try to mimic a head-to-head type of trial. Product A's manufacturer gets favorable results and wants to share those results with the general community, both on the payer and on the prescribing side. And they took great care, because of both good counsel and general risk aversion, they took great care to make sure that their study methodology followed as much as possible of the industry standard scientific approach that's considered best practice and to document that. And now they have gone out and disseminated that information, presumably to their advantage.

A year or two goes by and let's say the government comes forward and alleges that the company grew its market by promoting this product based on information obtained using this methodology that's not a head-to-head trial. When you then go to defend that company's actions, would you think that the fact that the company followed best practices and guidelines in undertaking this study would have weight in the discussion/negotiation environment in a government investigation?

Kalb I'd certainly like to think so, and it certainly should. Your question, however, raises another legal risk here, which is that to the extent that you are doing comparative analyses, the party whose product comes out less favorably may take umbrage at the results. They may complain to FDA or they may initiate party-to-party litigation.

Greenberg And have you seen much of that?

Kalb I don't know that I have, but it's certainly foreseeable.

Greenberg I ask because many years ago, a prominent lawyer in the Boston US Attorney's Office said that he expected to see private-sector litigation by one company against another where the competitor had engaged in off-label promotion and the first company had gone down the straight-and-narrow path of obtaining FDA approval for an expanded label.

Kalb We've not seen a lot of that. Perhaps it's been the "glass house" phenomenon, but, certainly in the antitrust world, for example, private enforcement is common, just as is government enforcement. And so, particularly if promotion moves from "here's what my product can do," to affirmatively adding "and it's better than yours," it is presumptively more likely to trigger a reaction from the aggrieved company. All of this just puts more and more pressure on the underlying question of "how good is your methodology and can it stand up to the legal standards"?

Pike Going back to things that might increase or decrease the risk of litigation: Paul provided a stylized example of basically taking two randomized controlled trials and blending them together to generate new data.

Another example is the use of observational data, like administrative claims data, that is becoming more readily available, or electronic medical health or medical record data. As data have evolved over time in the pharmaceutical industry, from a litigation context and in negotiations with the government, have you seen an evolution in terms of what types of data are considered more reliable or more acceptable in terms of being used for these types of studies?

Kalb The government does tend to rely on its own data sets in its investigations of company conduct. Certainly in negotiations concerning damages, the government has tended to accept the Medicare administrative claims database and the Medicaid claims data, or at least they understand analyses of those data sets. I suspect that more novel pharmacoeconomic or CER information may be a little harder for them to accept. But having said that, I can think of at least two examples in two different cases in which creative pharmacoeconomic analyses of retrospective data really hit the mark. In one, where the primary allegation concerned off-label promotion of a biological oncology therapy, we demonstrated the overall economic benefits associated with the relevant uses of that agent. In another, we defended allegations of off-label promotion by showing that the retrospective subset analyses performed by the company were truthful and non-misleading. In both cases the government really was forced to change its focus based on the novel studies that we performed defensively.

Greenberg And other than those two examples, have you seen – either in your own work or in a colleague's work – implementation of any other pharmacoeconomic types of investigations in litigation?

Kalb The other one that comes to mind is the work that Analysis Group did which analyzed prescribing patterns relative to the alleged conduct. There, you showed the absence of prescribing changes in response to alleged inducements. You performed very detailed data analyses, and that really resonated in our negotiations. So, yes, I would say analysis of prescribing patterns, if that falls within your definition of pharmacoeconomics, would be another example.

Greenberg And we certainly make a lot of use of clinical-trial data in cases where there is an allegation of product failure or product fraud, where there is a drug safety question. We often have to go back to the underlying clinical trial data and parse out in some detail what safety signals there were back when the clinical trials were first set up and whether what we're seeing in evidence in the ensuing years was hinted at already back in the pivotal clinical trials. So that's another place where – I wouldn't call that pharmacoeconomics so much as biostatistical or epidemiological type of work – that work would find its way meaningfully into a litigation environment.

Kalb Yes. So we've transitioned a little bit, and appropriately so, from the question of the litigation risk of communicating this information to the possible defensive uses. I think those really are related, but they are two different questions. In the defensive category is the kind of work that we did in those two matters to which I alluded earlier. These are really great examples of how one can bring thoughtful economic analysis to bear on the defendant's side even if it doesn't necessarily go to the core of the matter. And what I mean by that, for example, is we were not arguing about whether uses were on- or off-label, but rather our point was that even if the uses were off-label, communication of the off-label information actually benefitted the population as a whole, including the Medicare population, and therefore it was a really terrible case for the government to even contemplate prosecuting.

Greenberg It benefitted the population clinically, but it also benefitted the payer from a system-wide cost perspective.

Kalb Right, although we did get into these knotty ethical questions about whether the payers actually benefit when their beneficiaries live longer. I think it's worthwhile making the point that some of these arguments may go directly to the liability question. But some of them are more atmospheric. And some go to the prudential question of whether the government should bring cases in which the conduct, even if arguably illegal, was actually beneficial either purely from an efficacy standpoint or with respect to cost effectiveness.

Pike You talked about defensive uses for CER. Are there similar opportunities for plaintiffs to use this type of information to make their cases affirmatively?

Kalb I think that the government tends to start with a presumption that off-label use is somehow harmful. I've not yet seen anyone actually try to demonstrate that in any sophisticated way. Take, for example, the atypical antipsychotic litigation wars that went on for years, I don't think the government ever did a study to try to quantify its view of the risk created by the promotional conduct. Prosecutors certainly did point to anecdotal clinical studies which showed clinical downsides. But one could imagine that, particularly if the defense starts to generate these kinds of studies, the government or the relators' bar would focus on beefing up their own arguments by demonstrating the population efficacy impact or the aggregate cost effectiveness, although it can be both costly and time-consuming to do these types of studies.

Greenberg One other strand that we haven't talked about is speaker programs and the context that we've often grappled with together with respect to alleged kickbacks.

I have a scenario in my head of a physician who was involved in or even oversaw the kind of comparative effectiveness pseudo-trial, this sort of retrofit of the two different placebo-controlled trials to try to figure out what the head to head would have looked like had such a trial been undertaken. Suppose the lead author of that kind of study ends up being hired by a pharmaceutical company to be a speaker at a speaker program to talk about those results. On one reading I would say "well, that seems perfectly reasonable and advanced the state of knowledge and the state of science." On the other hand, given sensitivities that we know are out there, I could also see that kind of scenario unfolding in a very difficult way for the manufacturer.

Kalb Indeed. I see at least two levels of potential concern that this might raise. One is just the "payment of the author" issue. As you know, in the current environment, direct manufacturer support for research can be perceived by some as undermining the credibility of the research. But it's hard to get around that because who else is going to fund the research? That's a concern that obviously can be addressed through transparency rules, particularly at the journal level.

The other, though, is really just the means of communication. The dissemination of peer-reviewed information itself tends to be viewed as more credible than personal communication by an author, and particularly through a channel that feels as informal as a speaker program. So I think that the scenario that you just set up has some characteristics that could undermine the credibility of the message.

Let me close the loop here by pointing out that a form of this discussion has been going on for a long time. I wrote about an earlier version of it 25 years ago, focusing on the notion that it would be helpful to disseminate information about the cost and benefits of medical technologies. At the same time, I proposed a voluntary system involving three tiers of reimbursement based on the quality of evidence available in support of the medical intervention in question. The lowest tier would cover, essentially, a mandatory array of services, which would be those services that had been demonstrated to be safe and effective. And then I proposed progressively higher and more voluntary and more expensive tiers to cover products and service for which there was less proof of either efficacy or safety, so one could voluntarily obtain access to these progressively less proven technologies if one paid more.

At the heart of that proposal was the idea that the quality of scientific evidence matters, and the better it is, the more latitude that should confer on the manufacturer to share it with various kinds of audiences. That same insight applies today in the context of CER in a litigation context.

Part IV
Emerging Challenges, Methods and Applications of Comparative Effectiveness Research: Real-World and Big Data

Chapter 17
Application of Comparative Effectiveness Research to Promote Adherence to Clinical Practice Guidelines

Carl V. Asche, Stephen Hippler, and Dean Eurich

Abstract Comparative effectiveness research (CER) ideally should be used to inform clinical practice guidelines. The challenge is to effectively incorporate CER into guidelines and to promote adherence to these clinical practice guidelines. CER can serve as a tool for continuous quality improvement of clinical practice guidelines. Thousands of clinical practice guidelines exist, of varying quality, and often offer conflicting recommendations. In this chapter, we examine why clinicians do not follow agreed-upon evidence-based clinical practice guidelines. We discuss the advantages and disadvantages of using clinical practice guidelines and how they are adopted into practice. We discuss the measurement of adherence to clinical practice guidelines and the effect of multiple medical conditions and multiple clinical guidelines on adherence.

17.1 Introduction

Comparative effectiveness research (CER) by definition here is the comparison of the effectiveness of two or more different medical interventions within a defined population in a real-world clinical setting. CER is useful for informing clinical practice guidelines (CPGs) and in serving as a tool for continuous quality improvement

C.V. Asche, PhD (✉)
Center for Outcomes Research, University of Illinois College of Medicine at Peoria, Peoria, USA
e-mail: cva@uic.edu

S. Hippler, MD
OSF HealthCare, Peoria, USA
e-mail: Stephen.E.Hippler@osfhealthcare.org

D. Eurich, PhD
School of Public Health, University of Alberta, Alberta, Canada
e-mail: deurich@ualberta.ca

© Springer Nature Singapore Pte Ltd. 2017   221
H.G. Birnbaum, P.E. Greenberg (eds.), *Decision Making in a World of Comparative Effectiveness Research*, DOI 10.1007/978-981-10-3262-2_17

of CPGs. An important issue is to consider how to promote adherence to such CPGs. However, there are many thousands of CPGs in existence along with a multitude of systems designed to measure the strength of the evidence for making recommendations to improve health outcomes or health system efficiency.

In this chapter, we briefly describe the role of CER and provide a description of CPGs, followed by an examination of how one measures adherence to CPGs and their relative impact in an era where patients are affected by multiple medical conditions and where multiple CPGs may apply.

17.2 Comparative Effectiveness Research

To improve real-world outcomes, it is imperative that medical interventions are evaluated to properly gauge their respective potential harms and benefits. Generalizing such an evaluation to current medical practice is dependent upon the technique by which the medical intervention is evaluated. Evaluation techniques range from that of efficacy studies that are tightly controlled to that of an effectiveness study that occurs in a real-world condition to reflect how the medical intervention is really used. Historically, medical interventions have undergone evaluations on the basis of their efficacy, often in randomized controlled trials, where effectiveness studies serve to provide an evaluation of medical interventions as seen in actual practice. Effectiveness studies are beneficial in that they can apply to a much broader population than that of efficacy studies, thus offering increased utility for health-care decisions being made by a variety of stakeholders—including patients, service providers, and health-care insurance providers. However, these effectiveness studies are also more prone to potential bias often making them more difficult to evaluate and implement into front-line practice.

By definition, the Institute of Medicine (IOM) defines the role of CER as that of "assist[ing] consumers, clinicians, purchasers, and policy makers to allow them to make better informed decision that will ultimately serve to improve health care at both the patient and population levels" [1]. On this basis, the role of CER then becomes vital, as it helps facilitate decision making through the utilization of a host of research methods and tools, which are inclusive of clinical trials drawing head-to-head comparisons of one treatment versus another and retrospective and prospective observational studies utilizing data from electronic health records, registries, claims-based administrative records, and other databases.

The prevailing goal of CER is to determine which treatment works best for each patient or population and under what conditions does the treatment work best. The IOM utilizes the following characteristics to help describe the elements that define a CER study [2]:

1. CER results are described at the population and subgroup levels.
2. CER compares at least two alternative interventions, each with the potential to be considered "best practice."

3. CER employs methods and data sources appropriate for the decision of interest.
4. CER is conducted in settings that are similar to those in which the intervention will be used in practice.
5. CER measures outcomes—both benefits and harms—that are important to patients.
6. CER directly informs a specific clinical decision (patient perspective) or a health policy decision (population perspective).

17.3   Clinical Practice Guidelines

A CPG is a recommendation of the standardized workflow for decision making related to a specific clinical condition. These recommendations are intended to optimize patient care and should be informed by a systematic review of evidence and an assessment of the benefits and harms of alternative care options [2, 3].

CPGs are developed by governmental organizations, clinical specialty organizations, disease-specific organizations, and international organizations [2, 3].

In the United States, the main source for finding CPGs is the National Guideline Clearinghouse (http://www.guideline.gov), as multiple guidelines can exist for the same clinical situation. To help resolve this issue, the Journal of the Medical Association (JAMA) had introduced a series of "Clinical Guidelines Synopsis" articles, which serve to summarize guidelines recommendations in a format designed specifically for busy clinicians [4].

New CPGs are constantly either being introduced or serving as updates to existing guidelines in order to keep current with the often overwhelming amount of research that is continually being generated [5, 6]. Since there are multiple systems available for evaluating the strength of evidence of research studies and since there is no central body responsible for guideline development, it is often the case that multiple, conflicting guidelines exist for the same clinical condition, making implementation of those guidelines difficult. A classic example is prostate-specific antigen (PSA) testing for prostate cancer. Following the US Preventive Services Task Force (USPSTF) guideline statement against using the PSA test for screening in healthy men (Level D—not recommended), the American Urological Association quickly reiterated that they still support the use of the PSA test in screening of healthy men [7]. Similarly, the American Diabetes Association (ADA) recommends screening for diabetes in overweight or obese patients in the presence of other risk factors; however, the USPSTF recommends that irrespective of weight, otherwise healthy patients should not be screened [8, 9]. So how does the busy clinician or patient looking to be more informed manage these dilemmas?

There is also increasing awareness of the need to clearly identify the primary target audience of CPGs, which have been health-care providers. This by no means infers that caregivers and patients should not also be involved in discussions of CPGs with their providers. Indeed, in recent years there has been a large push within health care to more strongly engage patients in the decision-making process and CPG development. Patient and community experience is considered a strong component of quality care in today's health systems, and, as with CPGs, guidelines on how best to engage patients or communities in the decision-making process have also been developed [10]. In all practicality, there probably should be two types of CPGs developed: one for the health-care provider and another, more patient-friendly version, for the general public—as the more informed patients are in regard to their treatments and potential impacts on their health, the better placed they will be to engage in meaningful dialogue with their clinicians in the management of their conditions. However, the question of how best to elicit patient experience and value and implement this information into CPGs or into the shared decision-making process is still largely evolving, with evidence to suggest both intended and unintended consequences of patient engagement [11].

A trustworthy CPG should not be seen as replacing a health-care provider's role in clinical decision making but more so as support, as the CPG can be very challenging when one considers the patient's other medical conditions, values, and preferences [12]. Many parties outside the health-care providers, caregivers, and patients find utility in CPGs in generating metrics and care guides designed to reduce unnecessary variability and measure quality. These include health-care administrators, insurance plans, and those responsible for measuring successful practice. However, it should be noted that many CPGs are focused solely on the clinical evidence and often do not take a broader health system perspective. Moreover, what may be considered best practice from an evidence viewpoint may not fit with the values and references of patients. Although the transformation has been slow to take root in some people's minds, the inclusion of patients' values and preferences is becoming more prominent in today's CPGs. For example, the recent American Heart Association and American College of Cardiology guidelines recommend the use of statins to reduce cardiovascular risk in those at higher risk irrespective of low-density lipoprotein (LDL) levels [13]. Although many consider these guidelines not applicable to/in every clinical case, the guidelines do, in fact, make numerous references to the notion that any decision on the use of a statin must be a shared decision-making process with the patient. Unfortunately, the concept of patient preference is often overlooked when developing, interpreting, and implementing CPGs. As a result, CPGs can often introduce a disconnect between what is "best" based on the evidence and what is "best" for an individual patient or subpopulation. Furthermore, these recommendations may be completely at odds with what is feasible or affordable to a health system. Indeed, few CPGs to date include cost-effectiveness in their recommendations [14]. Although many developers of guidelines have provided the rationale behind their decision to not take into account cost-effectiveness in their CPGs [15], the reality is the cost of interventions, and the ability of the patient to pay for the intervention is one of the main barriers to the utilization of many therapies. Moreover, recent examples from Canada, and elsewhere, have clearly shown the disconnect between "evidence-based" CPGs vs. CPGs that include cost-effectiveness [16]. For example, CPGs for the management of diabetes strongly recommend routine self-monitoring of blood glucose by most adults with type 2 diabetes, irrespective of underlying therapy (oral vs. insulin therapies). However, when cost-effectiveness is considered, routine self-monitoring of blood glucose by most adults with type 2 diabetes using oral antidiabetes drugs is not recommended.

One of the major criticisms of CPGs is that we are often limited in terms of knowledge of what the correct answer is for either diagnosis or treatment of a particular condition. CPGs evolve over time as new evidence is developed and applied. However, initially much of the "evidence" in CPGs is often based on consensus recommendations as the research community requires time to generate the evidence which is considered the lowest grade of evidence by most evidence-grading systems [11]. Moreover, when there is insufficient evidence available to support a best practice recommendation, multiple CPGs for multiple medical conditions arise, as each organization provides their own recommendations. Indeed, it is not uncommon to see different recommendations at the institutional level, at the health maintenance organization level, and at the national or disease-specific organizational level. Another challenge occurs when one factors in not only multiple CPGs but also issues arising from contradictory medications and drug interactions, which may complicate further the appropriate use of CPGs.

The number of patients with multiple comorbidities will continue to grow as our population ages. Since CPGs are often developed by societies focused on one disease state, they often are developed from randomized controlled trials focused on one study condition, with confounding variables excluded. For example, it has been well documented in the literature that randomized controlled trials tend to exclude a high number of patients, due to age, sex, and comorbid conditions, among others, who could potentially benefit from the therapy. Indeed, systematic reviews of inclusion and exclusion criteria of randomized trials overwhelmingly show that common medical conditions form the basis of exclusion in the vast majority of trials [17]. Age is the next most common factor for exclusion (>70% of trials), which is concerning given the aging population and multimorbidity present within this population. In nearly half of all cases, the exclusion of these patients from the clinical trial was not well justified. Moreover, industry-sponsored trials, which account for a large portion of evidence, particularity in drug trials, tend to be even more likely to exclude patients based on age, concomitant medications, or comorbid conditions and were more likely to be poorly justified [18]. Even when patients with multimorbidity are included in the trials, limited information is available on these patients to provide meaningful information on how best to manage these specific patients [19].

This in turn leads to a trickle-down effect into the CPGs, where the guidelines are intended to be applied to patients with a specific disease but the focus disease rarely exists in isolation. The issue is compounded in today's health systems that often pay for performance based on guidance from CPGs, which tend to focus on very specific disease elements and often ignore the complexity observed in patients with multiple chronic conditions [20]. Not only can single-disease-focused CPGs result in suboptimal care for patients with multimorbidity, but care could result in unintended harms [21]. For example, viewing the patient from a single-disease viewpoint, where medications are often optimized to control a single disease without significant consideration of the presence of other diseases, may result in high medication burden to the patient and may also significantly increase the risk of adverse outcomes.

The disconnect between today's complex multimorbid patient and CPGs is substantial. Numerous systematic reviews have shown that the issue of multimorbidity is rarely included in CPGs, and when comorbidities are included, the CPG recommendation tends to focus on the index disease state and one additional comorbid condition. Although some organizations are beginning to recognize the multimorbid patient within guidelines [22, 23], currently few CPGs try to focus on the more complex patient (e.g., patients with two or more comorbid conditions). As a case example, take a typical patient with type 2 diabetes. Patients with type 2 diabetes tend to be older and have multisystem involvement. Indeed, it is the comorbid complications and the impact of diabetes on these comorbid conditions that often have the greatest negative impact on patients' health-related quality of life. Beyond the metabolic impact of diabetes, patients with type 2 diabetes are well known to have high rates of hypertension, myocardial infarction, heart failure, stroke, chronic kidney disease, neuropathies, mental health disorders, cancer, and significant functional limitations due to high rates of amputation and blindness. Yet, despite the substantial impact of comorbid conditions in this population, the evidence base for the management of type 2 diabetes is relatively unrepresentative of the typical type 2 diabetes patient. Despite over 40% of patients with type 2 diabetes having comorbid heart failure [24], not a single comparative effectiveness randomized controlled trial has ever been conducted on how best to manage type 2 diabetes in patients with comorbid diabetes and heart failure. Furthermore, heart failure is often an exclusion criterion in many trials of diabetes, and conversely diabetes is often an exclusion criterion in trials of heart failure. Although observational studies aim to provide some evidence to address these evidence gaps, the reality is that observational studies are always given a lower evidence level and appropriately so due to the risk of bias, within guidelines. As a result, there is less impact of observational studies in changing practice, and they are routinely discounted by front-line clinicians managing patients. Thus the evidence gap continues to exist, affecting millions of patients with type 2 diabetes, and has resulted in substantial debate within the clinical and research communities [20].

This limitation is just the tip of the iceberg, as the majority of clinical trials in type 2 diabetes patients have high exclusion rates (20% or more of potentially eligible populations) for patients with renal insufficiency, advanced disease (e.g., serious coexisting morbidity or requiring insulin therapy), or coronary artery disease [19]. The lack of information/recommendations with respect to multimorbid, complex, type 2 diabetes among diabetes CPGs is therefore not surprising. CPGs in diabetes tend to focus on diabetes and one additional comorbidity—such as hypertension or renal insufficiency [25]—although these are often clustered within an individual patient. Thus, the typical front-line clinician and patient are left in the difficult situation of dealing with competing conditions or competing disease-specific CPGs. Moreover, CPGs are even more limited with respect to those patients of advanced age and with multiple chronic conditions, providing minimal guidance to clinicians on appropriate care management in this increasingly growing population.

Indeed, it has been previously shown that major single-disease CPGs do not promote high-quality care in older patients with multiple chronic conditions, resulting in substantial treatment burden and risk for nonadherence, and they could result in unintended harms [21, 26].

The idea of developing CPGs to overcome the current limitations associated with a single-disease focus and expanding CPGs to be more inclusive of multiple chronic conditions is not new. Indeed, expert committee recommendations have been put forth in the past to provide a framework for development of such guidelines [27]. However, this is a slow process, since the evidence needed to develop such guidelines is truly lacking in most major chronic disease states.

As a result, a more fundamental shift in the generation of evidence in multimorbid patients will be required if the concept of CPGs for people with multiple chronic conditions is to become a reality. However, if randomized controlled trials are to include the multimorbid patient, these trials will likely have to be designed differently, and our expectations of these trials will need to be altered. Often patients not deemed eligible for clinical trials due to high levels of disease or multimorbidity have worse outcomes than patients enrolled in the trials. As a result, we may have to alter our expectations, as although these patients will benefit from therapy, it is possible that the benefit may not be as large as if the trial was conducted in the "perfect patient." It may also be difficult to find patients with substantial multimorbidity and foreshortened life expectancy to enroll in trials of therapies—for example, they may elect for palliative type of care or no treatment, as opposed to trying new aggressive therapies. Thus, careful consideration of how to effectively include multimorbid patients into randomized trials of therapies to optimize benefit for both the patient and the end users of these trials is urgently needed.

The benefits of using CPGs are multifaceted, as they serve to improve the quality of care and help practitioners and patients make informed decisions based on the best available research evidence [28, 29]. Still outstanding is the need for more research focused on the clinical outcomes. Although a great deal of time, money, and effort is put into the development of guidelines, the impact of guidelines on important clinical and patient-specific outcomes is less clear. Furthermore, although CPGs often provide strong support for the use of a therapy (e.g., use of an ACE inhibitor for risk reduction in heart failure), CPGs are often reluctant to recommend specific interventions or agents. For many front-line clinicians, the clinical dilemma is not centered around whether a therapy should or should not be prescribed but tends to be focused on which specific therapy should be prescribed to maximize benefit and minimize harm in their patients. This comparative effectiveness of therapies is rarely addressed in CPGs and if presented is usually given a lower grade of evidence within the CPGs, owing to the fact that comparative effectiveness research is rarely based on controlled clinical trials. However, when this evidence has been generated in well-conducted clinical trials, the benefits to both clinicians and patients can be substantial (e.g., the impact of the Antihypertensive and Lipid-Lowering Treatment to Prevent Heart Attack Trial [ALLHAT] in CPGs for the management of hypertension which equivalently showed older, cheaper drugs (i.e., thiazides) had similar benefits relative to newer, more expensive medications).

On the other side of the coin is the issue of CPG implementation and knowledge translation which has become a major focus for many organizations that develop guidelines. CPGs are complex and so is their successful implementation. There are many factors that affect the successful implementation of CPGs and the speed of the adoption of the CPG within clinical practice.

17.4   How Are CPGs Adopted?

A CPG is an innovation in the process of care. Factors affecting adoption of the innovation include characteristics of the innovation itself, communication channels, time, and the social system [30]. To better illustrate the concept, Greenhalgh and colleagues [31] conducted a systematic review of literature and developed a unifying conceptual model of innovation in health service delivery and organization (Fig. 17.1). The conceptual model was created "as a memory aide for considering the different aspects of a complex situation and their many interactions" [16]. Indeed, the conceptual model developed by Greenhalgh et al. is a comprehensive framework for understanding health service delivery innovations from a systems perspective.

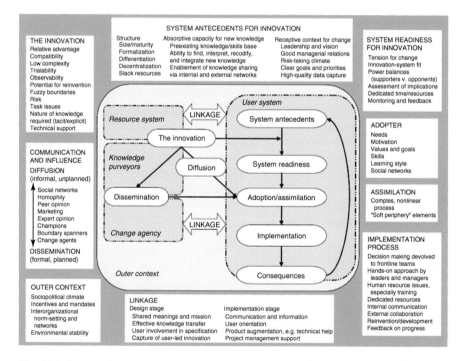

Fig. 17.1 Conceptual model for considering the determinants of diffusion, dissemination, and implementation of innovations in health service delivery and organizations by Greenhalgh et al. [31]

In a case where there is compelling evidence supporting an approach as superior to other approaches and where a trustworthy guideline has been developed, that guideline may be considered a "gold standard" or "best practice." We can understand why a clinical practice guideline is not followed when the recommendations are contested or when there is weak evidence for the recommendations.

17.5 What Are the Barriers to CPG Adoption?

What are the barriers when moving from a lack of awareness/lack of familiarity to integrating the innovation into practice [32, 33]? Cabana and colleagues conducted a systematic literature review to identify barriers to CPG adherence and organized their findings into a framework related to physician knowledge, attitudes, and behavior [34]. A qualitative study demonstrated the relevance of their framework: all of the key factors identified by Cabana and colleagues were perceived as important barriers to CPG use by general practitioners [34]. Individual provider use of CPG can be influenced by provider factors, such as personal differences in risk tolerance, the role of emotions in decision making, the role of self-perception and self-identity, and so on. These barriers are summarized by stakeholder perspective (provider factors, patient factors, guideline factors, and organizational factors) in Table 17.1. The barriers identified indicate poor fit between the CPG and the work environment.

Although the barriers to physician adherence to CPGs presented by Cabana and colleagues are a good starting point, one must understand that physicians are no longer "lone wolf" practitioners working in a cottage industry; the practice of modern medicine occurs in the context of the health-care system. It is important to

Table 17.1 Barriers to physician adherence to CPGs

Provider factors	Patient factors	Guideline construction	Environmental pressure
Lack of familiarity	Patient preference contradicts	Ambiguous guideline	Lack of time
Disagree with evidence	Patient without ability to follow guideline	Conflicting guidelines	Lack of resources
Lack of applicability		Comorbidity not addressed	System constraints
Loss of autonomy		Does not apply to typical patient	Misaligned financial incentives
Lack of outcome expectancy		Out of date	
Lack of motivation		Too complex	
		Evidence common sense mismatch	

Adapted from Cabana et al. [34]

consider the context in which the CPG is being used. There is some evidence that experts are less likely to adhere to a CPG when there are relevant patient-related factors that would "provide a reasonable basis for not following the CPG" [35].

In addition to the focus on adoption of CPGs by the health-care system and organizations within, we need to be specifically cognizant of the adoption of these innovative care ideas by the caregivers working within a complex adaptive system. Human providers have a choice within the context of a busy day whether or not to choose a new approach to care or default to the status quo. It has been well known that the diffusion of innovation is greatly enhanced by provider connectivity. Early work—over 50 years ago—showed that the adoption of a new antibiotic was related to the social connectedness of physicians, and this interconnectedness accelerated adoption. This phenomenon has also been shown with the adoption of hybrid seed corn in Iowa in the 1920s, where a close association with an early adopting colleague, in this case a farmer, increased the uptake of the innovative change. This further reinforces the idea that a CPG, although informative, is just a tool. In order for maximal benefit to reach patients and accrue to society at large, we must be cognizant of the social systems and structures that allow or foster the adoption of these changes.

Lastly and most importantly, patients are a critical element in this social network. Aside from implementing treatment recommendations from guidelines and standards, a physician's duty will always be to his or her patient. Each and every person is unique in terms of their diseases, comorbidities, social determinants, and goals of care. These all must be considered with a shared decision-making model to determine the best course of treatment for that one individual. When done at a high level, this is truly the art of medicine.

17.6   How CER Can Inform CPGs

An example that illustrates this point is ALLHAT study [36]. ALLHAT is regarded by experts in the field as the most robust primary prevention hypertension trial ever conducted. The major message from the ALLHAT study was that "thiazide-type diuretics should be used in drug treatment for most patients with uncomplicated hypertension…."—a finding that was subsequently included into the 2003 Seventh Report of the Joint National Committee on Prevention, Detection, Evaluation, and Treatment of High Blood Pressure Guidelines recommendations [37]. Another important example of the road from evidence building to integration into CPG recommendation was the fate of the findings from the comparison of warfarin and aspirin for symptomatic intracranial arterial stenosis (WASID) trial [38]. In 2006, CPGs only mentioned the trial's findings but carried no specific recommendations for use. In 2008, the guidelines were finally updated to recommend aspirin over warfarin on the basis of evidence from the WASID trial.

One of the lessons from both the ALLHAT and WASID studies was that "best" evidence may be immediately incorporated into CPGs, as in the case of ALLHAT, whereas other evidence might be delayed, as was the case of WASID. Furthermore, the incorporation of clinical trial findings into clinical practice, after inclusion into CPGs, still requires a concerted effort. The effort behind the implementation of evidence into practice includes the involvement of local opinion leaders, real-time clinical reminders, disease management strategies, critical pathways, and academic detailing. Furthermore, CPGs continually undergo revisions, and in subsequent iterations of CPGs the same evidence can be evaluated, weighted, and recommended differently, as was the case of the WASID trial over time. This can further add to the barriers of effectively implementing CPGs into clinical practice. Moreover, in some cases landmark clinical trials (e.g., RALES trial in heart failure) can fundamentally change clinical practice before the evidence can ever be incorporated into the CPGs [39].

The definitive proof that a particular piece of evidence indeed influences clinical practice would require conducting a study in which clinicians (and their patients) were randomly assigned to be exposed to the evidence. Furthermore, such an effort would require standardization of not only the evidence but also other influences such as pharmaceutical and direct-to-consumer marketing, etc. These types of trials have been attempted in the past, and although some have been deemed a success in changing behaviors and successfully implementing CPGs [40], others have shown less favorable results [41]. Collectively, the evidence indicates that no one approach is perfect in implementing CPGs and changing clinician behaviors and efforts must be customized for each unique setting and CPG.

17.7 Conclusions and Recommendations

Use of the "patient-centered" clinical practice guideline could be incorporated into the electronic health record, and this would become the quality measurement standard by which the clinical care would be measured. This would avoid the dangerous paradox created by following clinical guidelines where good clinicians are reprimanded for applying evidence-based management to individual patients.

References

1. Sox HC, Greenfield S (2009) Comparative effectiveness research: a report from the Institute of Medicine. Ann Intern Med 151(3):203–205 Epub 2009 Jun 30. PubMed PMID: 19567618
2. Institute of Medicine (US) (2009) Initial national priorities for comparative effectiveness research. National Academies Press (US), Washington, DC
3. Institute of Medicine (US) (2011) PubMed PMID: 22514810 In: Olsen LA, Saunders RS, JM MG (eds) Patients charting the course: citizen engagement and the learning health system: workshop summary. National Academies Press (US), Washington, DC
4. Cifu A, Davis A, Livingston E (2014) Introducing JAMA clinical guidelines synopsis. JAMA 312(12):1208–1209

5. Stange KC (2006) On track: medical research must consider context and complexity. Ann Fam Med 4:369–370. doi:10.1370/afm.613
6. Carl Heneghan. Carl Heneghan's blog. Retrieved 24 March 2016, from: http://blogs.trustthee-vidence.net/carl-heneghan/how-many-randomized-trials-are-published-each-year
7. Carter HB (2013) American Urological Association (AUA) guideline on prostate cancer detection: process and rationale. BJU Int 112(5):543–547
8. McCulloch DK, Hayward RA. Screening for type 2 diabetes mellitus. In: UpToDate. UpToDate. 2016. Retrieved 2 May 2016, from: http://www.uptodate.com/contents/screening-for-type-2-diabetes-mellitus
9. U.S. Preventive Services Task Force (2008) Screening for type 2 diabetes mellitus in adults: U.S. Preventive Services Task Force recommendation statement. Ann Intern Med 148(11):846–854
10. National Institute for Health and Care Excellence (NICE). Community engagement: improving health and wellbeing and reducing health inequalities. Guidance and guidelines; NICE guidelines [NG44]. National Institute for Health and Care Excellence. 2016. Retrieved 24 March 2016, from: https://www.nice.org.uk/guidance/NG44
11. Coulter A (2012) Patient engagement – what works? J Ambul Care Manage 35(2):80–89. doi:10.1097/JAC.0b013e318249e0fd Review. PubMedPMID: 22415281.
12. Djulbegovic B, Guyatt GH (2014) Evidence-based practice is not synonymous with delivery of uniform health care. JAMA 312(13):1293–1294. doi:10.1001/jama.2014.10713 PubMed PMID: 25268433
13. Smith SC, Benjamin EJ, Bonow RO, Braun LT, Creager MA, Franklin BA et al (2011) AHA/ACCF secondary prevention and risk reduction therapy for patients with coronary and other atherosclerotic vascular disease: 2011 update: a guideline from the American Heart Association and American College of Cardiology Foundation. Circulation 124(22):2458–2473
14. Koshman S, Johnson J (2005) Consensus, cost-effectiveness and clinical practice guidelines. Can J Diabetes 29:374–376
15. Harris S, McFarlane P, Lank C (2005) Consensus, cost-effectiveness and clinical practice guidelines: author's response. Can J Diabetes 29:376–378
16. Canadian Diabetes Association. Position statement and paper. Canadian Diabetes Association. 2011. Retrieved 24 March 2016, from: http://www.diabetes.ca/newsroom/search-news/position-statement-and-paper
17. Van Spall HG, Toren A, Kiss A, Fowler RA (2007) Eligibility criteria of randomized controlled trials published in high-impact general medical journals: a systematic sampling review. JAMA 297(11):1233–1240 Review. PubMed PMID: 17374817
18. Guthrie B, Payne K, Alderson P, McMurdo MET, Mercer SW (2012) Adapting clinical guidelines to take account of multimorbidity. BMJ 345(oct04_1):e6341
19. Boyd CM, Vollenweider D, Puhan MA (2012) Informing evidence-based decision-making for patients with comorbidity: availability of necessary information in clinical trials for chronic diseases. PLoS One 7(8):e41601. doi:10.1371/journal.pone.0041601 Epub 2012 Aug 3. Review. PubMed PMID: 22870234; PubMed Central PMCID: PMC3411714
20. Eurich DT, McAlister FA, Blackburn DF, Majumdar SR, Tsuyuki RT, Varney J, Johnson JA (2007) Benefits and harms of antidiabetic agents in patients with diabetes and heart failure: systemic review. BMJ 335:497–501. doi:10.1136/bmj.39314.620174.80
21. Tinetti ME, Bogardus ST Jr, Agostini JV (2004) Potential pitfalls of disease-specific guidelines for patients with multiple conditions. N Engl J Med 351(27):2870–2874 PubMed PMID: 15625341
22. Perk J, De Backer G, Gohlke H, Graham I, Reiner Z, Verschuren M et al (2012) European Guidelines on cardiovascular disease prevention in clinical practice (version 2012). The Fifth Joint Task Force of the European Society of Cardiology and Other Societies on Cardiovascular Disease Prevention in Clinical Practice (constituted by representatives of nine societies and by invited experts). Eur Heart J 33(13):1635–1701
23. Chamberlain JJ, Rhinehart AS, Shaefer CF, Neuman A (2016) Diagnosis and management of diabetes: synopsis of the 2016 American Diabetes Association Standards of Medical Care in Diabetes. Ann Intern Med 164(8):542–552

24. Bell DS (2003) Heart failure: the frequent, forgotten, and often fatal complication of diabetes. Diabetes Care 26(8):2433–2441 PMID:12882875
25. Standards of Medical Care in Diabetes – 2014. Diabetes Care 2014 Jan;37(Supplement 1): S14–S80. Https://doi.org/10.2337/dc14-s014
26. Boyd CM, Darer J, Boult C, Fried LP, Boult L, Wu AW (2005) Clinical practice guidelines and quality of care for older patients with multiple comorbid diseases: implications for pay for performance. JAMA 294(6):716–724 PubMed PMID: 16091574
27. Guiding principles for the care of older adults with multimorbidity: An approach for clinicians (2012) Guiding principles for the care of older adults with multimorbidity: an approach for clinicians: American geriatrics society expert panel on the care of older adults with multimorbidity. J Am Geriatr Soc 60(10):E1–E25. doi:10.1111/j.1532-5415.2012.04188.x
28. Sox HC (2014) Do clinical guidelines still make sense? Yes. Ann Fam Med 12(3):200–201. doi:10.1370/afm.1657 Erratum in: Ann Fam Med. 2014;12(4):301. PubMed PMID: 24821889; PubMed Central PMCID: PMC4018366
29. Woolf SH, Grol R, Hutchinson A, Eccles M, Grimshaw J (1999) Clinical guidelines: potential benefits, limitations, and harms of clinical guidelines. BMJ 318(7182):527–530 Review. PubMed PMID: 10024268; PubMed Central PMCID: PMC1114973
30. Rogers EM (2003) Diffusion of innovations, 5th edn. Free Press, New York
31. Greenhalgh T, Robert G, Macfarlane F, Bate P, Kyriakidou O (2004) Diffusion of innovations in service organizations: systematic review and recommendations. Milbank Q 82(4):581–629 Review. PubMed PMID: 15595944; PubMed Central PMCID: PMC2690184
32. Lugtenberg M, Zegers-van Schaick JM, Westert GP, Burgers JS (2009) Why don't physicians adhere to guideline recommendations in practice? An analysis of barriers among Dutch general practitioners. Implement Sci 4:54. doi:10.1186/1748-5908-4-54 PubMed PMID: 19674440; PubMed Central PMCID: PMC2734568
33. Casey DE Jr (2013) Why don't physicians (and patients) consistently follow clinical practice guidelines? JAMA Intern Med 173(17):1581–1583 PubMed PMID: 23897435
34. Cabana MD, Rand CS, Powe NR, Wu AW, Wilson MH, Abboud PA, Rubin HR (1999) Why don't physicians follow clinical practice guidelines? A framework for improvement. JAMA 282(15):1458–1465 Review. PubMed PMID: 10535437
35. Mercuri M, Sherbino J, Sedran RJ, Frank JR, Gafni A, Norman G. When guidelines don't guide: the effect of patient context on management decisions based on clinical practice guidelines. Acad Med. 2015 90(2):191–196. doi: 10.1097/ACM.0000000000000542. PubMed
36. ALLHAT Officers, Coordinators for the ALLHAT Collaborative Research Group (2002) The Antihypertensive and Lipid-Lowering Treatment to Prevent Heart Attack Trial. Major outcomes in high-risk hypertensive patients randomized to angiotensin-converting enzyme inhibitor or calcium channel blocker vs diuretic: The Antihypertensive and Lipid-Lowering Treatment to Prevent Heart Attack Trial (ALLHAT). JAMA 288(23):2981–2997 Erratum in: JAMA. 2004;291(18):2196. JAMA 2003;289(2):178. PubMed PMID: 12479763
37. Chobanian AV, Bakris GL, Black HR, Cushman WC, Green LA, Izzo JL Jr, Jones DW, Materson BJ, Oparil S, Wright JT Jr, Roccella EJ, Joint National Committee on Prevention, Detection, Evaluation, and Treatment of High Blood Pressure. National Heart, Lung, and Blood Institute, National High Blood Pressure Education Program Coordinating Committee (2003) Seventh report of the Joint National Committee on Prevention, Detection, Evaluation, and Treatment of High Blood Pressure. Hypertension 42(6):1206–1252 Epub 2003 Dec 1. PubMed PMID: 14656957
38. Chimowitz MI, Lynn MJ, Howlett-Smith H, Stern BJ, Hertzberg VS, Frankel MR, Levine SR, Chaturvedi S, Kasner SE, Benesch CG, Sila CA, Jovin TG, Romano JG, Warfarin-Aspirin Symptomatic Intracranial Disease Trial Investigators (2005) Comparison of warfarin and aspirin for symptomatic intracranial arterial stenosis. N Engl J Med 352(13):1305–1316 PubMed PMID: 15800226
39. Juurlink DN, Mamdani MM, Lee DS, Kopp A, Austin PC, Laupacis A, Redelmeier DA (2004) Rates of hyperkalemia after publication of the Randomized Aldactone Evaluation Study. N Engl J Med 351(6):543–551 PubMed PMID: 15295047

40. Schectman JM, Schroth WS, Verme D, Voss JD (2003) Randomized controlled trial of education and feedback for implementation of guidelines for acute low back pain. J Gen Intern Med 18(10):773–780 PubMed PMID: 14521638; PubMed Central PMCID: PMC1494929

41. Joseph AM, Arikian NJ, An LC, Nugent SM, Sloan RJ, Pieper CF, GIFT Research Group (2004) Results of a randomized controlled trial of intervention to implement smoking guidelines in Veterans Affairs medical centers: increased use of medications without cessation benefit. Med Care 42(11):1100–1110 PubMed PMID: 15586837

Chapter 18
Challenges in Developing and Assessing Comparative Effectiveness Evidence for Medical Technology

Richard Price and Genia Long

Abstract Medical devices exhibit characteristics distinct from pharmaceuticals, namely, that they are heterogeneous, subject to incremental and continuous innovation, and embedded in often complex patient-care processes, which affect comparative effectiveness assessments. Specifically, where procedures are performed, by whom, and when analysis is undertaken all are likely to affect results and therefore should be considered by payers and others in evidence reviews for coverage and patient access. In addition, there are special considerations for diagnostics and imaging technology, which provide clinicians with information, with the ultimate benefits dependent on the clinical actions taken.

While randomized controlled trials (RCTs) have been considered the "gold standard" for evidence, well-designed observational studies also provide relevant and valuable real-world evidence about therapies, particularly for medical technology, and there is increasing demand for high-quality observational studies. Unique device, diagnostic, and imaging technology characteristics should be considered in designing and evaluating evidence from both RCTs and observational studies. In light of growing demands for evidence for coverage and patient access decisions, evidence assessment tools and study strategies should reflect the clinical and economic characteristics of medical technology, rather than applying a "one-size-fits-all" approach.

R. Price
AdvaMed, Washington, DC, USA

G. Long (✉)
Analysis Group, Inc., Boston, MA, USA
e-mail: Genia.Long@analysisgroup.com

© Springer Nature Singapore Pte Ltd. 2017   235
H.G. Birnbaum, P.E. Greenberg (eds.), *Decision Making in a World of
Comparative Effectiveness Research*, DOI 10.1007/978-981-10-3262-2_18

18.1   Introduction

Whereas historically, medical technology innovators planned for safety and efficacy evidence requirements to support regulatory approval, increasingly they now also must plan for the evidence requirements associated with payer coverage, patient access, and physician utilization decisions. Increasingly, decision makers of various types require greater levels of evidence to support favorable decisions:

- By *payers* to support coverage decisions and guide utilization management approaches
- By *clinical professional societies* to support care guideline recommendations
- By *provider groups*, including accountable care organizations (ACOs), to support clinical protocol recommendations and technology acquisition and utilization decisions
- By *individual providers and patients* to support treatment decisions

While demands for evidence are increasing, there is also a growing recognition that randomized controlled trials (RCTs) alone will not meet all the information requirements of stakeholders and more high-quality real-world evidence will be required. While RCTs have been preferred historically as the "gold standard" for regulatory approval due to their randomized design and lack of bias, RCT results may not be possible or relevant to patients in "real-world" settings. There may be differences in rates of patient compliance, variation in treatment patterns, co-therapies administered by clinicians, and patient subtypes treated. In addition, other clinically and economically meaningful factors may be difficult to control and test in a traditional RCT (e.g., long-term health effects, interaction with rapidly changing treatment practices) [1].

Moreover, because therapeutic medical devices and diagnostic and imaging medical technology (together, medical technology) exhibit unique characteristics distinct from pharmaceuticals, innovators, payers, and other stakeholders face special challenges in designing and interpreting evidence for decision making, including evidence from both RCTs and observational studies.

There have been recent efforts to develop evidence frameworks and hierarchies that may more systematically guide the process of evidence review and assessment that reflect the characteristics of medical technology and to assist both innovators designing and producing evidence and those reviewing and evaluating evidence about medical technology.[1] However, challenges remain. As demands for decision-relevant, high-quality evidence increase, the impact of evidence-based decisions can be far-reaching for patients, physicians, payers, and manufacturers. Therefore, the evidence assessments supporting decision making should be appropriate and reflect best available practice for medical technology. In their absence, a mismatch can occur between the evidence demands and assessments of payers and other

---

[1] See, for example, results of the EU MedtecHTA project, which can be found at http://www.medtechta.eu/wps/wcm/connect/Site/MedtecHTA/Home/Publication/.

stakeholders and the evidence design strategies and investments of medical technology innovators, with potential ultimate impacts on patients.

This chapter summarizes key differences between medical technology and pharmaceuticals and the special considerations for innovators, payers, and other stakeholders when developing, reviewing, and assessing comparative effectiveness evidence for medical technology in the context of United States (US) market access-related decisions, with a focus on medical devices, but also considering implications for diagnostic and imaging technologies. Recent developments in best practices, remaining gaps, and implications for evidence practitioners and reviewers are identified.

18.2 Increasing Demands for Evidence to Guide Market Access and Other Decision Making

In the USA, commercial payers typically follow a two-step process for gathering, summarizing for review, and assessing evidence for coverage decisions. The first step, a *technology assessment*, is a structured scientific review of the available evidence and published literature; the second step, a *coverage decision*, reflects the results of the technology assessment and may also reflect other considerations, such as the past and expected future actions of other payers (both public and private), provider and plan member demand, and, in some cases, economic considerations.[2] As a result, coverage decisions vary across payers. For example, a recent study found that, for a sample of 47 medical devices considered in Centers for Medicare & Medicaid Services national coverage determinations, coverage policies by Medicare and private payers disagreed about half the time, with private payer coverage decisions being more restrictive about as often as they were less restrictive [3].

Structured technology assessments have become increasingly common among private and public payers in the context of setting new technology coverage and utilization management policies (such as prior authorization or other patient access requirements), as well as among large provider groups and ACOs, and clinical professional societies in defining clinical guidelines and care pathways. Some US payers conduct technology assessments with in-house resources. Others look to technology assessments and reviews conducted by various influential external bodies, such as the Agency for Healthcare Research and Quality (AHRQ), the Blue Cross Blue Shield Association Technology Evaluation Center (TEC), the Institute for Clinical and Economic Review (ICER), and the Cochrane Collaborative

---

[2] Previous payer research found that the stated importance and role of economic considerations in the coverage decision varied widely, with respondents reporting that "economic considerations never enter into technology coverage policy decisions," that they "always do—new technologies must demonstrate economic value as well as clinical value," and that they "sometimes" do, either "if the clinical assessment is a 'close call' or" if the economic impact is particularly large" [2].

("Cochrane Reviews") [4]. Some retain commercial third-party vendors to compile technology assessments for them.

To standardize the approach used, establish good practices, and improve the quality of evidence reviews and technology assessments, various evidence quality assessment frameworks for evaluating individual studies and for assessing overall bodies of evidence have been developed. They range from high-level principles and qualitative hierarchies to detailed numerical rating systems. However, some have focused on RCTs rather than observational studies and reflect the characteristics of drugs, either explicitly or implicitly.

In "gold standard" RCTs, patients are randomly assigned either to the therapy to be analyzed or to placebo (or, in some cases, another well-documented standard-of-care, active comparator therapy). In contrast, any study in which participants are not randomized or otherwise assigned by researchers to a treatment arm and the choice of treatments instead is up to patients and their physicians is an "observational" study. RCTs limit the potential for bias by addressing the possibility that characteristics of study participants are systematically related to whether they receive the treatment being analyzed.

While large-scale RCTs often are considered the evidence "gold standard" for drug approval and for certain medical devices, others have observed that RCTs are less likely to be conducted for devices (and their associated medical procedures) than for drugs, for a variety of reasons.[3] There may be practical barriers in applying standard randomization and double-blinding methods in which neither the patient nor the clinician is aware of whether the patient has been assigned to the study or the control group. There may also be ethical barriers (e.g., from exposing patients to sham surgeries from which they cannot benefit and with which there are significant risks). Moreover, they may be uneconomic for rapidly evolving medical technology innovations which represent incremental improvements over the previous standard of care, including such innovations as further miniaturization, battery life improvements, and improvements in feature ease of use by patients or providers. For typical devices with a life cycle of only 18–24 months, RCTs for each incremental improvement would be impractical, uneconomic, and untimely.

Device and diagnostic studies do include RCTs (albeit sometimes with smaller sample sizes), depending on the situation, and/or observational studies, in which treatment has been assigned on the basis of clinician judgment (rather than being randomly assigned). While as a result concerns have been raised about the potential for bias, others have noted that high-quality observational studies have a particularly important role to play in comparative effectiveness practice, as "they can address issues that are otherwise difficult or impossible to study. In addition, many clinical and policy decisions do not require the very high levels of certainty provided by large, rigorous randomized trials" [1]. In addition, there are also important benefits associated with well-designed observational studies, namely, lower cost (on a per capita basis), greater timeliness, potential for longer follow-up duration, and potential for greater fidelity to "real-world" factors such as clinical treatment

---

[3] See, for instance, an estimate that only 10% of the evidence base in surgery is from RCTs [5].

alternatives, demographic, and other factors, versus the highly controlled conditions in RCTs, which may be nonrepresentative of the real-world treatment environment.

Circumstances when observational studies are particularly useful include when they generate information that can only be provided through particularly large or long-term studies (e.g., the Framingham Heart or Nurses longitudinal studies that provide information on a wide variety of personal health behaviors over the span of many years); treatment adherence varies (e.g., assessment of one corticosteroid inhaler versus another); providers have different levels of training (e.g., comparison of outcomes of implantable cardioverter defibrillators by electrophysiologists and other physicians); and/or treatments are off-label (e.g., the potential for thrombosis among drug-eluting stent use was investigated by a combination of RCTs and observational studies).

In some cases, rather than *substituting* for RCTs, observational studies may *complement* RCTs. As noted by Dreyer et al., "[c]linical trials might not involve head-to-head comparisons of treatments or use the most relevant alternatives. Decision makers may want to fill this gap with observational studies – particularly when decisions must be made well before a formal trial could produce results" [1]. Other circumstances noted when observational studies can complement RCTs include evaluating the real-world applicability of RCT-generated evidence; studying patients or conditions not typically included in RCTs, or in circumstances when RCTs would be unethical or impractical; and providing information to better understand treatment practices in order to design future RCTs.

18.3   Differences Between Drugs and Medical Devices

There are a number of differences between drugs and medical technologies that have important implications for the economics of innovation in the two industry segments generally, as well as for the evidence environment and the review of evidence. These include among others differences in the regulatory process for approval, in the economics of innovation and in the pattern of clinical outcome improvement associated with new product introductions, and in underlying product design and features (Table 18.1).

With regard to the regulatory process for approval, in the USA, medical devices enter the market either through the Food and Drug Administration (FDA) premarket approval (PMA) or premarket notification (501(k)) processes. Medical devices fall into one of three classes, depending on the level of patient risk, with Class I devices representing the lowest level of patient risk and Class III (those that "support or sustain human life, are of substantial importance in preventing impairment of human health, or which present a potential, unreasonable risk of illness or injury," including many implantable devices) representing the highest level of patient risk. Class III devices must provide clinical and nonclinical data to support findings of safety and effectiveness and approval under the most stringent PMA process. For Class I (lowest risk) and Class II (low to moderate risk) devices, general and special controls

Table 18.1 Selected differences between drugs and medical devices

	Drugs	Medical Devices
Product features		
Product design and features	Molecule does not change; additional studies in different populations, for different indications, at different dosages and/or different delivery routes may be conducted	Product may evolve, with multiple successive product versions
Regulatory review		
FDA approval process	New Drug Application (NDA) or new Biologics License Applications (BLAs)	Premarket approval (PMA) or premarket notification (510(k))
Number of annual regulatory approvals	Dozens approved (41 new molecular entity NDAs or BLAs approved by Center for Drug Evaluation and Research (CDER) in 2014)	Dozens to thousands (25 original PMA approvals, 2,425 additional PMA supplemental approvals, and 3,217,510(k) approvals in 2014)
Evidence required for regulatory approval	At least two well-designed, randomized Phase III studies	PMA: valid scientific evidence providing reasonable assurance that the device is safe and effective for its intended use through well-controlled investigations, partially controlled studies, studies and objective trials without matched controls, well-documented case histories conducted by qualified experts, and reports of significant human experience with a marketed device, may differ by device type and conditions of use 510(k): FDA notification of equivalence to a predicate device (in one of three classification categories)
Economic factors		
Payment and reimbursement model	Typically a pharmacy benefit (managed by pharmacy director), with some exceptions	Typically a medical benefit (managed by medical director)
Economic life cycle	High R&D risk, cost, and duration (i.e., typically a dozen + years to reach market) Economic life is largely limited by patent and regulatory exclusivity – on average, market exclusivity period is approximately 12–13 years in the USA [6]	R&D risk, cost, and duration vary from similar to prescription drugs (e.g., stents) to lower-risk and cost frequent version updates (e.g., test equipment supplies, surgical trocars) Economic life may be limited by competitive alternatives before expiration of patent life

are considered sufficient to assure safety and effectiveness, and for Class II devices cleared under the FDA's 510(k) authority, substantial equivalence to a previously approved (predicate) device must be shown. Class I and II device market approval therefore does not generally involve RCTs.[4,5] For some devices, bench or mechanical evidence, rather than either RCTs or observational trials, may be important. For example, a new knee replacement may claim an effective life twice as long as an existing knee due to design or material improvements (e.g., 20 years versus 10 years). To support the claim, the manufacturer could conduct mechanical stimulation studies, in which laboratory conditions simulate the repeated stress of use over many years. Post-approval observational studies could also be conducted to monitor adverse events in actual use.

With regard to economic factors, the research and development (R&D) process and economically useful lives of drugs and devices tend to differ significantly. While some Class III devices may also require high levels of R&D investment over extended periods of time (drug-eluting stents being one example), Class I and II devices typically exhibit lower levels of development risk, duration, and cost, more rapid competitive market entry, and shorter useful economic lives.

An important related difference between medical devices and drugs is that a given medical technology typically evolves over time with successive product versions exhibiting continual and incremental improvements, while drugs correspond with fixed molecular structures.[6]

18.4 Resulting Special Challenges for Medical Technology Evidence Design and Reviews

These characteristics of medical technology and the economics of their development and market adoption can complicate the technology assessment process and create challenges in conducting high-quality comparative effectiveness research (and in defining appropriate evidence assessment guidelines). These challenges include the following [7]:

---

[4] See US Food and Drug Administration. Premarket Approval (PMA). Available at: http://www.fda.gov/MedicalDevices/DeviceRegulationandGuidance/HowtoMarketYourDevice/PremarketSubmissions/PremarketApprovalPMA/default.htm.

[5] Class I devices generally do not require premarket approval or clearance but must adhere to quality standards. Devices cleared under 510(k) authority must demonstrate they are at least as safe and effective as the predicate device and that technological changes do not "raise new question of safety and effectiveness." The type of evidence required to demonstrate that this standard is met varies with the device. For about 15% of 510(k) devices, clinical evidence of some type is required.

[6] Although some drugs may experience changes over time in molecular structure in order to yield improvements in safety or efficacy (such as new extended release or other formulations), new clinical evidence would be required.

- *Medical technologies are heterogeneous, creating challenges for a "one-size-fits-all" approach to evidence.*

- There is a wide range of medical device types (e.g., from implantable cardiovascular devices to tubing sets for ventilators) and medical device types vary widely in their complexity and in the degree and types of risks and benefits involved. Further, diagnostic technology (such as advanced imaging capital equipment) may be applied to a wide variety of clinical situations and patient types. A "one-size-fits-all" set of guideline principles or specific "checklists" encompassing such a broad range of technology instances presents potential for mismatch.

- *Medical technology innovation often proceeds incrementally and continuously, and as a result, devices and their associated procedures can be a "moving target" for analysis.*

- After devices come to market, improvements in them (and in their associated medical procedures) continue to accumulate over time, altering their clinical and cost-effectiveness. As a result, and also due to "operator learning curve" effects (rather than changes in device functionality or features), early assessments may tend to underestimate effectiveness and assessment conclusions may become out-of-date as devices and their use evolve, even in the same patient population(s).

- *Devices and their associated procedures are embedded in often complex processes of patient care, and patient, provider, and institutional factors can have important impacts on clinical and economic outcomes.*

- As opposed to drugs (where, assuming accurate dosing and patient compliance, operator skill is not an issue), medical device effectiveness is affected by how well they are deployed. Operator expertise and patient-care setting have been shown to affect surgical outcomes (e.g., laparoscopic surgery for GERD, gallbladder removal, and groin hernia repair), but appropriate methods taking them into consideration often are not incorporated into evaluations [8]. As a result, it can be difficult to separate multiple confounding effects from the measurement of clinical intervention and costs. In particular, the "learning curve" effect, in which the measured effectiveness of an intervention improves over time as a result of improving clinician proficiency (e.g., surgical skill), care delivery site institutional experience, and other patient-care model learning effects, can confound comparison between one intervention and another. For example, an analysis comparing a procedure involving a new, surgically implanted device and a traditional surgical approach without the new device could instead measure the difference in surgeons' expertise with the newer and more established procedures, which may narrow over time. Special analytic methods to compensate for such differences should be considered.

Other researchers have also highlighted these and related challenges to accurate comparative effectiveness evaluations and appropriate evidence guidelines and evaluation approaches. For example, a 2012 AHRQ-commissioned critical review of the quality of reporting in systematic reviews of implantable medical devices highlighted evi-

Table 18.2 Medical device evidence evaluation device-specific and operator-specific factors[a]

	Evidence factor	Examples
Device-specific factors	Differences in across-device characteristics	Different effects of bare-metal, paclitaxel-eluting, and sirolimus-eluting stents
	Differences in within-device characteristics	Differences in programming within implantable cardiac defibrillators
	Evolution of devices over time	Manufacturer modifications and enhancements to successive generations of devices
Operator-specific factors	Training/certification of operators (including lay users in case of home device use)	Electrophysiology-trained physicians versus cardiologists Lay users versus trained in-home care providers
	Ramp-up in provider expertise ("learning curve")	Number of procedures performed by a given surgeon
	Level of expertise in team/site	Lower rate of complications at centers with more stent graft procedures
	Practitioner variability	Different operators in different arms of a trial
	Differences in volume of procedures at different sites	High versus low experience with a given device/procedure combination at a given site

[a]Reflects eight evidence-related factors specific to medical devices identified in AHRQ Publication No. 12(13)-EHC116-EF [9].

dence-related factors specific to implantable medical devices, grouping the results into device-specific factors and operator-specific factors [9]. Device-specific factors include variation within and across types of devices and the phenomenon in which devices evolve over time as a result of clinician and manufacturer enhancements. Operator-specific factors include differences across clinicians, sites, and over time (Table 18.2).

Similarly, commentaries by Taylor and Iglesias [10] and Drummond, Griffin, and Tarricone [11] highlighted overlapping and related factors raising methodological challenges for evidence practitioners and reviewers, including the potential for device–operator interaction and incremental technological innovation over medical device lifetimes [10] and learning curve effects, impact of operator characteristics, importance of organizational context and capacity on comparative effectiveness findings, potential variation in clinical effect within a given technology class and less common "class effects," and differences in price trends over time and procurement and purchasing approaches [11].

The recent MedtecHTA project funded by the European Commission also highlights the importance of taking the learning curve and incremental technology innovation into consideration for both analysis and policy decisions. Nevertheless, in the context of health technology assessments (HTAs) conducted in Europe, they found that there is progress to be made: "It is well accepted that medical devices differ from drugs and other health technologies in a number of specific ways ... However, we found little evidence of differentiation in the methods used by HTA agencies to assess devices compared to non-device technologies [12].

18.5 Special Challenges for Diagnostic Medical Technology

In addition, there are special analytic challenges in the case of *diagnostic and imaging* technologies (rather than *therapeutic* medical technology), which merit discussion and consideration beyond the scope of this chapter. The core challenge is that the value of diagnostic technology lies in enabling improved clinical decision making and therapy selection, and this value is distinct from the value of the underlying therapy intervention itself [13]. As a result, evidence requirements, particularly by payers for coverage, should distinguish between therapeutics and diagnostics, and appropriate evidence frameworks specific to diagnostic technologies need to be developed and refined.

Advanced diagnostics (such as molecular diagnostics), which require especially high levels of up-front investment, may face particular evidence challenges from payers, together with the intellectual property protection, coding, and commercialization challenges they also face. For example, others have highlighted that although the guidelines for the molecular diagnostics program of one of Medicare's local administrative contractors provides for six possible levels of clinical utility evidence for molecular diagnostic tests, a diagnostic presenting evidence without at least one prospective study (i.e., "Prospective Observational Studies," "Prospective–Retrospective Trials," or "Randomized, Prospectively Controlled Trials") would be rejected without full clinical review [14].

18.6 Evidence Quality Guidelines and Observational Studies

Previous researchers have noted that one of the potentially problematic implications of an increased focus on evidence-based medicine is an inflexible set of assumptions with regard to the hierarchy of evidence quality for decision making, with Concato noting that "the popular belief that only randomized, controlled trials produce trustworthy results and that all observational studies are misleading do a disservice to patient care, clinical investigation, and the education of health care professionals." Study classifications "according to 'grades of evidence' on the basis of the research design" reserve the highest grade for "research involving 'at least one properly randomized controlled trial,' and the lowest grade is applied to descriptive studies (e.g., case series) and expert opinion; observational studies, both cohort studies and case-control studies, fall at intermediate levels." However, Concato's analysis of five clinical topic areas containing meta-analyses of both RCTs and observational studies finds that "there is evidence that observational studies can be designed with rigorous methods that mimic those of clinical trials and that well-designed observational studies do not consistently overestimate the effectiveness of therapeutic agents" [15].

For example, the tool developed by the Grading of Recommendations Assessment, Development and Evaluation (GRADE) Working Group generally assigns an initial "low" rating to all nonrandomized, observational studies. Ratings may be increased to "medium" ("upgraded observational studies") or possibly "high" ("double-upgraded observational studies"), based on several factors, including: observation of a large-magnitude effect, no obvious bias likely to be the cause of the observed large effect, and demonstration of a dose–response gradient (noting that GRADE distinguishes between the quality of evidence and the strength of recommendations) [16]. Other approaches incorporate similar evidence "grading" systems that generally characterize evidence generated by well-designed RCTs as having the top grade (e.g., "Level 1") and evidence generated by expert opinion alone as having the lowest grade.[7]

In response to the need for more nuanced assessments of evidence quality and appropriate methods for observational studies and noting that "doctors, patients, and other decision makers need access to the best available clinical evidence, which can come from systematic reviews, experimental trials, and observational research," Dreyer et al. reviewed a range of frameworks, standards, and principles designed to guide or evaluate the quality of these evidence sources [1]. They concluded that observational studies have an important and unique role to play (including for medical technology) and that, therefore, evidence assessment criteria and frameworks should address them. For useful guidance for observational study design, conduct, and analysis and reporting, the authors highlighted particularly the Good ReseArch for Comparative Effectiveness (GRACE) principles [18] and the guidance issued by the International Society for Pharmacoeconomics and Outcomes Research (ISPOR) [19].

The GRACE principles define "high-level concepts about good practice for nonrandomized comparative effectiveness research," rather than advocating a strict scoring approach (e.g., in the form of numerical scores, letter ratings, or high/medium/low/very low rankings) [18]. They were designed specifically to guide best practice in the design and evaluation of observational comparative effectiveness studies (e.g., studies of the relative effectiveness in the real world of one knee implant versus another over time in a particular defined patient population) and consist of a series of questions addressing the study plan, good study conduct and reporting practice, and appropriate inferences of the comparative effectiveness analysis to populations of interest. In recognition of the value of high-quality observational studies, the GRACE principles suggest "interpretation of these observational studies requires weighing of all available evidence, tempered by judg-

---

[7] See, for example, National Institute for Clinical Excellence and Guide to the Methods of Technology Appraisal [17] "Hierarchies typically grade studies as follows: from level 1 (RCTs), through level 2 (controlled observational studies, e.g., cohort studies, case–control studies), and level 3 (observational studies without control groups, e.g., case series), to level 4 (expert opinion based on pathophysiology, bench research, or consensus views)."

ment regarding the applicability of the studies to routine care" and "(n)o scoring system is provided or encouraged" [18]. A benefit of this flexible approach is that Dreyer et al. note the principles can be used in conjunction with other well-established standards – such as reviews of observational studies following the Cochrane principles, the AHRQ Methods Guide for Effectiveness and Comparative Effectiveness Reviews, and the AHRQ handbook on Registries for Evaluating Patient Outcomes [20], and are consistent with good pharmacoepidemiologic practice [19, 21, 22].

With the goal of accelerating acceptance into professional practice, researchers recently translated the GRACE principles into a concrete data and methods checklist, validating it against a dataset of published systematic review ratings using a panel of independent reviewers and against expert opinion concordance. The studies reviewed included those focused on drugs, medical devices, and clinical and surgical interventions [23].

18.7 Medical Device Study Special Challenges

The unique characteristics of medical technology outlined earlier lead to specific challenges in interpreting evidence from both observational studies and RCTs in determining coverage and patient access policies for devices. As a result, evidence development and review practices appropriate to drugs may be inappropriate without modification when applied to medical devices and other forms of medical technology.

Applying the GRACE evidence principle considerations relating to study planning, design, conduct, and interpretation and inference to the typical medical device characteristics identified earlier in this chapter (and summarized in Tables 18.1 and 18.2) highlights some important medical device-specific challenges in evidence planning, execution, and interpretation (Table 18.3).

Generally, these implications suggest that *where* procedures are performed, by *whom*, and *when* an analysis of effects is undertaken all are likely to have an impact on study results and therefore should be considered by payers, provider groups, and others in their reviews of and requirements for medical technology evidence for coverage, patient access, and utilization.

- *Who* **performs procedures may have an impact on study outcomes**.

- It is well established that clinical outcomes of procedures, particularly surgical procedures, reflect the proficiency of clinicians, which is typically related to such factors as specialized training, facility characteristics, and numbers of procedures performed. As a result, *who* conducts the procedure will also have an effect on study findings and conclusions about safety, efficacy, and relative effectiveness. For example, the characteristics of physicians taking part in early studies may differ from later adopters, due to study entry criteria and other factors.

- *Where* **procedures are performed also may have an impact on study outcomes**.

Table 18.3 Evidence design and evaluation challenges and implications due to medical technology characteristics

	Medical device evidence evaluation factors [9]	Evidence design and evaluation considerations for medical technology[a]	
		Study planning, study design, and conduct considerations	Evidence interpretation and inference challenges
Device-specific factors	Differences in across-device characteristics	Selection of model and comparators to be studied	Distinguishing between product and class-specific conclusions and restrictions
	Differences in within-device characteristics	Selection of features to be studied	Typicality of feature set
	Evolution of devices over time	Selection of model version/generation to be studied	Periodically revisiting restrictions based on earlier studies
Operator-specific factors	Training/certification of operators	Control for, or selection on the basis of, training or expertise	Extrapolating from RCTs to community practice settings with diverse practitioners
	Ramp-up in provider technique ("learning curve")	Control for, or selection on the basis of, experience with given procedure	Extrapolating from RCTs to community practice settings with diverse practitioners
	Level of expertise in team/site	Control for, or selection on the basis of, level of expertise or procedure volume (proxy for expertise)	Extrapolating from high-volume RCT sites to lower-volume community practice settings
	Practitioner variability	Consider in context of multi-arm study	Reliance on a single study
	Differences in volume of procedures at different sites	Control for, or selection on the basis of, procedure volume	Restrictions may be inappropriate if based on early adopters, volume outlier sites

[a]Reflects GRACE principle considerations relating to study planning, design, conduct, and interpretation and inference, as influenced by medical device characteristics.

- Moreover, *where* procedures are performed may have an impact on measured outcomes. The characteristics of early study centers may differ systematically from later practice locales (e.g., they may be more likely to be large, academic medical centers). Both factors may result in variation in outcomes, with implications for the generalizability of results.

- ***When* an analysis is performed also may have an impact on conclusions.**

- Due to the learning curve of clinical practitioners, *when* in the technology life cycle technology assessments and outcomes studies are conducted also is likely to have an effect on comparative effectiveness conclusions. Studies conducted early on in the life cycle may not reflect the beneficial effects of clinicians' increasing familiarity with the intervention and expertise in the associated procedure(s) and so underestimate true benefits in the target population by proficient clinicians. Conversely, to the degree that the technology diffuses to other patient subgroups (with either higher or lower underlying levels of risk and benefit) and/or to less proficient clinicians over time, there may be differences in findings relative to an initial study [24]. Moreover, as noted earlier, product design and features also may evolve over time, confounding the analysis further. In either event, earlier and later studies may show different results, and explicit plans may need to be formulated to revisit analyses over time. Conclusions about coverage or patient access that are reached through payers' comparative or cost-effectiveness analyses may need to be revisited over time.

- In addition, when outcomes are measured relative to the timing of the clinical intervention itself (i.e., the duration of the measurement and follow-up period) may have an important effect on study conclusions. For example, in the case of medical devices that are implanted surgically, because surgical procedures are typically associated with the potential for perioperative complications, measurements of benefits (for instance, impacts on mortality) will typically improve over time (e.g., survival curves cross over time when short-term outcomes are worse with surgery, but longer-term outcomes are better) [24]. In another example, in the case of orthopedic procedures such as knee, shoulder, and hip replacements, benefits such as increases in mobility and independence may take time to be fully realized as a result of rehabilitation activities.

18.8   Conclusions

Given continuing pressures on health-care spending, the increasing demand for evidence by various stakeholders, including both private- and public-sector payers, to support coverage, patient access, and therapy utilization decisions is unlikely to abate. While a number of structured frameworks to assess and compare the quality of evidence have been developed to support transparent and consistent scientific evidence review in the context of technology assessment and subsequent coverage decisions, historically they have reflected evidence expectations for drug therapies, rather than the unique and varied characteristics of medical technologies (including both therapeutic medical devices and diagnostics). As a result, there can be a mismatch between the evidence requirements of payers and other stakeholders and the clinically and economically appropriate evidence design strategies and investments by medical technology innovators, with potential ultimate impacts on patients. As noted, large-scale RCTs are likely to be uneconomic and impractical and therefore unlikely to be conducted for medical technology incremental improvements over the previous standard of care within highly similar devices. In addition, there may be challenges and limitations in designing and executing decision-relevant comparisons across sites and time periods with varying levels of operator expertise and when device design and features themselves are continually evolving.

Best practices for evidence assessment and decision frameworks for technology assessment should reflect the characteristics of medical devices and diagnostics, such as the frequently incremental and ongoing nature of technical improvement; the importance of evolving operator "learning curve" skill and variability of experience; the relationship between the device and a multi-dimensional, complex patient-care process; and the nature of imaging and diagnostic, rather than therapeutic, interventions. They should encompass both randomized controlled trials, where appropriate, and guidance for well-designed observational studies. Users who rely on strictly hierarchical methods in the context of medical device and other non-pharmaceutical interventions should be aware of the potential implications and decision biases that may result. In light of the high level of variation across medical technology types, a "one-size-fits-all" approach is unlikely to be appropriate.

There have been recent efforts to address gaps in evidence guidance for medical devices specifically. For example, in the context of European HTAs, a collaborative project (MedtecHTA) funded by the European Commission has been undertaken to "investigate improvement of HTA methods to allow for more comprehensive evaluation of medical devices by acknowledging complexities rising from their integration into clinical practice" [12]. Specific objectives include exploring current differences in methods used for HTA of medical devices across EU countries and developing improved methods for comparative effectiveness, economic evaluation, and organizational impact of medical devices. Initial findings include methodological recommendations such as that medical device interventions should be analyzed as complex interventions with user and context factors affecting results and that careful consideration should be given in particular to the impact of learning curves on clinical effects.

18.9 Takeaways for Practitioners

When developing and reviewing evidence for innovative medical technology to make coverage, patient access, and utilization decisions, stakeholders should consider that:

- While RCTs are the "gold standard" for pharmaceutical evidence, they may not always be possible (or even recommended) for medical devices, and under some circumstances, well-designed observational studies may be equally or more relevant components in the evidence environment.

- In designing and evaluating evidence from studies (whether RCTs or observational studies), the unique characteristics of medical devices should be taken into consideration and influence study design and evaluation and communication of results.

- In addition, special consideration should be given to diagnostics and imaging technologies, as their benefits derive from the value of additional information (such as additional imaging, genetic, or diagnostic data) provided as inputs to clinical decision makers, with the magnitude of the benefit being dependent on the subsequent actions taken by those decision makers.

References

1. Dreyer NA, Tunis SR, Berger M, Ollendorf D, Mattox P, Gliklich R (2010) Why observational studies should be among the tools used in comparative effectiveness research. Health Aff (Project Hope) 29(10):1818–1825. doi:10.1377/hlthaff.2010.0666
2. Long G, Mortimer R, Sanzenbacher G (2014) Evolving provider payment models and patient access to innovative medical technology. J Med Econ 17(12):883–893. doi:10.3111/13696998.2014.965255
3. Chambers JD, Chenoweth M, Thorat T, Neumann PJ (2015) Private payers disagree with Medicare over medical device coverage about half the time. Health Aff (Project Hope) 34(8):1376–1382. doi:10.1377/hlthaff.2015.0133
4. Feldman MD, Petersen AJ, Karliner LS, Tice JA (2008) Who is responsible for evaluating the safety and effectiveness of medical devices? The role of independent technology assessment. J Gen Intern Med 23(Suppl 1):57–63. doi:10.1007/s11606-007-0275-4
5. McCulloch P, Taylor I, Sasako M, Lovett B, Griffin D (2002) Randomised trials in surgery: problems and possible solutions. BMJ (Clin Res Ed) 324(7351):1448–1451
6. Grabowski H, Long G, Mortimer R (2014) Recent trends in brand-name and generic drug competition. J Med Econ 17(3):207–214. doi:10.3111/13696998.2013.873723
7. Raab GG, Parr DH (2006) From medical invention to clinical practice: the reimbursement challenge facing new device procedures and technology--part 1: issues in medical device assessment. J Am Coll Radiol JACR 3(9):694–702. doi:10.1016/j.jacr.2006.02.005
8. Ramsay CR, Grant AM, Wallace SA, Garthwaite PH, Monk AF, Russell IT (2001) Statistical assessment of the learning curves of health technologies. Health Technol Assess (Winch Eng) 5(12):1–79
9. Raman G, Gaylor JM, Rao M, Chan J, Earley A, Chang LKW, Salvi P, Lamont J, Lau J (2012) AHRQ methods for effective health care. In: Quality of reporting in systematic reviews of implantable medical devices (Prepared by Tufts Evidence-based Practice Center Under Contract No. 290-2007-10055-I), vol AHRQ Publication No. 12(13)-EHC116-EF. Agency for Healthcare Research and Quality (US), Rockville (MD)
10. Taylor RS, Iglesias CP (2009) Assessing the clinical and cost-effectiveness of medical devices and drugs: are they that different? Value Health J Int Soc Pharmacoecon Outcomes Res 12(4):404–406. doi:10.1111/j.1524-4733.2008.00476_2.x
11. Drummond M, Griffin A, Tarricone R (2009) Economic evaluation for devices and drugs – same or different? Value Health J Int Soc Pharmacoecon Outcomes Res 12(4):402–404. doi:10.1111/j.1524-4733.2008.00476_1.x
12. Project M. Review of International HTA Activities on Medical Devices Work Package 1 – Deliverable D 1.2, http://www.medtechta.eu/wps/wcm/connect/cff6d4bd-7ce6-40ea-8209-88e7528cdff4/MedtecHTA_D1.2.pdf?MOD=AJPERES

13. Pearson SD, Knudsen AB, Scherer RW, Weissberg J, Gazelle GS (2008) Assessing the comparative effectiveness of a diagnostic technology: CT colonography. Health Aff (Project Hope) 27(6):1503–1514. doi:10.1377/hlthaff.27.6.1503
14. E GBaT (2014) Advanced Diagnostics: Innovation, Reimbursement, And Coverage Challenges. In Vivo. https://www.pharmamedtechbi.com/publications/in-vivo?issue=Oct-28-2014. Accessed Last accessed 16 Nov 2015
15. Concato J, Shah N, Horwitz RI (2000) Randomized, controlled trials, observational studies, and the hierarchy of research designs. N Engl J Med 342(25):1887–1892. doi:10.1056/nejm200006223422507
16. Atkins D, Best D, Briss PA, Eccles M, Falck-Ytter Y, Flottorp S, Guyatt GH, Harbour RT, Haugh MC, Henry D, Hill S, Jaeschke R, Leng G, Liberati A, Magrini N, Mason J, Middleton P, Mrukowicz J, O'Connell D, Oxman AD, Phillips B, Schunemann HJ, Edejer T, Varonen H, Vist GE, Williams JW Jr, Zaza S (2004) Grading quality of evidence and strength of recommendations. BMJ Clin Res Ed 328(7454):1490. doi:10.1136/bmj.328.7454.1490
17. Excellence UKNIfC (2004) Guide to the methods of technology appraisal. https://www.gov.uk/government/uploads/system/uploads/attachment_data/file/191504/NICE_guide_to_the_methods_of_technology_appraisal.pdf. Accessed Last accessed 16 Nov 2015
18. Dreyer NA, Schneeweiss S, McNeil BJ, Berger ML, Walker AM, Ollendorf DA, Gliklich RE (2010) GRACE principles: recognizing high-quality observational studies of comparative effectiveness. Am J Manag Care 16(6):467–471
19. Berger ML, Mamdani M, Atkins D, Johnson ML (2009) Good research practices for comparative effectiveness research: defining, reporting and interpreting nonrandomized studies of treatment effects using secondary data sources: the ISPOR Good Research Practices for Retrospective Database Analysis Task Force Report – part I. Value Health J Int Soc Pharmacoecon Outcomes Res 12(8):1044–1052. doi:10.1111/j.1524-4733.2009.00600.x
20. AHRQ Methods for Effective Health Care (2014) In: Gliklich RE, Dreyer NA, Leavy MB (eds) Registries for evaluating patient outcomes: a user's guide. Agency for Healthcare Research and Quality (US), Rockville
21. Cox E, Martin BC, Van Staa T, Garbe E, Siebert U, Johnson ML (2009) Good research practices for comparative effectiveness research: approaches to mitigate bias and confounding in the design of nonrandomized studies of treatment effects using secondary data sources: the International Society for Pharmacoeconomics and Outcomes Research Good Research Practices for Retrospective Database Analysis Task Force Report – part II. Value Health J Int Soc Pharmacoecon Outcomes Res 12(8):1053–1061. doi:10.1111/j.1524-4733.2009.00601.x
22. Johnson ML, Crown W, Martin BC, Dormuth CR, Siebert U (2009) Good research practices for comparative effectiveness research: analytic methods to improve causal inference from nonrandomized studies of treatment effects using secondary data sources: the ISPOR Good Research Practices for Retrospective Database Analysis Task Force Report – part III. Value Health J Int Soc Pharmacoecon Outcomes Res 12(8):1062–1073. doi:10.1111/j.1524-4733.2009.00602.x
23. Dreyer NA, Velentgas P, Westrich K, Dubois R (2014) The GRACE checklist for rating the quality of observational studies of comparative effectiveness: a tale of hope and caution. J Manag Care Spec Pharm 20(3):301–308. doi:10.18553/jmcp.2014.20.3.301
24. Hartling L, McAlister FA, Rowe BH, Ezekowitz J, Friesen C, Klassen TP (2005) Challenges in systematic reviews of therapeutic devices and procedures. Ann Intern Med 142(12 Pt 2):1100–1111

Chapter 19
Evidence Generation Using Big Data: Challenges and Opportunities

Eberechukwu Onukwugha, Rahul Jain, and Husam Albarmawi

Abstract Big Data is defined as a large-volume dataset that is updated frequently and links data from a variety of sources, with the ability to derive unique and powerful insights from the linked datasets. Of note, a large sample size is not sufficient to characterize Big Data. With advancements in health information technology, Big Data provides opportunities for new insights regarding treatment effects. The use of Big Data in comparative effectiveness research (CER), including studies that examine heterogeneity of treatment effect, can expand our understanding of comparative effectiveness due to the availability of larger samples, with longer follow-up and richer measures. This chapter explores the advantages of using Big Data in CER, discusses the challenges related to analytics, and highlights the importance of translating evidence from CER. With appropriate attention to current lessons from CER, purposeful collection of theory-driven measures, and appropriate data linkages with minimal errors, the development and use of Big Data can support the conduct of CER for a diverse and evolving population.

19.1 Heterogeneity in Comparative Effectiveness Research and the Motivation for Big Data

Addressing heterogeneity is one of the key objectives in comparative effectiveness research (CER). According to the Institute of Medicine (IOM), the purpose of CER is to "… assist consumers, clinicians, purchasers, and policy makers to make informed decisions that will improve health care at both the individual and population levels" [1]. Additionally, one of the criteria that the National Institutes of Health

E. Onukwugha, PhD (✉) • H. Albarmawi, MS
Department of Pharmaceutical Health Services Research, University of Maryland School of Pharmacy, Baltimore, MD, USA
e-mail: eonukwug@rx.umaryland.edu

R. Jain, PhD
Boston Health Economics Inc, Waltham, MA, USA

© Springer Nature Singapore Pte Ltd. 2017
H.G. Birnbaum, P.E. Greenberg (eds.), *Decision Making in a World of Comparative Effectiveness Research*, DOI 10.1007/978-981-10-3262-2_19

(NIH) considers when prioritizing spending on CER studies is the emphasis on subgroups and population diversity [2]. This is because patient subgroups can exhibit different outcomes for the same treatment, and an estimate of the average treatment effect (ATE) for the whole population may not be sufficient as it will mask any variation in treatment effects across clinically important subgroups. Relying only on the ATE at the population level could also obscure evidence regarding qualitatively different treatment effects at the subgroup level and lead to policy decisions that prevent some patient subgroups from receiving the most effective health care or offer ineffective treatments to other subgroups [3, 4].

Heterogeneity of treatment effect (HTE) is a concept that refers to systematic variation in the direction or magnitude of outcomes among different patient subgroups receiving the same treatment [3–5]. Heterogeneity in outcomes may occur across subgroups defined by patients' demographics (such as age and gender), biologic factors (such as the presence of a genetic marker for a specific cancer), characteristics of the disease (acute vs. chronic disease), the presence of comorbidities, exposure to other treatments, geographic location, or other social and environmental factors [4–6].

Sources of heterogeneity in outcomes can also extend beyond those related to the patient and their physical environment, to include the health-care setting and providers' attributes [4]. For example, there is variation in the extent to which providers acknowledge the cultural and language barriers of patient populations and develop strategies to accommodate their needs. These strategies include enhancing staff awareness of cultural and language barriers and taking into consideration patient diversity when designing the physical space of the hospital [7]. To the extent that these considerations impact care delivery and patient engagement, they can lead to differences in observed process and health outcomes across patient groups defined at the provider level.

Subgroup analysis is a common method of addressing the heterogeneity of effect across patients in observational studies [5]. Using this approach, the researcher conducts the comparative effectiveness analysis in defined patient subgroups to determine whether the comparative results vary across the subgroups. In regression-based subgroup analysis, marginal analysis offers considerable flexibility to investigate and identify group differences [8–10]. Marginal analysis can be used to estimate the impact of a one-unit change in a covariate on the treatment outcome, while keeping all the other variables constant at predetermined values. This method reports the marginal effect (ME) in original units (absolute effect) as opposed to the relative effect (e.g., relative risks and odds ratios) estimated by other methods [8–11]. When the treatment effect is considered for distinct patient subgroups, using a correctly specified model (e.g., appropriate distribution, a correct specification, appropriate measures, etc.), the ME can be used to estimate the incremental difference in treatment effect between the patient subgroups. When the ME is utilized to account for sources of variation at the individual, geographical, and provider level, there are numerous intriguing opportunities to identify HTE across patient subgroups.

Traditionally, it has been challenging to address heterogeneity in CER studies because, on one hand, the sample sizes required to detect differences between heterogeneous subgroups are considerably larger than those needed to estimate the ATE [5] and, on the other hand, measures to identify and comprehensively investigate HTE typically are not available in datasets that are traditionally used for CER. Despite the strengths of marginal analysis for investigating HTE, these limitations restrict the application of marginal analysis in CER. This may change with the increasing availability of Big Data, which are large-volume datasets that are updated frequently and link data from a variety of sources.

Big Data can be defined using the 3V approach, which refers to volume, velocity, and variety [12]. Another "V" can be added to the definition, which refers to the value of new insight that is uniquely possible from Big Data [13, 14]. A large sample size is necessary but not sufficient to characterize a dataset as Big Data. Most observational CER studies typically use secondary data, where there is limited opportunity to define and then collect information about theorized factors of interest. As a result, information regarding potential sources of HTE may not be readily available. Big Data offer valuable advantages for generating evidence on CER by leveraging rich data sources that can provide patient demographic, clinical, lifestyle behavior (e.g., smoking status, physical activity metrics) and financial data (e.g., bankruptcy filing, employment status), among others. These linked datasets permit a holistic, contextual comparative analysis of health technologies.

19.2   Challenges and Opportunities with Big Data

Big Data presents many opportunities to consider HTE across patient-level and non-patient-level grouping factors. These opportunities, however, are not without potential challenges for decision makers; thus, we investigate both while highlighting the implications for CER.

19.2.1   New Opportunities with Big Data

To describe its potential, it is helpful to think of Big Data as a cuboid, where one axis may be considered as the number of people on which the information is available (long), the other as the length of time for which these individuals are followed (wide), and the third axis as different domains on which the data is available (deep).

A "long" dataset (i.e., large number of records) is often used to characterize Big Data; however, as was already noted, this is a necessary but not sufficient condition. Information on a larger set of individuals and/or patients provides flexibility in analyzing and presenting the data, e.g., administrative claims or registry data. It is important to emphasize that large sample size is not a substitute for correct model specifications (e.g., normality, a correct specification, appropriate measures, etc.). A large sample size does not reduce the bias introduced due to incorrect model specification, does not immunize the study against bias that arises from omitted relevant variables, and does not automatically increase the precision of the estimates.

Wider data, in our context, means following the same individual for a longer time. Depending on the study design, wide data provide either a longer patient history, follow-up, or both. Longer history provides flexibility in defining baseline health and the follow-up allows a longer-term evaluation of the impact of new interventions, which is especially important when analyzing patients with chronic diseases.

In our context, deeper data refers to the increasing number of domains along which measures are collected and available for research. Typically, in administrative claims data, any service or procedure that is covered by the payer is included in the data. Additionally, some basic demographic information is also available. Some of the administrative claims data from commercial plans do not include information on race and ethnicity that is available in public plans, e.g., Medicare and Medicaid. Typically, administrative claims data do not have information on clinical results (e.g., laboratory values), nor do they provide information on individuals' income, height and weight, or smoking history and individual health status (e.g., physical health, mental health, spiritual health, psychological health, nutrition) and context (e.g., physical and social environmental measures). Big Data, involving linkages across many domains, can extend the comparative effectiveness analysis beyond demographic and clinical factors to include behavioral, mental health, and other aspects of individual health that influence treatment and outcomes.

If Big Data is to support improved decision making, it must do more than provide large volume data; it must provide more appropriate, theory-driven, or conceptually grounded measures. It must allow researchers to develop models that substantively translate their conceptual frameworks into their empirical counterparts. Empirical analysis that ignores omitted variable bias is likely to have estimates that are biased (and the direction of the bias cannot be known a priori), and this bias does not shrink as the sample size increases [15].

Datasets with rich, contextual information are increasingly available through private sources, e.g., HealthCore's HIRD, which has the ability to link claims data with survey and electronic medical record (EMR) data, and HIRE—oncology which integrates clinical oncology with administrative claims data, Optum Labs' linked EMR and claims data, and IMS Health's suite of real-world evidence data products. The development of Big Data is supported through high-profile initiatives such as the National Institutes of Health Big Data to Knowledge (BD2K) program and the American Society of Clinical Oncology's CancerLINQ. As health information technology facilitates the continued integration of data platforms, linkable data will become more readily available from hospitals, outpatient physician offices, insurance companies, technology companies (e.g., those who sell wearable technology such as physical activity trackers), and a variety of information services companies that have access to vast amounts of client information through the various services they offer—including credit, consumer, and marketing services. Used appropriately, the linked datasets can provide new insights regarding CER and can facilitate comparisons of the estimated comparative benefit (or harm) across diverse patient subgroups [13].

As we consider these promising characteristics, it is important to keep in mind that Big Data do not necessarily overcome the challenges faced with datasets such as

EMRs, health-care claims, and disease registry data that are already available for conducting CER. To the extent that Big Data links these datasets to other (more or less reliable) sources of patient information, Big Data is not immune to missing, incomplete, and erroneous information. The next sections explore some of the key concerns with regard to the existing datasets and discuss their relevance for Big Data.

19.2.2   Missing Information

Missing information is a typical issue for Big Data [16]. Big Data involves linkages across different types of data, which may themselves exhibit distinct missing patterns. For example, a linked large-volume EMR dataset may have missing values because of human error, such as personnel forgetting to record patients' measurements. On the claims side, missing values will occur when there are coding errors or nonresponses. Moreover, the patterns of missing values over time may be complex in a way that is unique to Big Data (e.g., a linked large-volume, high-velocity EMR and claims dataset) due to the repeated and sometimes nonuniform updates. Missing data limit the outcomes and independent factors that can be studied and significantly reduce the population sample size [17].

Two main forms of missing information are relevant in the context of CER: unobserved measures and unobserved patients. Both factors will be examined in turn. Administrative claims data and EMRs are commonly used to compare the safety and effectiveness of health-care interventions [18, 19]. These are secondary sources of data, which means that the information is not collected for research purposes. Thus, information of interest to researchers, notably confounders, may be missing from these data. For example, smoking status and alcohol consumption, which are confounders of various exposures and outcomes, are almost always missing in administrative claims data [18, 19]. Other examples of missing measures include disease severity and over-the-counter medicine use [20]. Although linking claims data to EMRs enhances information related to patients' characteristics, relevant important information is still typically not available for all encounters [17]. In settings where multiple health-care providers exist, some patients may obtain health-care services from providers with independent systems, which precludes researchers from observing patients' full information [17]. When utilizing administrative claims data from a payer, it is important to recognize that information regarding services not covered by the payer, for which the patient pays entirely out of pocket, is also missing. For example, the use of over-the-counter medications is always missing from administrative claims data.

The other form of missing information is related to unobserved patients. Because secondary data are not generated unless patients utilize health-care services, individuals who do not access health-care services will be unobserved. These individuals can simply be insured nonusers or patients unable to access health-care services due to insurance restrictions or budget constraints [17]. Additionally, information regarding uninsured patients is typically unavailable in observational, secondary data sources. Baseline health status and patient demographics may differ between the categories (e.g., insured nonusers who do not need to access health services and insured nonusers who are not able to access health services) of unobserved patients in ways that could differentially impact treatment effectiveness or the generalizability of CER results. Big Data, linking additional information to these observational datasets, are not designed to address these limitations and may compound them if the linked datasets introduce a new category of unobserved patients (i.e., selection bias). For example, linking data from physical activity trackers excludes individuals who do not use the devices because they are unaware of these devices, unable to afford them, unable to use them due to activity restrictions, or otherwise unwilling to use a tracker. These missing individuals may differ in important ways from the non-users discussed above. The missing data may exclude specific populations from CER studies, which is likely to impact the generalizability of CER results.

19.2.3 *Erroneous Information*

Big Data are not immune to erroneous information or the inclusion of incorrect data. Data entry error is one source of errors in health-care datasets. In health-care systems, most data are entered manually, which makes the collected information vulnerable to human error. Data entry errors can result from direct mistakes in transferring the data to computers or from issues related to computer use [21]. For example, using different computer software packages leads to inconsistent data entry practices and scoring systems [21]. It has been suggested that electronic health-care data should be validated before its use in research [17] and this suggestion is no less applicable with Big Data that are used for CER.

Big Data includes large-volume, varied, and frequently updated datasets. Although linking these datasets can enrich the measures that are available for a CER study, it can be difficult to conduct these linkages. Compared to a one-time linkage between large-volume flat files (e.g., linking medical and pharmacy claims, linking claims and registry data), the linkages that are necessary with Big Data are considerably more complex due to the volume of data, variety of data sources, and the need to update the records. It will be important to understand how the accuracy and reliability of current data linkages evolve with the availability of Big Data. Ideally, the availability of Big Data will include concurrent attention to the reliability of data linkages across several data sources. Linking data related to the same individual from different sources provides integrated and more complete data about patients. However, it is often difficult to assess the success of the data linkage and, in particular, the impact of any non-matches on bias due to omission. A gold-standard method for assessing the success of data linkage allows for the comparison of the resulting linked data to an independent source that reports the outcome of interest [22]. Given the paucity of such sources of information, indirect methods, such as the "Duplicate Method," can be used to evaluate the accuracy of linkage [22, 23]. Several types of errors can occur during the linkage process, which leads to erroneous data and limits the utility of studies' results. These errors include failing to link data related to the same person or mistakenly linking unrelated data [24]. A systematic review conducted by Silveira and Artmann evaluated the accuracy of probabilistic record linkage as a process for linking health databases. The overall sensitivity and specificity of this linkage method ranged from 74% to 98% and 99% to 100%, respectively [22]. False negatives in linkage arise when records for the same individual are not matched due to missing or inaccurate records, and false positives arise when records for two separate individuals are linked as if they belonged to the same person [24]. In some cases, the linkage error is systematic.

Studies have illustrated the consequences of differential linkage error (e.g., differential linkage error by ethnic group or the systematic exclusion of vulnerable populations due to the use of poorly recorded identifiers) [25–29], where, for example, rankings of relative hospital performance were distorted due to differences in data quality between the sites that contributed study data [30]. Bohensky et al. (2010) identified several patient characteristics that are associated with incomplete data linkage, including patient age, gender, race, and socioeconomic and health status [24]. To the extent that these same factors are associated with health outcomes of CER studies, there is a potential for systematic bias in studies utilizing linked data, including Big Data. So, decision makers and stakeholders should be aware of the possibility of data linkage errors, the difficulty of assessing the accuracy of data linkage and the potential impact on omission. In the meantime, appropriate study design and attention to internal validity (e.g., measurement and construct validity) are still one of the best ways to reduce biases that can distort analyses (including subgroup analyses) designed to generate evidence on CER.

19.3   Translating Evidence for Decision Making

Drawing correct conclusions regarding the safety and effectiveness of health-care interventions requires data to be as accurate as possible. Moreover, systematic errors in data can lead to biased results, and decisions based on these results may lead to incorrect allocation of resources [31]. Big Data simplifies the process of evidence translation by providing substantially larger sample sizes and rich contextual data to generate evidence that is targeted to diverse patient populations. If the reliability and validity of data linkages and data quality improve with the availability of Big Data, these improvements in data quality generate confidence in the findings and will simplify the process of translating CER evidence. However, there is little guidance for decision makers on how to qualify data sources and evaluate their appropriateness for decisions that need to be made. As evident from the 2020 strategic plan of the International Society for Pharmacoeconomics and Outcomes Research (ISPOR), there is recognition of the need to provide guidance on qualifying data sources; thus, it is reasonable to expect specific guidance on this topic in the near future [32].

In the meantime, decision makers are faced with translating evidence into practice. In doing so, they first have to determine when the evidence is sufficient and then if/how to utilize the evidence. Case studies examined by the Office of the

Assistant Secretary for Planning and Evaluation (ASPE) led to the conclusion that CER studies might not produce clear-cut evidence even when they are well designed and properly conducted [33]. The absence of clear-cut evidence can lead to controversies when interpreting the results from CER studies [33].

Some organizations have taken joint steps to provide decision makers with guidance on how to assess the reliability and validity of CER studies. For example, ISPOR, the Academy of Managed Care Pharmacy, and the National Pharmaceutical Council formed a CER Collaborative to support greater uniformity and transparency among decision makers in the evaluation and use of CER evidence for decisions such as formulary coverage decisions. Among other tools, the CER Collaborative makes available online tools and training documents to help decision makers determine the reliability and validity of CER evidence for their specific decision setting. The CER Collaborative also provides an evidence rating matrix to guide the evaluation of comparative studies to support formulary decision making [34]. Two factors are included in the matrix: certainty of evidence and comparative net health benefit. Certainty of evidence is related to multiple factors including the amount of evidence, potential bias associated with study design or conduct, directness of evidence (e.g., surrogate outcomes, indirect comparisons), duration of study, and whether it allows for important benefits and adverse effects to be captured, as well as results' precision, consistency, and generalizability.

According to the ASPE report on the dissemination and adoption of CER findings [33], the translation of comparative evidence into practice occurs in five stages: (1) generation, which comprises CER study design and completion; (2) interpretation, which refers to giving meaning to the obtained results from the CER study; (3) formalization, which means converting the obtained results into guidelines for practice; (4) dissemination, which refers to actively conveying the results and the guidelines generated from CER studies to stakeholders; and (5) implementation, which refers to the application of the guidelines produced from the CER study in clinical practice. It is worth mentioning that these phases can overlap and that various stakeholders are involved in each of these stages.

The ASPE report provides real-world examples of the stages of CER evidence translation. One example is that of the Comparison of Medical Therapy, Pacing, and Defibrillation in Heart Failure (COMPANION) trial, which evaluated the effectiveness of resynchronizing the heart in patients with moderate to severe heart failure. The results of the trial indicated that heart synchronization in these patients improves survival and functional status [35]. In the generation phase of evidence translation, there were no major disagreements related to the study design and conduct. However, there were issues related to the generalizability of the results in the interpretation phase. Nevertheless, the findings were formalized quickly into guidelines. Different stakeholders participated in the dissemination of the results, including industry, continuing medical education (CME), registries, and, to a lesser extent, cardiology societies. Several factors affected the implementation of the guidelines, including reimbursement, referral to interventionists, and health-care setting. This case study, along with others in the ASPE report, can be used by decision makers to identify and anticipate some of the challenges that arise with translating CER evidence.

19.4 Conclusion

Big Data offer new opportunities to generate evidence on comparative effectiveness, explore heterogeneity of treatment effect across patient subgroups, and leverage existing measures such as the marginal effect to report identified subgroup effects. In looking forward to the widespread availability of Big Data, we should look back to the lessons learned from our experiences with generating CER using primarily administrative data, namely, the importance of a strong study design, the need to investigate subgroups, the advantages of useful and reliable measures, the difficulties that arise from missing or inaccurate data, and the continued need for guidance on translating evidence for decision making. We realize the full potential of Big Data for CER as we leverage each of these lessons from the past, anticipate challenges that Big Data introduces, and utilize its length (sample size), width (follow-up), and depth (relevant factors) to generate timely evidence for a diverse and evolving patient population.

References

1. Sox HC, Greenfield S (2009) Comparative effectiveness research: a report from the Institute of Medicine. Ann Intern Med 151(3):203–205
2. Lauer MS, Collins FS (2010) Using science to improve the nation's health system: NIH's commitment to comparative effectiveness research. JAMA 303(21):2182–2183
3. Kravitz RL, Duan N, Braslow J (2004) Evidence-based medicine, heterogeneity of treatment effects, and the trouble with averages. Milbank Q 82(4):661–687. doi:10.1111/j.0887-378X.2004.00327.x
4. Sox HC, Goodman SN (2012) The methods of comparative effectiveness research. Annu Rev Public Health 33:425–445. doi:10.1146/annurev-publhealth-031811-124610
5. Velentgas P, Dreyer NA, Nourjah P, Smith SR, Torchia MM, eds (2013) Developing a Protocol for Observational Comparative Effectiveness Research: A User's Guide. AHRQ Publication No. 12(13)-EHC099. Rockville, MD: Agency for Healthcare Research and Quality
6. Gomez SL, Shariff-Marco S, DeRouen M, Keegan TH, Yen IH, Mujahid M, Satariano WA, Glaser SL (2015) The impact of neighborhood social and built environment factors across the cancer continuum: current research, methodological considerations, and future directions. Cancer 121(14):2314–2330. doi:10.1002/cncr.29345
7. Wilson-Stronks A, Commission J (2008) One size does not fit all: meeting the health care needs of diverse populations. Joint Commission, Oakbrook Terrace
8. Karaca-Mandic P, Norton EC, Dowd B (2012) Interaction terms in nonlinear models. Health Serv Res 47(1 Pt 1):255–274. doi:10.1111/j.1475-6773.2011.01314.x
9. Ai C, Norton EC (2003) Interaction terms in logit and probit models. Econ Lett 80(1):123–129
10. Onukwugha E, Bergtold J, Jain R (2015) A primer on marginal effects–part I: theory and formulae. Pharmacoeconomics 33(1):25–30. doi:10.1007/s40273-014-0210-6

11. Verlinda JA (2006) A comparison of two common approaches for estimating marginal effects in binary choice models. Appl Econ Lett 13(2):77–80
12. Laney D (2001) 3D data management: controlling data volume, velocity and variety. META Group Res Note 6:70
13. Gray EA, Thorpe JH (2015) Comparative effectiveness research and big data: balancing potential with legal and ethical considerations. J Comp Eff Res 4(1):61–74. doi:10.2217/cer.14.51
14. Onukwugha E (2016) Big data and its role in health economics and outcomes research: a collection of perspectives on data sources, measurement, and analysis. Pharmacoeconomics 34(2):91–93. doi:10.1007/s40273-015-0378-4
15. Hill RC, Griffiths WE, Lim GC (2008) Principles of econometrics, vol 5. Wiley, Hoboken
16. Zhang Z (2015) Missing values in big data research: some basic skills. Annals of translational medicine 3(21): 323
17. Bayley KB, Belnap T, Savitz L, Masica AL, Shah N, Fleming NS (2013) Challenges in using electronic health record data for CER: experience of 4 learning organizations and solutions applied. Med Care 51:S80–S86
18. Strom BL, Kimmel SE, Hennessy S (2013) Textbook of pharmacoepidemiology, 2nd edn. Wiley Online Library, Chichester
19. Toh S, Garcia Rodriguez LA, Hernan MA (2012) Analyzing partially missing confounder information in comparative effectiveness and safety research of therapeutics. Pharmacoepidemiol Drug Saf 21 Suppl 2(S2):13–20. doi:10.1002/pds.3248
20. Suissa S, Garbe E (2007) Primer: administrative health databases in observational studies of drug effects–advantages and disadvantages. Nat Clin Pract Rheumatol 3(12):725–732. doi:10.1038/ncprheum0652
21. de Lusignan S, Liaw ST, Krause P, Curcin V, Vicente MT, Michalakidis G, Agreus L, Leysen P, Shaw N, Mendis K (2011) Key concepts to assess the readiness of data for international research: data quality, lineage and provenance, extraction and processing errors, traceability, and curation. Contribution of the IMIA Primary Health Care Informatics Working Group. Yearb Med Inform 6(1):112–120
22. Silveira DP, Artmann E (2009) Accuracy of probabilistic record linkage applied to health databases: systematic review. Rev Saude Publica 43(5):875–882
23. Blakely T, Salmond C (2002) Probabilistic record linkage and a method to calculate the positive predictive value. Int J Epidemiol 31(6):1246–1252
24. Bohensky MA, Jolley D, Sundararajan V, Evans S, Pilcher DV, Scott I, Brand CA (2010) Data linkage: a powerful research tool with potential problems. BMC Health Serv Res 10(1):346. doi:10.1186/1472-6963-10-346
25. Brenner H, Schmidtmann I, Stegmaier C (1997) Effects of record linkage errors on registry-based follow-up studies. Stat Med 16(23):2633–2643
26. Coeli CM, Barbosa FS, Brito AS, Pinheiro RS, Camargo KR Jr, Medronho RA, Bloch KV (2011) Estimativas de parâmetros no linkage entre os bancos de mortalidade e de hospitalização, segundo a qualidade do registro da causa básica do óbito. Cad Saude Publica 27(8):1654–1658. doi:10.1590/s0102-311x2011000800020
27. DuVall SL, Fraser AM, Rowe K, Thomas A, Mineau GP (2012) Evaluation of record linkage between a large healthcare provider and the Utah Population Database. J Am Med Inform Assoc 19(e1):e54–e59. doi:10.1136/amiajnl-2011-000335
28. Lariscy JT (2011) Differential record linkage by Hispanic ethnicity and age in linked mortality studies: implications for the epidemiologic paradox. J Aging Health 23(8):1263–1284. doi:10.1177/0898264311421369
29. Lawrence D, Christensen D, Mitrou F, Draper G, Davis G, McKeown S, McAullay D, Pearson G, Zubrick SR (2012) Adjusting for under-identification of Aboriginal and/or Torres Strait Islander births in time series produced from birth records: using record linkage of survey data and administrative data sources. BMC Med Res Methodol 12:90. doi:10.1186/1471-2288-12-90
30. Gibbs JL, Cunningham D, de Leval M, Monro J, Keogh B (2005) Paediatric cardiac surgical mortality after Bristol: paediatric cardiac hospital episode statistics are unreliable. BMJ (Clinical research ed) 330(7481):43–44 . doi:10.1136/bmj.330.7481.43-cauthor reply 44

31. Adler-Milstein J, Jha AK (2013) Healthcare's "big data" challenge. Am J Manag Care 19(7):537
32. Research ISFPO (2011) ISPOR VISION 2020. International Society for Pharmacoeconomics and Outcomes Research. http://www.ispor.org/vision2020.asp. Accessed 11 May 2016
33. Schneider EC, Timbie JW, Fox DS, Van Busum K, Caloyeras J (2011) Dissemination and adoption of comparative effectiveness research findings when findings challenge current practices. RAND Corporation, Santa Monica
34. Ollendorf D, Pearson SD (2013) ICER Evidence Rating Matrix: a user's guide. Available at: http://www.icer-review.org/wp-content/uploads/2008/03/Rating-Matrix-User-Guide-FINAL-v10-22-13.pdf. Accessed 11 May 2016
35. Bristow MR, Saxon LA, Boehmer J, Krueger S, Kass DA, De Marco T, Carson P, DiCarlo L, DeMets D, White BG, DeVries DW, Feldman AM, Comparison of Medical Therapy P, Defibrillation in Heart Failure I (2004) Cardiac-resynchronization therapy with or without an implantable defibrillator in advanced chronic heart failure. N Engl J Med 350(21):2140–2150. doi:10.1056/NEJMoa032423

Chapter 20
Indirect Comparisons: A Brief History and a Practical Look Forward

James Signorovitch and Jie Zhang

Abstract Indirect comparisons – the comparison of treatments across separate clinical trials – have become increasingly used over the past decade to inform healthcare decision making, especially for drug access and reimbursement. As a research tool, indirect comparisons have undergone a remarkable evolution. We provide a brief account of their development, highlight recent advances and long-standing challenges, and discuss how the use of indirect comparisons may continue to evolve in the near future.

20.1 Introduction

Healthcare decisions are ideally informed by comparative benefits, risks, and costs. For new drugs, the regulatory approval process focuses on measuring benefits and risks, usually in reference to a placebo or standard of care. Once a drug is approved, physicians and payers must answer broad questions about its clinical and economic value. Which of many eligible drug treatments should an individual patient receive? What price should be paid for a newly approved drug? The answers to these questions have significant consequences for individuals and populations. At the same time, the comparative evidence available to inform these decisions is often limited.

The gold standard for comparative evidence is the randomized controlled trial (RCT). While most new drug treatments will have been compared with a placebo or standard of care in randomized trials, randomized comparisons are rarely available for all pairs of approved treatments. Standards of care may also vary among geographic regions and across time. Furthermore, many innovative and urgently needed treatments, especially for rare diseases and late-stage oncology indications, receive

J. Signorovitch (*)
Analysis Group, Inc., Boston, MA, USA
e-mail: James.Signorovitch@analysisgroup.com

J. Zhang
Novartis, East Hanover, NJ, USA
e-mail: Jie.Zhang@novartis.com

© Springer Nature Singapore Pte Ltd. 2017
H.G. Birnbaum, P.E. Greenberg (eds.), *Decision Making in a World of Comparative Effectiveness Research*, DOI 10.1007/978-981-10-3262-2_20

accelerated approvals without any randomized comparison. For all of these reasons, payers and physicians are frequently left without gold-standard, randomized evidence for large numbers of decisions and must strive to make timely decisions based on the best available evidence.

Payers have taken multiple approaches to making reimbursement decisions with limited evidence. One collection of approaches seeks to condition reimbursement, in one way or another, on future evidence. The broad role of risk-sharing schemes and outcome-based contracts has been reviewed by Nazareth et al. [1]. The granting of reimbursement conditional on ongoing real-world evidence generation, as enacted for cancer drugs by the National Health Service in the United Kingdom (UK), is a recent example [2].

Whether or not future evidence generation is formally required for reimbursement, there is a need to maximize the value of currently available evidence. In this chapter, we consider so-called indirect comparisons, a collection of research methods that seek to fill gaps in direct comparative evidence from randomized trials by comparing treatments indirectly across separate clinical trials. These methods have become increasingly used over the past decade, as indicated by exponential growth in the number of peer-reviewed publications per year (Fig. 20.1). Indirect comparisons are now a standard component of the evidence developed to inform access and reimbursement discussions for pharmaceutical products in many countries. A total of 28 jurisdictions across the globe were identified as accepting of indirect evidence in a recent review [3]. As a means of understanding the current environment for indirect comparisons, we take a look back at how use of these methods has evolved over time, along with their practical strengths and limitations. We then survey recent developments and expectations for the future role of indirect evidence in decision making.

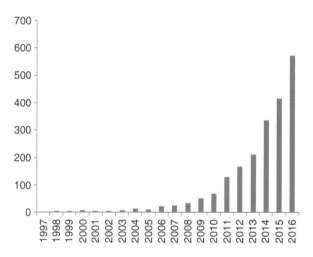

Fig. 20.1 Published indirect comparisons by year (Based on a PubMed search for "indirect comparison" or "mixed treatment comparison" or "network meta-analysis" up to November 2016. The count for 2016 was extrapolated to cover the full year)

20.2   The Evolution of Indirect Comparisons

Since indirect comparisons involve comparisons between nonrandomized treatment groups – i.e., a departure from the gold standard – they must be approached with caution. The indirect comparisons that are conducted today have a long prehistory. An early consideration of the potential value – and the potential perils – of indirect comparisons was made in a 1976 article by Stuart Pocock, which presented criteria for evaluating comparisons to external controls (essentially indirect comparisons) including similarities in control group treatments, patient characteristics, inclusion criteria, methods for outcome evaluation, time periods, study centers, and investigators [4]. These criteria remain central to all indirect comparisons, regardless of statistical methodologies. When trials differ on these criteria, cross trial comparisons of outcomes are at risk of bias. By categorizing these sources of potential bias, Pocock's criteria help guide the assessment of whether a comparison across separate trials would be apples to apples or apples to oranges. Expanding on Pocock's 1976 criteria, a non-exhaustive list of trial characteristics that need to be evaluated to assess suitability for indirect comparisons is presented in Table 20.1.

Table 20.1 Areas for evaluating trial similarity and suitability for indirect comparisons

Category	Examples
Trial design and conduct	Blinding and concealment Duration of follow-up Run-in periods Dropout procedures Allowance for crossovers Study managers and investigators
Trial settings	Diagnostic criteria Year(s) of trial Geography Health system Standards of care Study centers
Inclusion criteria and patient characteristics	Demographics Disease severity and duration Biomarkers Treatment history Concomitant therapies Comorbidities Inclusion/exclusion criteria
Treatments	Dose and schedule Duration Background therapies Subsequent therapies
Outcome measures	Definition Assessment criteria Assessment frequency Handling of missing data

20.2.1 Anchor-Based Approaches

A significant development in the evolution of indirect comparisons was the advent of "anchor-based" approaches, as described in the seminal paper of Bucher et al. (1997) [5]. These approaches introduce an assumption that, if reasonable, allows a relaxation of Pocock's criteria. Instead of requiring that trial designs and patient populations are nearly identical, an anchor-based approach makes a weaker requirement: cross-trial differences are allowed as long as they do not impact relative treatment effects measured against a common comparator treatment (i.e., the anchor). For example, if one trial has compared drug A vs. placebo and another trial has compared drug B vs. placebo, an anchor-based analysis would compare effects measured relative to placebo between the trials to estimate the effect of drug A vs. drug B (Fig. 20.2). The trials may differ, for example, in baseline disease severity, but as long as the effect of drug A vs. placebo and the effect of drug B vs. placebo are each invariant to baseline disease severity, that baseline difference will not bias the anchor-based comparison of drugs A and B. Bucher et al. (1997) [5] describe the theoretical basis for such anchor-based comparisons, the assumptions under which they can provide unbiased comparisons, and the statistical formulae needed for measuring uncertainty. These tools have enabled broader scientific use of indirect comparisons and laid the groundwork for additional development. Assessing cross-trial similarity remains a critical step in conducting indirect comparisons, but cross-trial differences can be tolerated if there is reason to believe that they do not modify relative treatment effects. Even when trials appear similar, however, the potential for confounding due to unobserved differences must be acknowledged.

20.2.2 Evidence Networks

The next round of advancement began in the early 2000s and enabled a dramatic expansion in use of indirect comparisons. Strong demand from decision makers and the global development of health technology assessment (HTA) authorities [6] were important drivers of this growth. Methodological advances also improved the utility of indirect evidence by enabling synthesis of all available evidence, from multiple treatments into a single, unified analysis [7–11]. In brief, these new developments

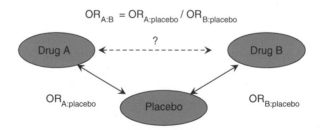

$$OR_{A:B} = OR_{A:placebo} / OR_{B:placebo}$$

Fig. 20.2 An anchor-based indirect comparison (the anchor-based indirect OR comparing A vs. B is estimated as the ratio of ORs comparing each drug to placebo. *OR* odds ratio)

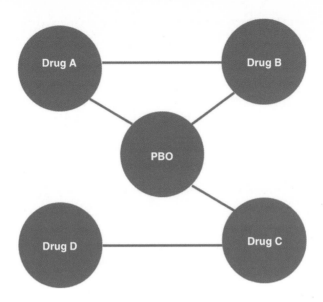

Fig. 20.3 An evidence network (links between treatments indicate the presence of head-to-head comparisons in a randomized trial) (*PBO* placebo)

extended the anchor-based framework, which links two treatments through a single common comparator, to larger "evidence networks" in which multiple treatments are connected through one or more common comparator treatments, i.e., treatments that are shared across head-to-head randomized trials. For example, if we had randomized trials of drug A vs. placebo, B vs. placebo, and C vs. placebo, treatments A, B, and C can be linked through placebo as the common anchor. A head-to-head randomized trial of A vs. B could also be added to the network, with links to drug A and drug B arms in the placebo-controlled trials. A trial of drug C vs. drug D could also be included (Fig. 20.3). With a connected network of treatments, the relative effects of any pair of included treatments can be estimated in a "network meta-analysis." The key assumption underlying network meta-analysis is that the included trials are sufficiently similar, such that any cross-trial differences do not impact the relative treatment effects. As with all anchor-based comparisons, assessment of cross-trial similarity, including the types of factors tabulated in Table 20.1, is an important first step. With greater numbers of trials and treatments included in an evidence network, the potential for cross-trial heterogeneity may increase and needs to be carefully considered. Approaches for mitigating cross-trial differences are referenced in subsequent sections.

20.2.3   Guidance and Evaluations

The development of multiple review articles and guidance documents has been extremely valuable to the advancement of indirect comparisons. The guidance developed by the International Society for Pharmacoeconomics and Outcomes

Research (ISPOR) has been valuable in disseminating consensus on good practices for conduct and reporting of indirect comparisons [12, 13]. The guidance issued or commissioned by HTAs, such as the National Authority for Health [14] in France, the Pharmaceutical Benefits Advisory Committee [15] in Australia, the Scottish Medicines Consortium [16] in Scotland, the Institute for Quality and Efficiency in Health Care [17] in Germany, Canadian Agency for Drugs and Technologies in Health [18] in Canada, the Health Insurance Review and Assessment Service [19] in South Korea, as well as guidance issued by the European Network for Health Technology Assessment [20], has been valuable for the preparation of submissions. The methodological guidance developed by the UK's National Institute for Health and Care Excellence Decision Support Unit is especially comprehensive and detailed [21–23].

A wealth of reviews has also been published, surveying and critiquing published indirect comparisons [24, 25] and explaining the role of indirect evidence and different methodologies to broader audiences [26, 27]. Readers interested in a deeper exposition of indirect comparison methodologies and guidelines for use and interpretation have many resources to draw on.

In an illuminating set of studies, Song et al. have taken an empirical approach to evaluating the reliability of indirect comparisons [28, 29]. By constructing over 100 sets of comparisons that included both indirect evidence and direct evidence from randomized trials, Song et al. estimated that 14% of indirect comparisons exhibit significant divergence from the available randomized comparative evidence. Analyses that included smaller numbers of trials subjectively assessed outcomes, and significant treatment effects were at higher risk of divergence.

20.2.4   Adjusting for Cross-Trial Differences

While anchor-based indirect comparisons and network meta-analyses are robust to cross-trial differences that do not impact relative treatment effects, one cannot always be certain whether a particular cross-trial difference is a potential source of bias. Several approaches have been developed to adjust indirect comparisons for cross-trial differences. One approach is meta-regression, which adjusts for average differences in baseline characteristics within a network meta-analysis [23]. When individual patient data are available from at least some of the trials, meta-regression adjustment can be combined with regression adjustments that utilize the individual patient data [30–32]. This can enable more precise adjustment and is preferred when possible, but access to individual patient data is often limited. Simulated treatment comparisons (STCs) constitute a special case of network meta-analyses with regression adjustment that uses available patient-level data [33].

Another approach to adjusting for cross-trial differences is the matching-adjusted indirect comparison (MAIC) [34, 35]. This approach uses individual patient data from trials of one treatment, but does not require individual patient data from all treatments. The trials with individual patient data are adjusted via propensity score weighting to match the average baseline characteristics reported for comparator patients. Treatment outcomes can then be compared across balanced populations. MAICs have been used in multiple HTA submissions, either to complement unadjusted anchor-based comparisons or to provide evidence when anchor-based comparisons were not feasible [36]. MAICs have been particularly useful for comparisons of time-to-event outcomes [33, 37, 38] and binary response outcomes, which can present complications for regression-based approaches that combine patient-level and population-level evidence. The strengths and limitations of MAIC and STC have recently been reviewed [33] and parallel the strengths and limitations of general regression-based and propensity score-based approaches to adjusted analyses. Both approaches assume that unobserved cross-trial differences have not led to bias.

20.2.5 Single-Arm Trials

While RCTs are the gold standard for comparative evidence, they are not always feasible or ethical. Nonrandomized, single-arm trials are increasingly common, especially for accelerated approvals in rare diseases and late-stage oncology indications. Over 50% of the Food and Drug Administration accelerated approvals and 20% of European Medicines Agency oncology approvals (in the past 20 years) have been based on single-arm trials [39, 40]. These settings present significant challenges for indirect comparisons, since conventional anchor-based analyses cannot be applied to single-arm studies. This leaves what are often urgently needed therapies that address areas of high unmet need with a less clear pathway for generating the necessary evidence to support access and reimbursement. Over the last few years, non-anchor-based indirect comparisons have been increasingly applied in these settings. Rather than relying on relative effect measures to account for cross-trial differences, these approaches rely on extensive adjustment for patient baseline characteristics. In this respect, they are more akin to traditional observational studies (e.g., with adjustment based on multivariable regression or propensity scores) and to comparisons with historical or external control groups than to meta-analyses. MAIC in particular has been frequently used in this setting, for example, for indirect comparisons of single-arm oncology trials [37] and for breakthrough treatments for hepatitis C virus [41]. MAIC has also been used to link single-arm trials into a broader evidence network which can then be analyzed using network meta-analysis.

The absence of a common comparator arm presents important limitations for indirect comparisons, but should not be disqualifying. The quality of any particular analysis must be assessed based on the cross-trial similarity and the adequacy of adjustment for observed cross-trial differences, and the potential for confounding due to unobserved differences – essentially returning to an evaluation of Pocock's (1976) criteria.

20.3   Moving Forward

Indirect comparisons have continued to evolve. Below we summarize a few areas of active development and potential future growth.

20.3.1   *Comparing Outcomes in Real-World Practice vs. Clinical Trials*

An important limitation of indirect comparisons based on clinical trials is that, like the underlying clinical trials themselves, they may not reflect real-world outcomes. This can decrease the relevance of indirect comparisons for new healthcare technologies for which benefits and risks can differ between real-world practice and protocol-driven settings. Technologies that improve adherence are an important example and include long-acting formulations, smart pill bottles, ingestible microchips, and implantable drug dispensation devices. In this volume the value of pragmatic trials for measuring real-world benefits, and the process for conducting such trials, is reviewed by Hung et al. [42]. The growing information technology capabilities of healthcare payers and providers can lower the barriers for pragmatic trials and could increase the role of such evidence in reimbursement decisions. Best practices for indirect comparisons and network meta-analyses that incorporate pragmatic trials have yet to be developed. While the existing theory for indirect comparisons will be largely applicable, the potential for new sources of heterogeneity among pragmatic trials is likely to present important challenges. Given the importance of real-world evidence for the ultimate value of new healthcare technologies, the incorporation of pragmatic trials into indirect comparisons, if possible, will be a welcome development.

20.3.2   *The Role of Indirect Evidence in the United States (USA)*

While HTA authorities outside the USA have been documenting and systematizing preferred approaches to indirect comparisons, the response to indirect evidence in the USA has been more ad hoc. With the recent growth in interest in value-based frameworks and the increasing activity of groups that systematically conduct relative-value assessments in the USA (e.g., the cancer DrugAbacus [43] and the Institute for Clinical and Economic Review), there is the potential for further increases in expectations for indirect evidence among decision makers in the USA. The extent to which a systematic appraisal process develops in the USA, whether driven by payers themselves or by third-party organizations, will be interesting to observe in the coming years.

20.3.3 Assessing the Consistency and Reliability of Indirect Evidence

An important development that we would expect over the next several years is a systematic evaluation of past indirect comparisons. With over 10 years of published indirect analyses by now, the opportunities to systematically validate published indirect comparisons against data from head-to-head randomized trials should increase. Since indirect comparisons are often used to inform value assessments and reimbursement, a quantitative assessment of the degree of similarities and differences between direct and indirect evidence will be important, beyond testing for statistically significant differences. These types of evaluations will be tremendously important for benchmarking the overall reliability of indirect evidence and for identifying features that predict risk of bias.

20.3.4 Broader Access to Patient-Level Data

With data sharing initiatives gaining traction, it is possible that an increasing number of indirect comparisons will have the opportunity to access individual patient data [44, 45]. In this environment, further development of methods for incorporating individual-level data into indirect comparisons will be valuable. Just as important will be the development of understanding and guidance on when and how individual patient data should be used. Though individual patient data is generally preferred (it is difficult to argue against having more data), the more complicated modeling required for using individual-level data can significantly increase the time and resources required for analyses and can decrease the transparency of the results to reviewer if all data are not accessible to them. Potential biases may also arise from cross-trial differences in the availability of individual-level data and of specific data elements. In the context of appraisals, this presents important practical and scientific challenges.

20.4 Summary

Indirect comparisons have achieved rapid growth in use, driven by intense demand for comparative evidence, the development of well-suited methodologies, and the extensive and timely development of guidance. Expertise in the conduct and evaluation of indirect evidence has become essential for all stakeholders involved in drug access and reimbursement negotiations, especially for HTA jurisdictions that issue explicit guidance and requirements. With the continued development of new treatments and continued interest in managing drug spending, the need for indirect evidence is expected to continue to grow, particularly if more markets develop systematized approaches to drug evaluation. Broader availability of patient-level data in coming years will present interesting opportunities and challenges for indirect comparisons – but overall access to more data should, if used appropriately, help improve the quality of indirect evidence.

The opinions expressed in this article are those of the contributors, who are solely responsible for its content.

References

1. Nazareth T, Ko J, Frois C, Carpenter S, Demean S, Wu E, Sasane R, Navarro R (2015) Outcomes-based pricing and reimbursement arrangements for pharmaceutical products in the U.S. and EU-5: payer and manufacturer experience and outlook. Poster presented at the AMCP NEXUS, Orlando. 26–29 Oct 2015
2. https://www.england.nhs.uk/wp-content/uploads/2013/04/cdf-sop.pdf
3. Kleijnen S, George E, Goulden S, d'Andon A, Vitre P, Osinska B, Rdzany R, Thirstrup S, Corbacho B, Nagy BZ, Leufkens HG, de Boer A, Goettsch WG (2012) Relative effectiveness assessment of pharmaceuticals: similarities and differences in 29 jurisdictions. Value Health J Int Soc Pharmacoecon Outcomes Res 15(6):954–960. doi:10.1016/j.jval.2012.04.010
4. Pocock SJ (1976) The combination of randomized and historical controls in clinical trials. J Chronic Dis 29(3):175–188
5. Bucher HC, Guyatt GH, Griffith LE, Walter SD (1997) The results of direct and indirect treatment comparisons in meta-analysis of randomized controlled trials. J Clin Epidemiol 50(6):683–691
6. Xie J, Chalkidou K, Kamae I, Dittrich RE, Mahbub R, Vasan A, Metallo C (2017) Policy considerations: Ex-U.S. Payers and regulators. In: Birnbaum HG, Greenberg PE (eds) Decision making in a World of comparative effectiveness research. Springer, Singapore
7. Lumley T (2002) Network meta-analysis for indirect treatment comparisons. Stat Med 21(16):2313–2324. doi:10.1002/sim.1201
8. O'Regan C, Ghement I, Eyawo O, Guyatt GH, Mills EJ (2009) Incorporating multiple interventions in meta-analysis: an evaluation of the mixed treatment comparison with the adjusted indirect comparison. Trials 10:86. doi:10.1186/1745-6215-10-86
9. Glenny AM, Altman DG, Song F, Sakarovitch C, Deeks JJ, D'Amico R, Bradburn M, Eastwood AJ (2005) Indirect comparisons of competing interventions. Health Technol Assess (Winchester, England) 9(26):1–134 iii–iv
10. Lu G, Ades AE (2004) Combination of direct and indirect evidence in mixed treatment comparisons. Stat Med 23(20):3105–3124. doi:10.1002/sim.1875
11. Lu G, Ades AE (2006) Assessing evidence inconsistency in mixed treatment comparisons. J Am Stat Assoc 101(474):447–459. doi:10.1198/016214505000001302 http://dx.doi.org/10.1198/016214505000001302
12. Hoaglin DC, Hawkins N, Jansen JP, Scott DA, Itzler R, Cappelleri JC, Boersma C, Thompson D, Larholt KM, Diaz M, Barrett A (2011) Conducting indirect-treatment-comparison and network-meta-analysis studies: report of the ISPOR Task Force on Indirect Treatment Comparisons Good Research Practices: part 2. Value Health J Int Soc Pharmacoecon Outcomes Res 14(4):429–437. doi:10.1016/j.jval.2011.01.011
13. Jansen JP, Fleurence R, Devine B, Itzler R, Barrett A, Hawkins N, Lee K, Boersma C, Annemans L, Cappelleri JC (2011) Interpreting indirect treatment comparisons and network meta-analysis for health-care decision making: report of the ISPOR Task Force on Indirect Treatment Comparisons Good Research Practices: part 1. Value Health J Int Soc Pharmacoecon Outcomes Res 14(4):417–428. doi:10.1016/j.jval.2011.04.002

14. http://www.has-sante.fr/portail/upload/docs/application/pdf/2011-02/summary_report__indi-rect_comparisons_methods_and_validity_january_2011_2.pdf
15. http://www.pbs.gov.au/industry/useful-resources/pbac-technical-working-groups-archive/indirect-comparisons-working-group-report-2008.pdf
16. https://www.scottishmedicines.org.uk/files/about/Indirect_comparison_and_mixed_treat-ment_comparison_checklist_final_V4_021013.doc
17. https://www.iqwig.de/download/Joint_Statement_Indirect_Comparisons.pdf
18. https://www.cadth.ca/sites/default/files/pdf/MH0003_Guidance_on_IDC_Reporting.pdf
19. Bae S, Lee S, Bae EY, Jang S (2013) Korean guidelines for pharmacoeconomic evaluation (second and updated version): consensus and compromise. Pharmacoeconomics 31(4):257–267. doi:10.1007/s40273-012-0021-6
20. Guideline: comparators & comparisons: direct and indirect comparisons. EUnetHTA network. (November 2015)
21. Dias S, Welton NJ, Sutton AJ, Ades AE (2014) NICE decision support unit technical support documents. In: A generalised linear modelling framework for pairwise and network meta-analysis of randomised controlled trials. National Institute for Health and Care Excellence (NICE) unless otherwise stated. All rights reserved., London
22. Dias S, Welton NJ, Sutton AJ, Caldwell DM, Lu G, Ades AE (2013) Evidence synthesis for decision making 4: inconsistency in networks of evidence based on randomized controlled trials. Med Decis MakingInt J Soc Med Decis Mak 33(5):641–656. doi:10.1177/0272989x12455847
23. Dias S, Sutton AJ, Welton NJ, Ades A (2013) Evidence synthesis for decision making 3 heterogeneity—subgroups, meta-regression, bias, and bias-adjustment. Med Decis Making 33(5):618–640
24. Kim H, Gurrin L, Ademi Z, Liew D (2014) Overview of methods for comparing the efficacies of drugs in the absence of head-to-head clinical trial data. Br J Clin Pharmacol 77(1):116–121. doi:10.1111/bcp.12150
25. Song F, Loke YK, Walsh T, Glenny AM, Eastwood AJ, Altman DG (2009) Methodological problems in the use of indirect comparisons for evaluating healthcare interventions: survey of published systematic reviews. BMJ (Clinical research ed) 338:b1147. doi:10.1136/bmj.b1147
26. Mills EJ, Thorlund K, Ioannidis JP (2013) Demystifying trial networks and network meta-analysis. BMJ (Clinical research ed) 346:f2914. doi:10.1136/bmj.f2914
27. Mills EJ, Ioannidis JP, Thorlund K, Schunemann HJ, Puhan MA, Guyatt GH (2012) How to use an article reporting a multiple treatment comparison meta-analysis. JAMA 308(12):1246–1253. doi:10.1001/2012.jama.11228
28. Song F, Xiong T, Parekh-Bhurke S, Loke YK, Sutton AJ, Eastwood AJ, Holland R, Chen YF, Glenny AM, Deeks JJ, Altman DG (2011) Inconsistency between direct and indirect comparisons of competing interventions: meta-epidemiological study. BMJ (Clinical research ed) 343:d4909. doi:10.1136/bmj.d4909
29. Song F, Altman DG, Glenny AM, Deeks JJ (2003) Validity of indirect comparison for estimating efficacy of competing interventions: empirical evidence from published meta-analyses. BMJ (Clinical research ed) 326(7387):472. doi:10.1136/bmj.326.7387.472
30. Saramago P, Sutton AJ, Cooper NJ, Manca A (2012) Mixed treatment comparisons using aggregate and individual participant level data. Stat Med 31(28):3516–3536. doi:10.1002/sim.5442
31. Jansen JP (2012) Network meta-analysis of individual and aggregate level data. Res Synth Methods 3(2):177–190. doi:10.1002/jrsm.1048
32. Veroniki AA, Straus SE, Ashoor HM, Hamid JS, Hemmelgarn BR, Holroyd-Leduc J, Majumdar SR, McAuley G, Tricco AC (2016) Comparative safety and effectiveness of cognitive enhancers for Alzheimer's dementia: protocol for a systematic review and individual patient data network meta-analysis. BMJ Open 6(1):e010251. doi:10.1136/bmjopen-2015-010251
33. Ishak KJ, Proskorovsky I, Benedict A (2015) Simulation and matching-based approaches for indirect comparison of treatments. Pharmacoeconomics 33(6):537–549. doi:10.1007/s40273-015-0271-1

34. Signorovitch JE, Sikirica V, Erder MH, Xie J, Lu M, Hodgkins PS, Betts KA, Wu EQ (2012) Matching-adjusted indirect comparisons: a new tool for timely comparative effectiveness research. Value Health J Int Soc Pharmacoecon Outcomes Res 15(6):940–947. doi:10.1016/j.jval.2012.05.004

35. Signorovitch JE, Wu EQ, Yu AP, Gerrits CM, Kantor E, Bao Y, Gupta SR, Mulani PM (2010) Comparative effectiveness without head-to-head trials: a method for matching-adjusted indirect comparisons applied to psoriasis treatment with adalimumab or etanercept. Pharmacoeconomics 28(10):935–945. doi:10.2165/11538370-000000000-00000

36. Thom H, Jugl S, Palaka E, Jawla S (2016) Matching Adjusted Indirect Comparisons to assess comparative effectiveness of therapies: usage in scientific literature and Health Technology Appraisals. Poster Presentation at ISPOR 21st Annual International Meeting, Washington, DC, 21–25 May 2016

37. Tan DS, Araujo A, Zhang J, Signorovitch J, Zhou ZY, Cai X, Liu G (2016) Comparative efficacy of ceritinib and crizotinib as initial ALK-targeted therapies in previously treated advanced NSCLC: an adjusted comparison with external controls. J Thorac Oncol 11(9):1550–1557. doi:10.1016/j.jtho.2016.05.029

38. Signorovitch J, Swallow E, Kantor E, Wang X, Klimovsky J, Haas T, Devine B, Metrakos P (2013) Everolimus and sunitinib for advanced pancreatic neuroendocrine tumors: a matching-adjusted indirect comparison. Exp Hematol Oncol 2(1):32. doi:10.1186/2162-3619-2-32

39. http://www.fda.gov/downloads/AdvisoryCommittees/CommitteesMeetingMaterials/Drugs/OncologicDrugsAdvisoryCommittee/UCM242009.pdf

40. Martinalbo J, Camarero J, Delgado-Charro B, Démolis P, Ersbøll J, Foggi P, Jonsson B, O'Connor D, Pignatti F (2016) Single-arm trials for cancer drug approval and patient access. Ann Oncol 27(Suppl 6). doi:10.1093/annonc/mdw435.49. http://annonc.oxfordjournals.org/content/27/suppl_6/1362O_PR.short

41. Swallow E, Song J, Yuan Y, Kalsekar A, Kelley C, Peeples M, Mu F, Ackerman P, Signorovitch J (2016) Daclatasvir and sofosbuvir versus sofosbuvir and ribavirin in patients with chronic hepatitis C coinfected with HIV: a matching-adjusted indirect comparison. Clin Ther 38(2):404–412. doi:10.1016/j.clinthera.2015.12.017

42. Hung A, Baas C, Bekelman J, Fitz-Randolph M, Mullins CD (2017) Patient and stakeholder engagement in designing pragmatic clinical trials. In: Birnbaum HG, Greenberg PE (eds) Decision making in a World of comparative effectiveness research. Springer, Singapore

43. http://www.drugabacus.org/

44. Sydes MR, Johnson AL, Meredith SK, Rauchenberger M, South A, Parmar MK (2015) Sharing data from clinical trials: the rationale for a controlled access approach. Trials 16:104. doi:10.1186/s13063-015-0604-6

45. Strom BL, Buyse M, Hughes J, Knoppers BM (2014) Data sharing, year 1–access to data from industry-sponsored clinical trials. N Engl J Med 371(22):2052–2054. doi:10.1056/NEJMp1411794

Chapter 21
Decision Making with Machine Learning in Our Modern, Data-Rich Health-Care Industry

Nick Dadson, Lisa Pinheiro, and Jimmy Royer

Abstract Recent innovation in the health-care industry has given us an abundance of data with which we can compare the efficacy of alternative treatments, drugs, and other health interventions. Machine learning has proven to be particularly adept at finding intricate relationships within large datasets. In this chapter we emphasize the potential for machine learning to help us digest and use health-care data effectively. We first provide an introduction to machine learning algorithms, particularly neural network and ensemble algorithms. We then discuss machine learning applications in three areas of the health-care industry. Learning algorithms have been used within the lab as a method of automation to complement problem solving and decision making in the workplace. They have been used to compare the effectiveness of alternative interventions, such as drugs taken together. Given the rise in genomic data, they have been used to develop new treatments and drugs. Taken together, these trends suggest there is vast potential for the expanded application of these algorithms in health care.

21.1 Introduction

In August 2015, a 62-year-old builder in Alberta, Canada, purchased an Apple Watch. At work two weeks later, feeling unwell, he took a break after lunch. While he felt like he was simply having a bad run of the flu, the watch heart rate sensor indicated he was actually in the process of having a heart attack. He immediately called 911, was rushed to the emergency room, and survived [1, 2].

This anecdote is an example of the close relationship between health and technological innovation today. Health-care technology rapidly improved in the early years of the twenty-first century. Cases in point include the finite precision at which

N. Dadson • L. Pinheiro (✉) • J. Royer
Analysis Group, Inc., Montreal, QC, Canada
e-mail: Lisa.Pinheiro@analysisgroup.com

© Springer Nature Singapore Pte Ltd. 2017
H.G. Birnbaum, P.E. Greenberg (eds.), *Decision Making in a World of Comparative Effectiveness Research*, DOI 10.1007/978-981-10-3262-2_21

we can now measure brain processes and the ability to examine characteristics of individual heartbeats; both were impossible a decade ago. These technological improvements have dramatically increased the quantity of data available. In addition to our heart rate, wearable technology can now track the distances we travel, the calories we burn, and the hours we sleep.[1] In genomics, the introduction of next-generation genome sequencing machines in the mid-2000s increased throughput per machine 500,000-fold [5]. In 2015, less than a decade later, a team of biologists and computer scientists expressed concern over the inability of their discipline to deal with the anticipated flood of genome data [6].

If health-care technology improves without corresponding advances in our ability to digest, interpret, and use its data, these technological innovations will be wasted. This is particularly important in comparative effectiveness research (CER), in which alternative health-care technologies are evaluated by their efficacy for particular populations under a range of specific conditions. In this chapter we emphasize the potential for machine learning to ensure these recent innovations do not go to waste.

In the next section, we provide an introduction to machine learning. We pay particular attention to neural network and ensemble algorithms, which have been applied often in recent health-care applications. We then discuss recent health-care applications, which can be read independently of our description of these algorithms.

21.2   Machine Learning Algorithms

The classic approach to data science has been to assume explicit functional relationships between input and response variables. Machine learning, in contrast, begins with fewer assumptions on the forms of these relationships. Computer algorithms are designed to learn these relationships from the data itself, automatically improving their own performance through experience [7].

Whether the task involves classification or relating various inputs to expected outcomes, algorithms are evaluated based on how well they perform when given new data, i.e., "out of sample" performance. The algorithm is first trained on a sample of inputs for which the target output responses are known. Training enables the algorithm to learn highly complex and intricate relationships in high-dimensional data, rather than pre-imposing assumptions on how inputs and outcomes are related. Algorithm performance is then assessed on a different sample to determine how it would perform in the real world.

---

[1] In the future, biometric and wearable patient identification devices will potentially automate patient identification and data entry [3]. In addition, wearable devices are being developed that will provide information on patient and consumer vital signs, weight, glucose levels, and respiratory function [4].

21.2.1 Neural Networks

Inspired by the architecture of the human brain, artificial neural network models consist of interconnected units, referred to as "neurons." Neurons take multiple inputs and produce a single numeric output; the output of one neuron may then become the input to many other neurons. Continuing the biological metaphor, connections between neurons are referred to as "synapses," and the corresponding connection strengths are referred to as "synaptic weights."

Neural network structure is often represented by a directed graph, organized in layers of neurons connected by synapses. Figure 21.1 provides an example of a very simple network with three layers: an input layer with three neurons, a "hidden" layer with five neurons, and an output layer with a single neuron. We show the example of a network that takes three symptoms as inputs, e.g., blood pressure, age, and the presence of a headache. The second layer is hidden in the sense that it is not directly observable from the input or output of the network. Each hidden neuron combines the inputs using a nonlinear function and passes the result to the output neuron. The output neuron then uses the information passed from the intermediate layer and returns the response of the entire network: the probability that the patient will benefit from the treatment.

Synaptic weights indicate the effect or importance of the information coming from the neuron in the previous layer. "Learning" in a neural network context is the procedure by which the machine computes the numeric values of the synaptic weights. In other words, the algorithm learns how much weight to give each

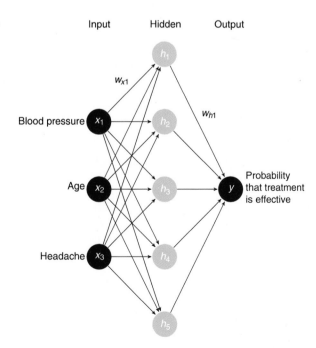

Fig. 21.1 Simple artificial neural network

combination of inputs in order to best predict the correct output. The most common form of learning is "supervised," in which the machine is trained on a set of examples. Each training example consists of inputs and corresponding target responses— i.e., what the appropriate output should be. In our patient example above, training would repeat the following steps:

1. The machine is given a random training example, corresponding to a particular combination of symptoms. Given the inputs and current weights,[2] the machine produces output as a single score between 0 and 1—the likelihood the patient will benefit from treatment.
2. An error function measures the distance between the output score and the target score associated with the training example, e.g., the target score would be 1 if the example patient actually did benefit from treatment.
3. The machine modifies the weights to reduce this distance.

The training procedure stops when the error function in step 2 stops decreasing, which suggests the potential for errors in treatment has been minimized. The training procedure also implicitly incorporates a penalty for each additional weight or variable used (called the regularization process) to avoid overfitting and to limit the variables used to those that are key to prediction performance.

"Deep" networks involve multiple hidden layers, which allow the model to discover intricate and unexpected relationships between input and output.[3] The brain appears to process information through a particularly deep architecture [12]. Recent research applications have shown the impressive performance of two deep network architectures in particular: convolutional neural networks and recurrent neural networks [13]. We describe these two algorithms and some of their typical applications below.

Image recognition involves the classification of images into particular categories. We illustrate a simple image recognition example in Fig. 21.2: what number does the image contain?[4] Images are split up into a grid of pixels, and the numeric value of each pixel corresponds to its light intensity (grayscale or color). In theory, we could apply a network structure similar to Fig. 21.1 to an image recognition problem, each input neuron encoding the intensity of a pixel. This structure would, however, ignore where pixels are located relative to one another; pixels close to one another in an image tend to be more strongly correlated than pixels more distant. Convolutional neural networks (ConvNets) use "local receptive fields" to improve recognition. Panel (a) in Fig. 21.2 illustrates the local receptive field for a single neuron in the hidden "convolutional layer." It is a 2×2 subregion of pixels, its small size ensuring it extracts only local information. Each set of neurons in the convolutional layer, called a "feature map," uses the same synaptic weights for its receptive

---

[2] At the outset, the weights are usually initialized with random values drawn from a probability distribution [8].

[3] Deep neural networks were not particularly popular until successful attempts were reported in the mid-2000s, when deep belief networks were introduced and related deep learning algorithms proposed in 2006 [9–11].

[4] Modern applications of image recognition are far more impressive than the number recognition example here. For example, Chen et al. use a deep learning algorithm for automated glaucoma classification, and Gao et al. use a convolutional-recursive network to grade cataracts [14, 15].

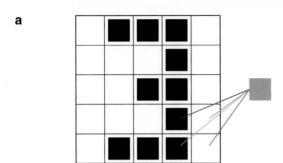

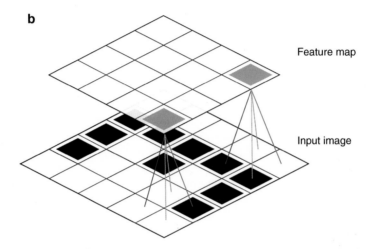

Fig. 21.2 (**a**, **b**) Convolutional neural network

field (colors in panel (b) in Fig. 21.2). Like the receptive fields of neurons in our visual cortex, these fields overlap to cover the entire input image.[5] Feature maps are followed by a "pooling layer," which condenses the information it receives from a feature map—often by taking the maximum of groups of feature map neurons. ConvNets alternate between convolutional and pooling layers until a fully connected output layer, in which every neuron in the last pooling layer is connected to each output neuron.

The idea of feature maps followed by pooling was inspired by the interaction between simple and complex cells found by Hubel and Wiesel in the visual cortex of cats [16, 17]. Pooling allows neurons in subsequent layers to work on larger patches of data. Lower network levels focus on fine details, their output is merged and passed to higher levels, and higher levels focus on the bigger picture [18]. We can also think of this organization as broadly resembling the hierarchical organization of a firm. Lower-level employees focus on specific customer and production issues. Higher-level

---

[5] Hidden neurons sharing the same weights are collectively called a feature map. The repeated application of the same set of weights across the input image is, mathematically speaking, a convolution. This gives these networks their name.

Fig. 21.3 Recurrent neural
network

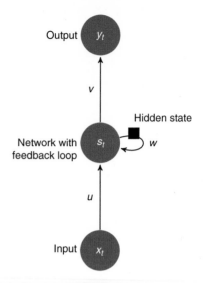

executives focus on bigger picture issues and business strategy [19]. Some posit that hierarchies such as these are nature's solution to the trade-off between invariance, the ability to recognize objects despite small differences in exact appearance, and selectivity, the ability to discriminate between different objects. Hierarchies address this trade-off by decomposing a complex task, e.g., invariant object recognition, into a hierarchy of simpler ones at each stage of processing [20].

The network architectures described above are "feedforward" networks: information flows in one direction only, inputs to outputs. The sequential order of input images is unimportant since the network only considers the current input. Recurrent neural networks (RNNs), in contrast, are explicitly designed to recognize patterns in sequences of data. Based on features of our nervous system, their architecture includes some form of "feedback loop": the output of a neuron in the system influences, in part, its own input.[6] Feedback loops in RNNs allow past information to persist and inform current decision making. They therefore effectively use two input sources, the present input and a memory of recent past inputs. Important information is often embedded in the sequence of inputs, over and above the inputs themselves. Without the ability to put current information in the proper context, feedforward networks perform much worse than RNNs in applications with sequential inputs, such as language recognition.

Figure 21.3 presents an example of an RNN, which processes one element of an input sequence at time t. The central node s represents a potentially large number of

---

[6] Feedback is present in almost all parts of the nervous system (Freeman 1975). The number of feedback connections between different areas in the brain is at least as large as the number of feedforward connections [21]. For example, the primary visual cortex receives (feedforward) signals from the retina through the lateral geniculate nucleus (LGN). The number of signals in the opposite direction, from V1 to the LGN, is approximately ten times as large [17]. Visual cortex area V2 also sends signals back to V1 and may even play a role during immediate recognition [12].

hidden neurons forming a neural network. These neurons receive the first input x_1 and produce output h_1. They also record information, the "hidden state," to be fed back to the network at time $t = 2$. The hidden state is the sequential memory of the network and is updated at each time. At time t, the network uses the information in its memory together with the current input x_t to produce output and record information for $t + 1$. These networks are "recurrent" in the sense that they repeat this same process for each input in the sequence. They are essentially a collection of multiple copies of the same neural network—with feedback loops between copies.

21.2.2   Ensemble Methods

There are many machine learning algorithms; neural networks are simply among the most common [22]. Ensemble methods train several different models separately, which possibly use different algorithms, and aggregate their results. Intuitively, ensemble methods work because different models will not make the same errors. Machine learning algorithms can benefit substantially from the aggregation of results at the price of increased computational costs.[7] For this reason, ensemble techniques often use fast algorithms such as decision trees.

To explain decision tree learning, we revisit our previous classification problem of whether a patient will benefit from treatment. Decision trees work a lot like a doctor asking a patient a series of questions about his or her symptoms until deciding on a final treatment plan. The doctor wants to correctly diagnose the patient, ideally with the smallest number of questions. At the top node of the tree in Fig. 21.4, the doctor must decide which question to ask. Each question is evaluated using a statistical test to determine how well it alone classifies the training examples into treatment or no treatment. The best question is selected (e.g., blood pressure) and used as the decision test at the initial node. Descendant nodes are created for high and low blood pressure, and the training examples are sorted into to the two appropriate groups. The process is repeated at each descendant node with the associated training examples used to select which among the remaining attributes alone best classifies the training examples [7].

One limitation of the above decision tree learning algorithm is that it never looks back to reconsider earlier choices. The random forest algorithm, on the other hand, combines decision trees with ensemble methods to achieve superior results. The algorithm features randomization at two steps. First, random samples of the training data are used to create several different decision trees. The result of each decision tree is given equal weight in the aggregation (averaging) of the results. Second, instead of selecting the best question out of all remaining questions at each node, the

---

[7]"Bagging," for example, involves constructing k different datasets, each the size of the original dataset. Each of these different datasets is constructed by sampling with replacement from the original dataset. Model i is then trained on dataset i. The differences between which examples are included in each dataset result in differences between the trained models [23].

Fig. 21.4 Decision tree

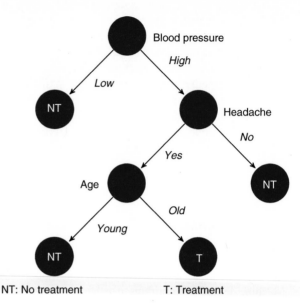

NT: No treatment   T: Treatment

best is chosen out of a random subset of the remaining questions at that node. While this second randomization may seem counterintuitive, this ensemble method performs very well when benchmarked against other classifiers [24] because randomization allows it to find combinations of factors that perform better than those selected sequentially.

21.3 Machine Learning Applications

Machine learning algorithms are useful in at least three areas of the health-care industry. They have been used within the lab as a method of automation; they have been used to compare the effectiveness of alternative interventions; and they have been used to develop new treatments, drugs, and methods. We discuss these three areas in turn below.

21.3.1 Complementary Tool

Enabling algorithms to discover relationships in the data has led to significant improvements in performance. The algorithms generally have the potential to replace costly manual, repetitive tasks and to complement problem solving and decision making in the workplace. Moreover, they are less prone to human subjectivity and error.

In lung cancer screening, for example, conventional computer-aided diagnosis (CAD) requires many manual intermediate image processing and pattern recognition steps [25]. Hua et al. use a deep learning algorithm to classify computed tomography (CT) images of lung nodules as malignant or benign. Their algorithm automatically uncovers features from training images and learns their interaction, simultaneously circumventing the need for intermediate processing steps and outperforming conventional feature computing CAD methods [26]. Ciompi et al. use an ensemble of 2D classifiers to classify 3D images as pulmonary peri-fissural nodules (PFNs) or non-PFN; classifying nonmalignant PFNs can reduce the need for additional examinations to investigate suspicious nodules. They find their 3D ensemble classification method outperforms the 2D classifiers. They also find a recent 3D classifier had similar performance to human experts, even without problem-specific tuning [27].

21.3.2   Comparative Effectiveness

The suite of machine learning algorithms is not only useful as a means to improve productivity within the lab, however. They also provide tools to compare the effectiveness of alternative pharmaceuticals, devices, procedures, and other interventions. While we have seen rapid innovation in health-care technology recently, in most cases these technologies essentially perform tasks already carried out by conventional systems; they just perform them faster, more precisely, or more safely [28].

Understanding the trade-offs associated with choosing one alternative over another is not a simple task. Comparisons of competing technologies and methods were given greater prominence in the past decade with the American Recovery and Reinvestment Act of 2009 (ARRA). In the midst of rising national medical costs, ARRA allocated $1.1 billion to support research comparing the effectiveness of procedures and services used in the prevention, diagnosis, or treatment of various health conditions [29]. Most government intervention up to that point was intended to establish whether a given treatment was safe and effective, not whether it was better than alternatives [30].[8]

Emerging health-care technologies have given us an abundance of data to use for comparative effectiveness research. Recent innovations in medical imaging, for example, carry the promise of more three-dimensional data. Unlike conventional CT imaging, which uses X-rays, proton CT relies on proton beams. A proton CT can produce a detailed three-dimensional image of the body, allowing radiologists to target treatments more precisely and reduce the amount of health tissue exposed to radiation [28].

In genomics, the costs of sequencing DNA continue to drop markedly, producing a corresponding marked increase in the quantity of genomic data available. This increase has been so large that some researchers have warned the genomic computing demands could surpass those of other "big data" generators, such as YouTube and Twitter [6]. The central repository GenBank had one billion base pairs in 1997;

---

[8] For example, note that the US Food and Drug Administration enabling law (FDC Act, as amended in 1962) does not require an assessment of comparative effectiveness to support its decisions [31].

they now have more than 300 billion[9] [34]. Recent genomics applications have used genome data to predict drug resistance for a given patient, which has numerous implications for treatments that are costly or have many side effects [35].

Genomic data is an important piece of the evidence base for a new approach in health care: "precision medicine." The goal of precision medicine is identifying which treatments will be effective for which patients based on individuals' genes, environments, and lifestyles [36].[10] Despite the advances that have been made in precision medicine, it is presently used very little. In the 2015 State of the Union Address, President Obama announced the Precision Medicine Initiative. The Initiative is intended to extend precision medicine to all diseases and to improve the chances individuals remain healthy throughout their lives [37]. To do so, a cohort program will be established that will include clinical data from electronic health record systems, patient-reported data, and genomic data [32].

The tremendous size and complexity of the health data now available is both very exciting and somewhat daunting. Although these data may not fit squarely within the definition of randomized controlled trials (RCTs), which are designed to minimize confounding, many clinically relevant questions cannot be answered with an RCT. The retrospective analyses that are possible with these new data sources can provide results faster than clinical trials [38]. Machine learning algorithms can be combined with the dramatic improvements we have seen in computation speed, e.g., the use of graphical processing units (GPUs), to provide exceptional performance even with very large datasets.[11] In addition, as stated earlier, the algorithms are able to discover intricate relationships between inputs and outputs that may be impossible to anticipate in advance.

As an example, consider drug-drug interactions (DDIs), i.e., when two or more drugs taken together alter the way one drug interacts with the body and lead to unexpected side effects. Using reports in the FDA's Adverse Event Reporting System, Tatonetti, Fernald, and Altman trained a learning algorithm to identify DDIs while addressing the primary limitation of conventional detection methods: the underreporting of adverse events. They used electronic medical records to validate their model, which identified 171 new drug interactions [40].[12] In another study, Liu et al.

---

[9] If the two examples above did not provide you with sufficient context to understand the size of the genomic data, consider another example. It has been estimated that the entire printed collection of the Library of Congress is approximately 10 terabytes (terabyte = 10^{12} bytes); the raw genomic data corresponding to a single cohort of one million patients would require approximately 5700 terabytes [32, 33].

[10] The older term "personalized medicine" is sometimes used interchangeably with precision medicine. While some do, we do not make a distinction here between the two terms.

[11] Many modern machine learning algorithms can be trained in parallel, i.e., across multiple processors simultaneously. The first major application of deep belief networks, in speech recognition, was possible because fast and easy to program GPUs allowed researchers to train the networks up to 20 times faster. Similarly, the recent success of ConvNets can partly be attributed to the efficient use of GPUs. Whereas training deep ConvNet architectures with 10–20 layers would have taken weeks two years ago, advances in hardware, software, and parallelization have reduced this time to a few hours [39].

[12] White et al. also examined adverse effects due to drug pairing, but used data drawn from queries entered into search engines, e.g., Google [41]. Their analysis of this large quantity of data revealed

use a ConvNet to detect DDIs in unstructured text and classify them into predefined categories. This can be used to find and aggregate DDIs reported in the sea of scientific articles, books, and technical reports, which can otherwise be difficult to stay up to date with on a continuous basis. Their experimental results demonstrate their network outperforms the best alternative learning algorithms [42].

In the future, big data and machine learning could potentially support joint decision making by patients and their providers regarding the best treatment plan given the patient's characteristics—lifestyle, environment, and genetics. They might also support the creation and revision of treatment guidelines, providing a deeper understanding of which genetic markers are associated with which side effects, or how patients who followed the treatment guidelines fared relative to those who did not [4].

21.3.3   Creation of New Interventions

Learning algorithms can also play an important role in the design of new drugs. A deeper understanding of the relationships between phenotypes, e.g., disease risk, and genotypes is important for the development of targeted drug therapies. In this vein, Leung et al. suggest that learning algorithms be trained to learn the relationship between genotypes and intermediate "molecular phenotypes," such as the concentrations of proteins, and that these can then be linked to phenotype. Since these molecular phenotypes are intermediate biochemically active quantities, they are good targets for therapies [43].

Jamali et al. used machine learning algorithms to predict whether a protein interacts with a drug or nutraceutical molecule (is "druggable"). Their neural network classifier had an accuracy of 90% and provided novel drug targets [44]. Relatedly, Prachayasittikul et al. used machine learning algorithms to classify compounds that interact with P-glycoprotein (Pgp) as inhibitors or non-inhibitors. Given its expression in several types of cancer and its association with multidrug resistance, inhibition of Pgp has been considered a strategy to combat multidrug-resistant cancers. Their models generally yielded good predictive performance, e.g., the decision tree had an accuracy of 87% [45].

21.4   Concluding Remarks

The health-care industry is quickly becoming characterized by its large quantity of data. Machine learning is a valuable tool that decision makers can combine with new data, old data, and existing resources to improve productivity, evaluate alternative interventions, and develop new treatments. Given the pace of innovation in health-care technology and machine learning research, there is vast potential for the application of these algorithms in the modern health-care industry.

---

prescription drug side effects before they were found by the US FDA's warning system. Although White et al. did not use them, machine learning algorithms could be used in this application as they were in Tatonetti et al. [40].

References

1. Morris I (2016) Apple watch saves man's life. Forbes. Available from: http://www.forbes.com/sites/ianmorris/2016/03/28/apple-watch-saves-mans-life/#7eda2e275783. Accessed 18 May 2016
2. Snowdon W (2016) Apple watch saved Alberta man's life, makes international headlines. CBC News. Available from: http://www.cbc.ca/news/canada/edmonton/apple-watch-saved-alberta-man-s-life-makes-international-headlines-1.3495397. Accessed 18 May 2016
3. Thrall JH (2012) Look ahead: the future of medical imaging. RSNA News 25(8):4–6
4. Berger ML, Doban V (2014) Big data, advanced analytics and the future of comparative effectiveness research. J Comp Eff Res 3(2):167–176
5. Baker M (2010) Next-generation sequencing: adjusting to data overload. Nat Methods 7(7): 495–499
6. Stephens ZD, Lee SY, Faghri F, Campbell RH, Zhai C, Efron MJ et al (2015) Big data: astronomical or genomical? PLoS Biol 13(7):e1002195
7. Mitchell TM (1997) Machine learning. Machine learning. McGraw-Hill, New York
8. Günther F, Fritsch S (2010) Neuralnet: training of neural networks. R J 2(1):30–38
9. Hinton GE, Osindero S, Teh Y-W (2006) A fast learning algorithm for deep belief nets. Neural Comput 18(7):1527–1554
10. Bengio Y, Lamblin P, Popovici D, Larochelle H (2007) Greedy layer-wise training of deep networks. In: Schölkopf B, Platt J, Hoffman T (eds)., Advances in Neural Information Processing Systems 19 (NIPS'06), p. 153–60. Available from: http://www.iro.umontreal.ca/~lisa/pointeurs/BengioNips2006All.pdf
11. Ranzato MA, Poultney C, Chopra S, LeCun Y (2006) Efficient learning of sparse representations with an energy-based model. Nips 1:1137–1144
12. Serre T, Kreiman G, Kouh M, Cadieu C, Knoblich U, Poggio T (2007) A quantitative theory of immediate visual recognition. Prog Brain Res 165:33–56
13. Nielsen MA (2015) Neural networks and deep learning. Determination Press
14. Chen X, Xu Y, Wong DWK, Wong TY, Liu J (2015) Glaucoma detection based on deep convolutional neural network. Conf Proc Annu Int Conf IEEE Eng Med Biol Soc IEEE Eng Med Biol Soc Annu Conf 2015:715–718
15. Gao X, Lin S, Wong TY (2015) Automatic feature learning to grade nuclear cataracts based on deep learning. IEEE Trans Biomed Eng 62(11):2693–2701
16. Hubel DH, Wiesel TN (1962) Receptive fields, binocular interaction and functional architecture in the cat's visual cortex. J Physiol 160:106–154
17. Haykin SS (2009) Neural networks and learning machines. Pearson Education, Upper Saddle River
18. Bishop CM (2006) Pattern recognition and machine learning (information science and statistics). Springer-Verlag New York, Inc., Secaucus
19. Phillips J, Gully SM (2013) Organizational behavior. Tools for success, 2nd edn. South-Western Cengage Learning, Mason, p xxvii, 574
20. Serre T (2015) Hierarchical models of the visual system. Encycl Comput Neurosci 1309–18
21. Churchland PS, Sejnowski TJ (1992) The computational brain. MIT Press, Cambridge, MA
22. Encyclopædia Britannica Inc. Machine learning|Artificial intelligence. In: Britannica.com. Encyclopædia Britannica, Inc. (2016). Available from: http://www.britannica.com/technology/machine-learning. Accessed 19 May 2016
23. Goodfellow I, Bengio Y, Courville A. Deep learning. In: book in preparation for MIT Press. MIT Press. 2016. Available from: http://www.deeplearningbook.org/. Accessed 19 May 2016
24. Liaw A, Wiener M (2002) Classification and regression by randomForest. R News 2(3):18–22
25. Cheng J-Z, Chou Y-H, Huang C-S, Chang Y-C, Tiu C-M, Chen K-W et al (2010) Computer-aided US diagnosis of breast lesions by using cell-based contour grouping. Radiology 255(3):746–754

26. Hua K-L, Hsu C-H, Hidayati SC, Cheng W-H, Chen Y-J (2015) Computer-aided classification of lung nodules on computed tomography images via deep learning technique. Onco Targets Ther 8:2015–2022

27. Ciompi F, de Hoop B, van Riel SJ, Chung K, Scholten ET, Oudkerk M et al (2015) Automatic classification of pulmonary peri-fissural nodules in computed tomography using an ensemble of 2D views and a convolutional neural network out-of-the-box. Med Image Anal 26(1): 195–202

28. Gwynne P (2013) Next-generation scans: seeing into the future. Nature 502(7473):S96–S97

29. American Recovery and Reinvestment Act of 2009 (2009) Available from: https://www.gpo.gov/fdsys/pkg/PLAW-111publ5/html/PLAW-111publ5.htm. Accessed 19 May 2016

30. Pear R (2009) U.S. to study effectiveness of treatments. The New York Times. A1. Available from: http://www.nytimes.com/2009/02/16/health/policy/16health.html?_r=0. Accessed 19 May 2016

31. IJzerman M, Manca A, Keizer J, Ramsey S (2015) Implementation of comparative effectiveness research in personalized medicine applications in oncology: current and future perspectives. Comp Eff Res 5:65

32. Huser V, Cimino JJ (2015) Impending challenges for the use of big data. Int J Radiat Oncol Biol Phys. doi:10.1016/j.ijrobp.2015.10.060

33. Bunn J (2012) How big is a petabyte, exabyte, zettabyte, or a yottabyte? In: High scalability. Todd Hoff. Available from: http://highscalability.com/blog/2012/9/11/how-big-is-a-petabyte-exabyte-zettabyte-or-a-yottabyte.html. Accessed 19 May 2016

34. National Research Council (US) Committee on A Framework for Developing a New Taxonomy of Disease (2011) Toward precision medicine: building a knowledge network for biomedical research and a new taxonomy of disease. National Academies Press (US), Washington, DC

35. Farhat MR, Sultana R, Iartchouk O, Bozeman S, Galagan J, Sisk P, Stolte C, Nebenzahl-Guimaraes H, Jacobson K, Sloutsky A, Kaur D, Posey J, Kreiswirth BN, Kurepina N, Rigouts L, Streicher EM, Victor TC, Warren RM, van Soolingen D, Murray M (2016) Genetic determinants of drug resistance in mycobacterium tuberculosis and their diagnostic value. Am J Respir Crit Care Med. 194(5):621–630. doi: 10.1164/rccm.201510-2091OC

36. The White House (2015) Precision medicine initiative. The White House. Available from: https://www.whitehouse.gov/precision-medicine. Accessed 19 May 2016

37. National Institutes of Health (NIH) (2015) Precision Medicine Initiative. National Institutes of Health. U.S. Department of Health and Human Services. Available from: https://www.nih.gov/precision-medicine-initiative-cohort-program. Accessed 19 May 2016

38. Chen RC, Gabriel PE, Kavanagh BD, McNutt TR (2015) How will big data impact clinical decision making and precision medicine in radiation therapy? Int J Radiat Oncol Biol Phys. doi:10.1016/j.ijrobp.2015.10.052

39. LeCun Y, Bengio Y, Hinton G (2015) Deep learning. Nature 521(7553):436–444

40. Tatonetti NP, Fernald GH, Altman RB (2012) A novel signal detection algorithm for identifying hidden drug-drug interactions in adverse event reports. J Am Med Inform Assoc 19(1): 79–85

41. White RW, Tatonetti NP, Shah NH, Altman RB, Horvitz E (2013) Web-scale pharmacovigilance: listening to signals from the crowd. J Am Med Inform Assoc 20(3):404–408

42. Liu S, Tang B, Chen Q, Wang X (2016) Drug-drug interaction extraction via convolutional neural networks. Comput Math Methods Med 2016:1–8

43. Leung MKK, Delong A, Alipanahi B, Frey BJ (2016) Machine learning in genomic medicine: a review of computational problems and data sets. Proc IEEE 104(1):176–197

44. Jamali AA, Ferdousi R, Razzaghi S, Li J, Safdari R, Ebrahimie E (2016) DrugMiner: comparative analysis of machine learning algorithms for prediction of potential druggable proteins. Drug Discov Today. doi:10.1016/j.drudis.2016.01.007

45. Prachayasittikul V, Worachartcheewan A, Shoombuatong W, Prachayasittikul V, Nantasenamat C (2015) Classification of P-glycoprotein-interacting compounds using machine learning methods. EXCLI J 14:958–970

Printed by Printforce, the Netherlands